Combinatorics: The Rota Way

Gian-Carlo Rota was one of the most original and colorful mathematicians of the twentieth century. His work on the foundations of combinatorics focused on revealing the algebraic structures that lie behind diverse combinatorial areas and created a new area of algebraic combinatorics. His graduate courses influenced generations of students.

Written by two of his former students, this book is based on notes from his courses and on personal discussions with him. Topics include sets and valuations, partially ordered sets, distributive lattices, partitions and entropy, matching theory, free matrices, doubly stochastic matrices, Möbius functions, chains and antichains, Sperner theory, commuting equivalence relations and linear lattices, modular and geometric lattices, valuation rings, generating functions, umbral calculus, symmetric functions, Baxter algebras, unimodality of sequences, and location of zeros of polynomials. Many exercises and research problems are included and unexplored areas of possible research are discussed.

This book should be on the shelf of all students and researchers in combinatorics and related areas.

JOSEPH P. S. KUNG is a professor of mathematics at the University of North Texas. He is currently an editor-in-chief of *Advances in Applied Mathematics*.

GIAN-CARLO ROTA (1932–1999) was a professor of applied mathematics and natural philosophy at the Massachusetts Institute of Technology. He was a member of the National Academy of Science. He was awarded the 1988 Steele Prize of the American Mathematical Society for his 1964 paper "On the Foundations of Combinatorial Theory I. Theory of Möbius Functions." He was a founding editor of *Journal of Combinatorial Theory*.

CATHERINE H. YAN is a professor of mathematics at Texas A&M University. Prior to that, she was a Courant Instructor at New York University and a Sloan Fellow.

Cambridge Mathematical Library

Cambridge University Press has a long and honorable history of publishing in mathematics and counts many classics of the mathematical literature within its list. Some of these titles have been out of print for many years now and yet the methods they espouse are still of considerable relevance today.

The Cambridge Mathematical Library will provide an inexpensive edition of these titles in a durable paperback format and at a price that will make the books attractive to individuals wishing to add them to their personal libraries. It is intended that certain volumes in the series will have forewords, written by leading experts in the subject, which will place the title in its historical and mathematical context.

Gian-Carlo Rota, Circa 1970
Pencil drawing by Eleanor Blair

Combinatorics: The Rota Way

JOSEPH P. S. KUNG
University of North Texas

GIAN-CARLO ROTA
CATHERINE H. YAN
Texas A&M University

CAMBRIDGE
UNIVERSITY PRESS

CAMBRIDGE
UNIVERSITY PRESS

University Printing House, Cambridge CB2 8BS, United Kingdom

One Liberty Plaza, 20th Floor, New York, NY 10006, USA

477 Williamstown Road, Port Melbourne, VIC 3207, Australia

314-321, 3rd Floor, Plot 3, Splendor Forum, Jasola District Centre, New Delhi - 110025, India

79 Anson Road, #06-04/06, Singapore 079906

Cambridge University Press is part of the University of Cambridge.

It furthers the University's mission by disseminating knowledge in the pursuit of education, learning and research at the highest international levels of excellence.

www.cambridge.org
Information on this title: www.cambridge.org/9780521737944

First published 2009

A catalogue record for this publication is available from the British Library

Library of Congress Cataloging in Publication data
Kung, Joseph P. S.
Combinatorics : the Rota way / Joseph P. S. Kung, Gian-Carlo Rota, Catherine H. Yan.
p. cm.
Includes bibliographical references and index.
ISBN 978-0-521-88389-4 (hardback)
1. Rota, Gian-Carlo, 1932–1999. 2. Combinatorial analysis. I. Rota, Gian-Carlo, 1932–1999. II. Yan, Catherine H. III. Title.
QA164.K86 2009
511′.6 – dc22 2008037803

ISBN 978-0-521-88389-4 Hardback
ISBN 978-0-521-73794-4 Paperback

Contents

Preface

The working title of this book was "Combinatorics 18.315." In the private language of the Massachusetts Institute of Technology, Course 18 is Mathematics, and 18.315 is the beginning graduate course in combinatorial theory. From the 1960s to the 1990s, 18.315 was taught primarily by the three permanent faculty in combinatorics, Gian-Carlo Rota, Daniel Kleitman, and Richard Stanley. Kleitman is a problem solver, with a prior career as a theoretical physicist. His way of teaching 18.315 was intuitive and humorous. With Kleitman, mathematics is fun. The experience of a Kleitman lecture can be gleaned from the transcripts of two talks.[1] Stanley's way is the opposite of Kleitman. His lectures are careful, methodical, and packed with information. He does not waste words. The experience of a Stanley lecture is captured in the two books *Enumerative Combinatorics I* and *II*, now universally known as *EC1* and *EC2*. Stanley's work is a major factor in making algebraic combinatorics a respectable flourishing mainstream area.

It is difficult to convey the experience of a Rota lecture. Rota once said that the secret to successful teaching is to reveal the material so that at the end, the idea – and there should be only one per lecture – is obvious, ready for the audience to "take home." We must confess that we have failed to pull this off in this book. The immediacy of a lecture cannot (and should not) be frozen in the textuality of a book. Instead, we have tried to convey the method behind Rota's research. Although he would object to it being stated in such stark simplistic terms, mathematical research is not about *solving* problems; it is about *finding* the right problems. One way of finding the right problems is to look for ideas common to subjects, ranging from, say, category theory to statistics. What is shared may be the implicit algebraic structures that hide behind the technicalities, in which case finding the structure is part

[1] Kleitman (1979, 2000).

of "applied universal algebra." The famous paper *Foundations I*, which revealed the role of partially ordered sets in combinatorics, is a product of this point of view. To convey Rota's thinking, which involves all of mathematics, one must go against an *idée reçue* of textbook writing: the prerequisites for this book are, in a sense, all mathematics. However, it is the ideas, not the technical details, that matter. Thus, in a different sense, there are no prerequisites to this book: we intend that a minimum of technical knowledge is needed to seriously appreciate the text of this book. Those parts where special technical knowledge is needed, usually in the exercises, can be skimmed over.

Rota taught his courses with different topics and for different audiences. The chapters in this book reflect this. Chapter 1 is about sets, functions, relations, valuations, and entropy. Chapter 2 is mostly a survey of matching theory. It provides a case study of Rota's advice to read on the history of a subject before tackling its problems. The aim of Chapter 2 is to find what results one should expect when one extends matching theory to higher dimensions. Possible paths are suggested in Section 2.8. The third chapter offers a mixture of topics in partially ordered sets. The first section is about Möbius functions. After the mid-1960s, Möbius functions were never the focus of a Rota course; his feeling was that he had made his contribution. However, a book on Rota's combinatorics would be incomplete without Möbius functions. Other topics in Chapter 3 are Dilworth's chain partition theorem; Sperner theory; modular, linear, and geometric lattices; and valuation rings. Linear lattices, or lattices represented by commuting equivalence relations, lie at the intersection of geometric invariant theory and the foundations of probability theory. Chapter 4 is about generating functions, polynomial sequences of binomial type, and the umbral calculus. These subjects have been intensively studied and the chapter merely opens the door to this area. Chapter 5 is about symmetric functions. We define them by distribution and occupancy and apply them to the study of Baxter algebras. This chapter ends with a section on symmetric functions over finite fields. The sixth chapter is on polynomials and their zeros. The topic is motivated, in part, by unimodality conjectures in combinatorics and was the last topic Rota taught regularly. Sadly, we did not have the opportunity to discuss this topic in detail with him.

There is a comprehensive bibliography. Items in the bibliography are referenced in the text by author and year of publication. In a few cases when two items by the same authors are published in the same year, suffixes *a* and *b* are appended according to the order in which the items are listed. Exceptions are several papers by Rota and the two volumes of his selected papers; these are referenced by short titles. Our convention is explained in the beginning of the bibliography.

We should now explain the authorship and the title of this book. Gian-Carlo Rota passed away unexpectedly in 1999, a week before his 67th birthday. This book was physically written by the two authors signing this preface. We will refer to the third author simply as Rota. As for the title, we wanted one that is not boring. The word "way" is not meant to be prescriptive, in the sense of "my way or the highway." Rather, it comes from the core of the cultures of the three authors. The word "way" resonates with the word "cammin" in the first line of Dante's *Divina commedia*, "Nel mezzo del cammin di nostra vita." It also resonates as the character "tao" in Chinese. In both senses, the way has to be struggled for and sought individually. This is best expressed in Chinese:

道可道，非常道

Inadequately translated into rectilinear English, this says "a way which can be *wayed* (that is, taught or followed) cannot be a way." Rota's way is but one way of doing combinatorics. After "seeing through" Rota's way, the reader will seek his or her own way.

It is our duty and pleasure to thank the many friends who have contributed, knowingly or unknowingly, to the writing of this book. There are several sets of notes from Rota's courses. We have specifically made use of our own notes (1976, 1977, 1994, and 1995), and more crucially, our recollection of many conversations we had with Rota. Norton Starr provided us with his notes from 1964. These offer a useful pre-foundations perspective. We have also consulted notes by Miklós Bona, Gabor Hetyei, Richard Ehrenborg, Matteo Mainetti, Brian Taylor, and Lizhao Zhang from the early 1990s. We have benefited from discussions with Ottavio D'Antona, Wendy Chan, and Dan Klain. John Guidi generously provided us with his verbatim transcript from 1998, the last time Rota taught 18.315. Section 1.4 is based partly on notes of Kenneth Baclawski, Sara Billey, Graham Sterling, and Carlo Mereghetti. Section 2.8 originated in discussions with Jay Sulzberger in the 1970s. Sections 3.4 and 3.5 were much improved by a discussion with J. B. Nation. William Y. C. Chen and his students at the Center for Combinatorics at Nankai University (Tianjin, China) – Thomas Britz, Dimitrije Kostic, Svetlana Poznanovik, and Susan Y. Wu – carefully read various sections of this book and saved us from innumerable errors. We also thank Ester Rota Gasperoni, Gian-Carlo's sister, for her encouragement of this project.

Finally, Joseph Kung was supported by a University of North Texas faculty development leave. Catherine Yan was supported by the National Science Foundation and a faculty development leave funded by the Association of Former Students at Texas A&M University.

May 2008

Joseph P. S. Kung
Catherine H. Yan

1

Sets, Functions, and Relations

1.1 Sets, Valuations, and Boolean Algebras

We shall usually work with finite sets. If A is a finite set, let $|A|$ be the number of elements in A. The function $|\cdot|$ satisfies the functional equation

$$|A \cup B| + |A \cap B| = |A| + |B|.$$

The function $|\cdot|$ is one of many functions measuring the "size" of a set. Let v be a function from a collection $\mathcal{C}$ of sets to an algebraic structure $\mathbb{A}$ (such as an Abelian group or the nonnegative real numbers) on which a commutative binary operation analogous to addition is defined. Then v is a *valuation* if for sets A and B in $\mathcal{C}$,

$$v(A \cup B) + v(A \cap B) = v(A) + v(B),$$

whenever the union $A \cup B$ and the intersection $A \cap B$ are in $\mathcal{C}$.

Sets can be combined algebraically and sometimes two sets can be compared with each other. The operations of union $\cup$ and intersection $\cap$ are two basic algebraic binary operations on sets. In addition, if we fix a universal set S containing all the sets we will consider, then we have the unary operation A^c of *complementation,* defined by

$$A^c = S \backslash A = \{a\colon a \in S \text{ and } a \notin A\}.$$

Sets are partially ordered by containment. A collection $\mathcal{C}$ of subsets is a *ring of sets* if $\mathcal{C}$ is closed under unions and intersections. If, in addition, all the sets in $\mathcal{C}$ are subsets of a universal set and $\mathcal{C}$ is closed under complementation, then $\mathcal{C}$ is a *field of sets*. The collection 2^S of all subsets of the set S is a field of sets.

Boolean algebras capture the algebraic and order structure of fields of sets. The axioms of a Boolean algebra abstract the properties of union, intersection, and complementation, without any mention of elements or points. As John von Neumann put it, the theory of Boolean algebras is "pointless" set theory.

A *Boolean algebra* P is a set with two binary operations, the *join* $\vee$ and the *meet* $\wedge$; a unary operation, *complementation* $\cdot^c$; and two nullary operations or constants, the *minimum* $\hat{0}$ and the *maximum* $\hat{1}$. The binary operations $\vee$ and $\wedge$ satisfy the *lattice axioms*:

L1. Idempotency: $x \vee x = x, \quad x \wedge x = x.$
L2. Commutativity: $x \vee y = y \vee x, \quad x \wedge y = y \wedge x.$
L3. Associativity: $x \vee (y \vee z) = (x \vee y) \vee z, \quad x \wedge (y \wedge z) = (x \wedge y) \wedge z.$
L4. Absorption: $x \wedge (x \vee y) = x, \quad x \vee (x \wedge y) = x.$

Joins and meets also satisfy the *distributive axioms*

$$x \vee (y \wedge z) = (x \vee y) \wedge (x \vee z), \quad x \wedge (y \vee z) = (x \wedge y) \vee (x \wedge z).$$

In addition, the five operations satisfy the *De Morgan laws*

$$(x \vee y)^c = x^c \wedge y^c, \quad (x \wedge y)^c = x^c \vee y^c,$$

two pairs of rules concerning complementation

$$x \vee x^c = \hat{1}, \quad x \wedge x^c = \hat{0}$$

and

$$\hat{0} \neq \hat{1}, \ \hat{1}^c = \hat{0}, \ \hat{0}^c = \hat{1}.$$

It follows from the axioms that complementation is an involution; that is, $(x^c)^c = x$. The smallest Boolean algebra is the algebra $\underline{2}$ with two elements $\hat{0}$ and $\hat{1}$, thought of as the *truth values* "false" and "true." The axioms are, more or less, those given by George Boole. Boole, perhaps the greatest simplifier in history, called these axioms "the laws of thought."[1] He may be right, at least for silicon-based intelligence.

A *lattice* is a set L with two binary operations $\vee$ and $\wedge$ satisfying axioms L1–L4. A *partially ordered set* or *poset* is a set P with a relation $\leq$ (or $\leq_P$ when we need to be clear which partial order is under discussion) satisfying three axioms:

PO1. Reflexivity: $x \leq x$.
PO2. Transitivity: $x \leq y$ and $y \leq z$ imply $x \leq z$.
PO3. Antisymmetry: $x \leq y$ and $y \leq x$ imply $x = y$.

[1] Boole (1854). For careful historical studies, see, for example, Hailperin (1986) and Smith (1982).

The *order-dual* $P^{\downarrow}$ is the partial order obtained from P by inverting the order; that is,

$$x \leq_{P^{\downarrow}} y \quad \text{if and only if} \quad y \leq_P x.$$

Sets are partially ordered by containment. This order relation is not explicit in a Boolean algebra, but can be defined by using the meet or the join. More generally, in a lattice L, we can define a partial order $\leq_L$ compatible with the lattice operations on L by $x \leq_L y$ if and only if $x \wedge y = x$. Using the absorption axiom L4, it is easy to prove that $x \wedge y = x$ if and only if $x \vee y = y$; thus, the following three conditions are equivalent:

$$x \leq_L y, \quad x \wedge y = x, \quad x \vee y = y.$$

The join $x \vee y$ is the *supremum* or *least upper bound* of x and y in the partial order $\leq_L$; that is, $x \vee y \geq_L x$, $x \vee y \geq_L y$, and if $z \geq_L x$ and $z \geq_L y$, then $z \geq_L x \vee y$. The meet $x \wedge y$ is the *infimum* or *greatest lower bound* of x and y. Supremums and infimums can be defined for arbitrary sets in partial orders, but they need not exist, even when the partial order is defined from a lattice. However, supremums and infimums of finite sets always exist in lattices.

By the De Morgan laws, the complementation map $x \mapsto x^c$ from a Boolean algebra P to itself exchanges the operations $\vee$ and $\wedge$. This gives an (order) duality: if a statement P about Boolean algebra holds for all Boolean algebras, then the statement $P^{\downarrow}$, obtained from P by the exchanges $x \leftrightarrow x^c$, $\wedge \leftrightarrow \vee$, $\leq \leftrightarrow \geq$, $\hat{0} \leftrightarrow \hat{1}$, is also valid over all Boolean algebras. A similar duality principle holds for statements about lattices.

We end this section with representation theorems for Boolean algebras as fields of subsets. Let P and Q be Boolean algebras. A function $\phi : P \to Q$ is a *Boolean homomorphism* or *morphism* if

$$\begin{aligned} \phi(x \vee y) &= \phi(x) \vee \phi(y) \\ \phi(x \wedge y) &= \phi(x) \wedge \phi(y) \\ \phi(x^c) &= (\phi(x))^c. \end{aligned}$$

1.1.1. Theorem. A finite Boolean algebra P is isomorphic to the Boolean algebra 2^S of all subsets of a finite set S.

Proof. An *atom* a in P is an element covering the minimum $\hat{0}$; that is, $a > \hat{0}$ and if $a \geq b > \hat{0}$, then $b = a$. Atoms correspond to one-element subsets. Let S be the set of atoms of B and $\psi : P \to 2^S$, $\phi : 2^S \to P$ be the functions defined by

$$\psi(x) = \{a\colon a \in S, a \leq x\}, \quad \phi(A) = \bigvee_{a \in A} a.$$

It is routine to check that both compositions $\psi\phi$ and $\phi\psi$ are identity functions and that ϕ and ψ are Boolean morphisms. □

The theorem is false if finiteness is not assumed. Two properties implied by finiteness are needed in the proof. A Boolean algebra P is *complete* if the supremum and infimum (with respect to the partial order $\leq$ defined by the lattice operations) exist for every subset (of any cardinality) of elements in P. It is *atomic* if every element x in P is a supremum of atoms. The proof of Theorem 1.1.1 yields the following result.

1.1.2. Theorem. A Boolean algebra P is isomorphic to a Boolean algebra 2^S of *all* subsets of a set if and only if P is complete and atomic.

Theorem 1.1.2 says that not all Boolean algebras are of the form 2^S for some set S. For a specific example, let S be an infinite set. A subset in S is *cofinite* if its complement is finite. The *finite–cofinite Boolean algebra* on the set S is the Boolean algebra formed by the collection of all finite or cofinite subsets of S. The finite–cofinite algebra on an infinite set is atomic but not complete. Another example comes from analysis. The algebra of measurable sets of the real line, modulo the sets of measure zero, is a nonatomic Boolean algebra in which unions and intersections of countable families of equivalence classes of sets exist.

One might hope to represent a Boolean algebra as a field of subsets constructed from a topological space. The collection of open sets is a natural choice. However, because complements exist and complements of open sets are closed, we need to consider *clopen* sets, that is, sets that are both closed and open.

1.1.3. Lemma. The collection of clopen sets of a topological space is a field of subsets (and forms a Boolean algebra).

Since meets and joins are finitary operations, it is natural to require the topological space to be compact. A space X is *totally disconnected* if the only connected subspaces in X are single points. If we assume that X is compact and Hausdorff, then being totally connected is equivalent to each of the two conditions: (a) every open set is the union of clopen sets, or (b) if p and q are two points in X, then there exists a clopen set containing p but not q. A *Stone space* is a totally disconnected compact Hausdorff space.

1.1.4. The Stone representation theorem.[2] Every Boolean algebra can be represented as the field of clopen sets of a Stone space.

[2] Stone (1936).

There are two ways, topological or algebraic, to prove the Stone representation theorem. In both, the key step is to construct a Stone space X from a Boolean algebra P. A $\underline{2}$-*morphism* of P is a Boolean morphism from P onto the two-element Boolean algebra $\underline{2}$. Let X be the set of $\underline{2}$-morphisms of P. Regarding X as a (closed) subset of the space $\underline{2}^P$ of all functions from P into $\underline{2}$ with the product topology, we obtain a Stone space. Each element x in P defines a continuous function $X \to \underline{2}$, $f \mapsto f(x)$. Using this, we obtain a Boolean morphism from P into the Boolean algebra of clopen sets of X.

The algebraic approach regards a Boolean algebra P as a commutative ring, with addition defined by $x + y = (x \wedge y^c) \vee (x^c \wedge y)$ and multiplication defined by $xy = x \wedge y$. (Addition is an abstract version of symmetric difference of subsets.) Then the set of prime ideals Spec(P) of P is a topological space under the *Zariski topology*: the closed sets are the order filters in Spec(P) under set-containment. The order filters are also open, and hence clopen. Then the Boolean algebra P is isomorphic to the Boolean algebra of clopen sets of Spec(P). Note that in a ring constructed from a Boolean algebra, $2x = x + x = 0$ for all x. In such a ring, every prime ideal is maximal. Maximal ideals are in bijection with $\underline{2}$-morphisms and so Spec(P) and X are the same set (and less obviously, the same topological space).[3]

The Boolean operations on a field P of subsets of a universal set S can be modeled by addition and multiplication over a ring $\mathbb{A}$ using indicator (or characteristic) functions. If S is a universal set and $A \subseteq S$, then the *indicator function* χ_A *of* A is the function $S \to \mathbb{A}$ defined by

$$\chi_A(a) = \begin{cases} 1 & \text{if } a \in A, \\ 0 & \text{if } a \notin A. \end{cases}$$

The indicator function satisfies

$$\chi_{A \cap B}(a) = \chi_A(a)\chi_B(a),$$
$$\chi_{A \cup B}(a) = \chi_A(a) + \chi_B(a) - \chi_A(a)\chi_B(a).$$

When $\mathbb{A}$ is GF(2), the (algebraic) field of integers modulo 2, then the indicator function gives an injection from P to the vector space $\mathrm{GF}(2)^S$ of dimension $|S|$ with coordinates labeled by S. Since GF(2) is the Boolean algebra $\underline{2}$ as a ring, indicator functions also give an injection into the Boolean algebra $\underline{2}^{|S|}$. Indicator functions give another way to prove Theorem 1.1.1.

It will be useful to have the notion of a multiset. Informally, a multiset is a set in which elements can occur in multiple copies. For example, $\{a, a, b, a, b, c\}$

[3] See Halmos (1974) for the topological approach. A no-nonsense account of the algebraic approach is in Atiyah and MacDonald (1969, p. 14). See also Johnstone (1982).

is a multiset in which the element a occurs with multiplicity 3. One way to define multisets formally is to generalize indicator functions. If S is a universal set and $A \subseteq S$, then a *multiset* M is defined by a *multiplicity function* $\chi_M : S \to \mathbb{N}$ (where $\mathbb{N}$ is the set of nonnegative integers). The *support* of M is the subset $\{a \in S\colon \chi_M(a) > 0\}$. Unions and intersections of multisets are defined by

$$\chi_{A\cap B}(a) = \min\{\chi_A(a), \chi_B(a)\},$$
$$\chi_{A\cup B}(a) = \max\{\chi_A(a), \chi_B(a)\}.$$

We have defined union so that it coincides with set-union when both multisets are sets. We also have the notion of the *sum* of two multisets, defined by

$$\chi_{A+B}(a) = \chi_A(a) + \chi_B(a).$$

This sum is an analog of disjoint union for sets.

Exercises

1.1.1. *Distributive and shearing inequalities.*

Let L be a lattice. Prove that for all $x, y, z \in L$,

$$(x \wedge y) \vee (x \wedge z) \leq x \wedge (y \vee z)$$

and

$$(x \wedge y) \vee (x \wedge z) \leq x \wedge (y \vee (x \wedge z)).$$

1.1.2. *Sublattices forbidden by the distributive axioms.*[4]

A *sublattice* of a lattice L is a subset of elements of L closed under meets and joins. Show that a lattice L is distributive if and only if L does not contain the *diamond* M_5 and the *pentagon* N_5 as a sublattice (see Figure 1.1).

1.1.3. *More on the distributive axioms.*

(a) Assuming the lattice axioms, show that the two identities in the distributive axioms imply each other. Show that each identity is equivalent to the self-dual identity

$$(x \vee y) \wedge (y \vee z) \wedge (z \vee x) = (x \wedge y) \vee (y \wedge z) \vee (z \wedge x).$$

(b) Show that a lattice L is distributive if and only if for all $a, x, y \in L$, $a \vee x = a \vee y$ and $a \wedge x = a \wedge y$ imply $x = y$.

[4] Birkhoff (1934).

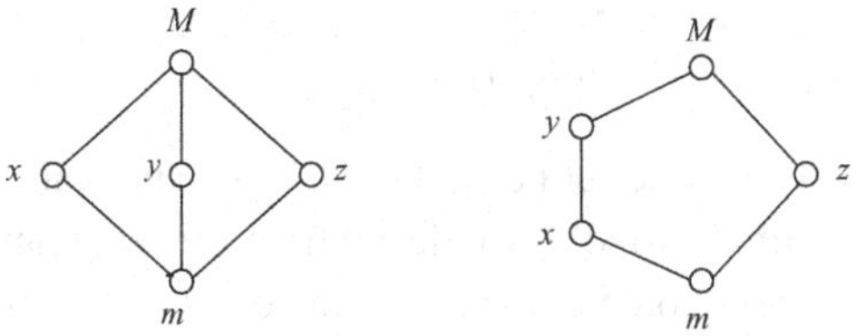

Figure 1.1 The diamond and the pentagon.

1.1.4. *Implication.*

Define the binary operation $\rightarrow$ of *implication* on a Boolean algebra P by

$$x \rightarrow y = x^c \vee y.$$

Show that the binary operation $\rightarrow$ and the constant $\hat{0}$ generate the operations $\vee, \wedge, \cdot^c$ and the constant $\hat{1}$. Give a set of axioms using $\rightarrow$ and $\hat{0}$.

1.1.5. *Conditional disjunction.*

Define the ternary operation $[x, y, z]$ of *conditional disjunction* by

$$[x, y, z] = (x \wedge y) \vee (y \wedge z) \vee (z \wedge x).$$

Note that $[x, y, z]$ is invariant under permutations of the variables. Show that $\vee$ and $\wedge$ can be defined using conditional disjunction and the constants $\hat{1}$ and $\hat{0}$. Find an elegant set of axioms for Boolean algebras using conditional disjunction and complementation.

1.1.6. *Huntington's axiom.*[5]

Show that a Boolean algebra P can be defined as a nonempty set with a binary operation $\vee$ and a unary operation $\cdot^c$ satisfying the following three axioms:

H1. $\vee$ is associative.

H2. $\vee$ is commutative.

H3. *Huntington's axiom:* For all x and y,

$$(x^c \vee y^c)^c \vee (x^c \vee y)^c = x.$$

1.1.7. *The Sheffer stroke.*[6]

Show that a Boolean algebra P can be defined as a set P with at least two elements with single binary operation $|$ satisfying the axioms:

Sh1. $(a|a)|(a|a) = a$.

[5] Huntington (1933). [6] Sheffer (1913).

Sh2. $a|(b|(b|b)) = a|a$.
Sh3. $(a|(b|c))|(a|(b|c)) = ((b|b)|a)|((c|c)|a)$.

1.1.8. Let S be the countable set $\{1/n\colon 1 \leq n < \infty\}$ and consider the topological space $S \cup \{0\}$ with the topology induced from the real numbers. Show that the finite–cofinite algebra on S is the collection of open sets of $S \cup \{0\}$.

1.1.9. (a) Let $\mathcal{H}$ be the collection of all unions of a finite number of subsets of rational numbers of the following form:

$$\{r\colon r < b\},\ \{r\colon a \leq r < b\},\ \text{or}\ \{r\colon a \leq r\}.$$

Show that $\mathcal{H}$ is a countable Boolean algebra (under set-containment) with no atoms.

(b) Show that any two countable Boolean algebras with no atoms are isomorphic.

1.1.10. Is there a natural description of the Stone space of the Boolean algebra of measurable sets of real numbers modulo sets of measure zero?

1.1.11. *Infinite distributive axioms.*

The infinite distributive axioms for the lattice operations say

$$\bigwedge_{i:i\in I}\ \bigvee_{j:\,j\in J} x_{ij} = \bigvee_{f:f:I\to J}\ \bigwedge_{i:i\in I} x_{i,f(i)}, \qquad \bigvee_{i:i\in I}\ \bigwedge_{j:\,j\in J} x_{ij} = \bigwedge_{f:f:I\to J}\ \bigvee_{i:i\in I} x_{i,f(i)},$$

with f ranging over all functions from I to J. To see that this is the correct infinite extension, interpret $\wedge$ as multiplication and $\vee$ as addition. Then formally

$$(x_{11}+x_{12}+x_{13}+\cdots)(x_{21}+x_{22}+x_{23}+\cdots)(x_{31}+x_{32}+x_{33}+\cdots)\cdots$$
$$= \sum_{f:\,f:I\to J} x_{1,f(1)}x_{2,f(2)}x_{3,f(3)}\cdots.$$

Prove the following theorem of Tarski.[7] Let P be a Boolean algebra. Then the following conditions are equivalent:

1. P is complete and satisfies the infinite distributivity axioms.
2. P is complete and atomic.
3. P is the Boolean algebra of all subsets of a set.

1.1.12. *Universal valuations for finite sets.*

Let S be a finite set, $\{x_a\colon a \in S\}$ be a set of variables, one for each element of S, x_0 be another variable, and $\mathbb{A}[\underline{x}]$ be the ring of polynomials in the

[7] Tarski (1929).

set of variables $\{x_a : a \in S\} \cup \{x_0\}$ with coefficients in a ring $\mathbb{A}$. Show that $v : 2^S \to \mathbb{A}[\underline{x}]$ defined by

$$v(A) = x_0 + \sum_{a:\, a \in A} x_a$$

is a valuation taking values in $\mathbb{A}[\underline{x}]$ and every valuation taking values in $\mathbb{A}$ can be obtained by assigning a value in $\mathbb{A}$ to each variable in $\{x_a : a \in S\} \cup \{x_0\}$.

1.2 Partially Ordered Sets

Let P be a partially ordered set. An element x *covers* the element y in the partially ordered set if $x > y$ and there is no element z in P such that $x > z > y$. An element m is *minimal* in the partial order P if there are no elements y in P such that $y < m$. A *maximal* element is a minimal element in the dual $P^{\downarrow}$.

Two elements x and y in P are *comparable* if $x \le y$ or $y \le x$; they are *incomparable* if neither $x \le y$ nor $y \le x$. A subset $C \subseteq P$ is a *chain* if any two elements in C are comparable. A subset $A \subseteq P$ is an *antichain* if any two elements in A are incomparable. If C is a finite chain and $|C| = n + 1$, then the elements in C can be *linearly ordered,* so that

$$x_0 < x_1 < x_2 < \cdots < x_n.$$

The *length* of the chain C is n, 1 less than the number of elements in C. A chain $x_0 < x_1 < \cdots < x_n$ in the partial order P is *maximal* or *saturated* if x_{i+1} covers x_i for $1 \le i \le n$. A function r defined from P to the nonnegative integers is a *rank function* if $r(x) = 0$ for every minimal element and $r(y) = r(x) + 1$ whenever y covers x. The partial order P is *ranked* if there exists a rank function on P. The *rank* of the entire partially ordered set P is the maximum $\max\{r(x) : x \in P\}$. If $x \le y$ in P, the *interval* $[x, y]$ is the set $\{z : x \le z \le y\}$.

If P is finite, then we draw a picture of P by assigning a vertex or dot to each element of P and putting a directed edge or arrow from y to x if x covers y. Thinking of the arrows as flexible, we can draw the picture so that if $x > y$, then x is above y. It is not required that the edges do not cross each other. Helmut Hasse drew such pictures for field extensions. For this reason, pictures of partial orders are often called *Hasse diagrams.*

Let P and Q be partially ordered sets. A function $f : P \to Q$ is *order-preserving* if for elements x and y in P, $x \le_P y$ implies $f(x) \le_Q f(y)$. A function f is *order-reversing* if $x \le_P y$ implies $f(x) \ge_Q f(y)$.

A subset $I \subseteq P$ is an *(order) ideal* of P if it is "down-closed;" that is, $y \le x$ and $x \in I$ imply $y \in I$. Note that we do not require ideals to be closed under joins if P is a lattice. The union and intersection of an arbitrary collection of ideals are ideals. There is a bijection between ideals and antichains: an ideal I is associated with the antichain $A(I)$ of maximal elements in I. If a is an element of P, then the set $I(a)$ defined by

$$I(a) = \{x\colon x \le a\}$$

is an ideal. An ideal is *principal* if it has this form or, equivalently, if it has exactly one maximal element a. The element a *generates* the principal ideal $I(a)$.

If A is a set of elements of P, then the *ideal* $I(A)$ *generated by* A is the ideal defined, in two equivalent ways, by

$$I(A) = \{x\colon x \le a \text{ for some } a \in A\}$$

or

$$I(A) = \bigcup_{a\colon a \in A} I(a).$$

Ideals are also in bijection with order-preserving functions from P to the Boolean algebra $\underline{2}$: the ideal I corresponds to the function $f : P \to \underline{2}$ defined by $f(x) = \hat{0}$ if $x \in I$ and $f(x) = \hat{1}$ otherwise.

Filters are "up-closed;" in other words, filters are ideals in the order-dual $P^{\downarrow}$. The set complement $P \backslash I$ of an ideal is a filter. Any statement about ideals inverts to a statement about filters. In particular, the map sending a filter to the antichain of its minimal elements is a bijection. Hence, there is a bijection between the ideals and the filters of a partially ordered set. If A is a set of elements of P, then the *filter* $F(A)$ *generated by* A is the filter defined by

$$F(A) = \{x\colon x \ge a \text{ for some } a \in A\}.$$

When A is a single-element set $\{a\}$, the filter $F(\{a\})$, written $F(a)$, is the *principal filter* generated by a.

Let P be a partial order and Q be a partial order on the same set P. The partial order Q is an *extension* of P if $x \le_P y$ implies $x \le_Q y$ or, equivalently, as a subset of the Cartesian product $P \times P$, the relation $\le_P$ is contained in $\le_Q$. If Q is a chain, then it is a *linear extension* of P.

1.2.1. Lemma.[8] Let P be a finite partially ordered set. If x is incomparable with y, then there is a linear extension L of P such that $x <_L y$.

Proof. We can construct a linear extension in the following way: let $\min(P)$ be the set of minimal elements of P. Then choose an element x_1 from $\min(P)$,

[8] Dushnik and Miller (1941).

an element x_2 from $\min(P\backslash\{x_1\})$, an element x_3 from $\min(P\backslash\{x_1, x_2\})$, and so on. This gives a linear extension in which $x_1 < x_2 < x_3 < \cdots$. Intuitively, this can be done by drawing the Hasse diagram and tilting it "slightly" so that the partial order is preserved and no two elements lie at the same height. Then we can read off the linear extension from bottom to top.

Now suppose that x is incomparable with y. Then in the construction, always choose an element other than x or y. This is possible unless, at some stage, the set $\min(P\backslash\{x_1, x_2, \ldots, x_i\})$ is $\{x, y\}$. (The set of minimal elements cannot be the one-element set $\{x\}$ or $\{y\}$; otherwise x and y would be comparable.) Choosing x before y, we obtain a linear extension L in which $x \leq_L y$.

Another proof uses the simple but useful result: if a relation in $P \times P$ contains the diagonal and has no directed cycles of positive length, then its transitive closure is a partial order. Consider the relation $P \cup \{(x, y)\}$, where $(x, y) \notin P$. This relation has no directed cycle and so its transitive closure P_x^y is a partial order. A linear extension of P_x^y is a linear extension of P in which $x \leq_L y$. □

Lemma 1.2.1 is the finite case of Szpilrajn's lemma:[9] every partially ordered set has a linear extension. In full generality, Szpilrajn's lemma is equivalent to the axiom of choice.

It is routine to show that if P and Q are two partial orders on the same set, then the intersection of the order relations P and Q, as subsets of the partial order, is a partial order (on the same set). The *order dimension* $\dim(P)$ of a partially ordered set P is the minimum number d such that there exist d linear extensions of P such that

$$P = \bigcap_{i=1}^{d} L_i.$$

For example, chains have order dimension 1 and antichains have order dimension 2. By Lemma 1.2.1, P is the intersection of all its linear extensions. Hence, the order dimension of a finite partially ordered set P is at most the number of linear extensions of P; in particular, the order dimension exists.

The *(Cartesian) product* $P \times Q$ of two partial orders P and Q is the order defined by

$$(x, u) \leq_{P \times Q} (y, v) \text{ if } x \leq_P y \text{ and } u \leq_P v.$$

If $L_1, L_2, \ldots, L_d$ are d linear extensions intersecting at P, then the diagonal map $P \to L_1 \times L_2 \times \cdots \times L_d$, $a \mapsto (a, a, \ldots, a)$, gives an isomorphism

[9] Szpilrajn (1930).

of P onto the image $\{(a, a, \ldots, a) : a \in P\}$. Conversely, given an injective function $a \mapsto (f_1(a), f_2(a), \ldots, f_d(a))$ from P to a product $K_1 \times K_2 \times \cdots \times K_d$ of d chains such that the image is isomorphic to P under the inherited order, then one can obtain d linear extensions in the following way: for each index i, choose a "default" linear extension L and define the linear extension L_i by

$$x < y \text{ if } \begin{cases} f_i(x) < f_i(y) \\ f_i(x) = f_i(y) \text{ and } x < y \text{ in } L. \end{cases}$$

Then $P = \bigcap L_i$. We conclude that the order dimension is the smallest number d such that there exists an injective order-preserving function from P into a product of d chains (of any size). Thinking of a chain as a one-dimensional line, this gives a geometric interpretation of the order dimension.

Exercises

1.2.1. *Finite partial orders and topological spaces.*

A *topological space* on a set S may be defined by a collection of *open sets* containing S and $\emptyset$ are closed under arbitrary unions and finite intersections. A topology is T_0 if for any two elements x and y, there is an open set containing x but not y or an open set containing y but not x. Show that if P is a finite set, a T_0-topology defines a partial order and, conversely, a partial order defines a T_0-topology.

1.2.2. *Standard examples for order dimension.*[10]

(a) Let $A_i = \{i\}$ and $B_i = \{1, 2, \ldots, i-1, i+1, \ldots, n\}$, ordered by set-containment, so that $A_i \leq B_j$ whenever $i \neq j$ and A_i and B_i are incomparable. The *standard example* S_n is the rank-2 partially ordered set on the $2n$ sets A_i and B_i ordered by set-containment. Show that S_n has order dimension n.

(b) Show that if Q is a suborder of P, then $\dim(Q) \leq \dim(P)$.

(c) Show that if P is a collection of subsets of a set S ordered by set-containment, then $\dim(P) \leq |S|$. Using (a) and (b), conclude that the Boolean algebra $2^{\{1,2,\ldots,n\}}$ has order dimension n.

1.2.3. *Order dimension of Cartesian products.*[11]

(a) Let P and Q be partially ordered sets, each having a maximum, a minimum, and size at least 2. Then

$$\dim(P \times Q) = \dim(P) + \dim(Q).$$

[10] Dushnik and Miller (1941).

[11] See the exposition and references in Trotter (1992, chapters 1 and 2).

In particular, if C_i are chains of positive length, then

$$\dim(C_1 \times C_2 \times \cdots \times C_n) = n.$$

In general, all that can be proved is the inequality

$$\dim(P \times Q) \leq \dim(P) + \dim(Q).$$

This inequality can be strict.

(b) Let S_n be the standard example defined in Exercise 1.2.2. Show that

$$\dim(S_n \times S_n) = 2n - 2.$$

1.2.4. *The order polynomial.*[12]

Let P be a finite partially ordered set and $\langle n \rangle$ be a chain with n elements. Let $\Omega(P; n)$ be the number of order-preserving functions $P \to \langle n \rangle$.

(a) Show that

$$\Omega(P; n) = \sum_{s=1}^{|P|} e_s \binom{n}{s},$$

where e_s is the number of surjective order-preserving functions from P to the chain $\langle s \rangle$. In particular, $\Omega(P; n)$ is a polynomial in n, called the *order polynomial* of P.

(b) Let x and y be any two incomparable elements of P. Let ${P_y}^x$ be the partial order obtained by taking the transitive closure of $P \cup \{(y, x)\}$ (defined in the proof of Lemma 1.2.1). Let P_{xy} be the partially ordered set obtained by identifying x and y. Show that $\Omega(P; n)$ satisfies the relation

$$\Omega(P; n) = \Omega({P_x}^y; n) + \Omega({P_y}^x; n) - \Omega(P_{xy}; n).$$

An order-preserving map $f : P \to Q$ is *strict* if $x < y$ implies $f(x) < f(y)$. Let $\bar{\Omega}(P; n)$ be the number of strict order-preserving functions $P \to \langle n \rangle$.

(c) Show that $\bar{\Omega}(P; n) = (-1)^n \Omega(P; -n)$. This is a simple example of a "combinatorial reciprocity theorem."

(d) Show that the order polynomial satisfies the convolution identity

$$\Omega(P; m + n) = \sum_{I} \Omega(I; m) \Omega(P \backslash I; n),$$

where the sum ranges over all order ideals I of P.

[12] Johnson (1971) and Stanley (1973, 1974).

1.2.5. *Well-quasi-orders.*[13]

Rota observed that

> an infinite class of finite sets is no longer a finite set, and infinity has a way of getting into the most finite of considerations. Nowhere more than in combinatorial theory do we see the fallacy of Kronecker's well-known saying that "God created the integers; everything else is man-made." A more accurate description might be: "God created infinity, and man, unable to understand infinity, had to invent finite sets."[14]

Rota might have added that many infinite classes of finite objects can be "finitely generated." The theory of well-quasi-orders offers a combinatorial foundation for studying finite generation of infinite classes.

A *quasi-order* is a set Q with a relation $\leq$ satisfying the reflexivity and transitivity axioms (PO1) and (PO2), but not necessarily the antisymmetry axiom (PO3). Almost all the notions in the theory of partial orders extend to quasi-orders with minor adjustments.

(a) Show that if Q is finite, a quasi-order defines a topology and, conversely, a topology defines a quasi-order.

(b) Show that the relation $x \sim y$ if $x \leq y$ and $y \leq x$ is an equivalence relation on Q. Define a natural partial order on the equivalence classes from $\leq$.

A quasi-order Q is a *well-quasi-order* (often abbreviated to *wqo*) if there are no infinite antichains or infinite descending chains. The property of being well-quasi-ordered is equivalent to four properties:

The infinite-nondecreasing-subsequence condition. Every infinite sequence $(x_i)_{1 \leq i < \infty}$ of elements in Q contains an infinite nondecreasing subsequence.

No-bad-sequences condition. If $(x_i)_{1 \leq i < \infty}$ is an infinite sequence of elements in Q, then there exist indices i and j such that $i < j$ and $x_i \leq x_j$.

Finite basis property for order filters. If F is an order filter of Q, then there exists a finite set B (called a *basis* for F) such that for each element $a \in F$, there exists an element b in B such that $b \leq a$.

Ascending chain condition for order filters. There is no strictly increasing chain (under the subset order) of order filters.

[13] Dickson (1913), Gordan (1885, 1887), Higman (1952), Kruskal (1960, 1972), and Nash-Williams (1965, 1967).

[14] Rota (1969a, pp. 197–208).

(c) Show that each of the four properties is equivalent to being a well-quasi-order.

(d) Let Q be a well-quasi-order. Show that if $P \subseteq Q$, then P (with quasi-order inherited from Q) is a well-quasi-order. Show that if P is a quasi-order and there exists a quasi-order-preserving map $Q \to P$, then P is a well-quasi-order.

(e) *An induction principle.* For an element a in a quasi-order set Q, let $F(a) = \{x \colon x \geq a\}$, the principal filter generated by a. Show that if the complements $Q \backslash F(a)$ are well-quasi-orders for all $a \in Q$, then Q itself is a well-quasi-order.

(f) Show that if P and Q are well-quasi-orders, then the Cartesian product $P \times Q$ is a well-quasi-order.

The set $\mathbb{N}$ of nonnegative integers, ordered so that $0 < 1 < 2 < 3 < \cdots$, is a well-quasi-order. From (f), one derives immediately *Gordan's lemma:* the Cartesian product $\mathbb{N} \times \mathbb{N} \times \cdots \times \mathbb{N}$ of finitely many copies of $\mathbb{N}$ is a well-quasi-order.

Gordan proved his lemma in the following form:

Let

$$\begin{aligned} a_{11}X_1 + a_{12}X_2 + \cdots + a_{1m}X_m &= 0, \\ &\vdots \\ a_{k1}X_1 + a_{k2}X_2 + \cdots + a_{km}X_m &= 0 \end{aligned}$$

be a system of linear diophantine equations, where the coefficients a_{ij} are integers. Let $\mathcal{S}$ be the set of solutions $(s_1, s_2, \ldots, s_m)$, where the coordinates s_i are nonnegative integers. Then there exists a finite set $\{\underline{b}_1, \underline{b}_2, \ldots, \underline{b}_t\}$ of vectors in $\mathcal{S}$ such that every solution in $\mathcal{S}$ can be written as a linear combination

$$\sum_{i=1}^{t} c_i \underline{b}_i,$$

where the coefficients c_i are nonnegative integers.

Consider the ring $\mathbb{F}[x_1, x_2, \ldots, x_n]$ of polynomials in the variables $x_1, x_2, \ldots, x_n$ with coefficients in a field $\mathbb{F}$. An ideal I in $\mathbb{F}[x_1, x_2, \ldots, x_n]$ is a *monomial ideal* if it can be generated by monomials. *Dickson's lemma,* that every monomial ideal can be generated by a finite set of monomials, follows immediately from Gordan's lemma. *Hilbert's basis theorem* says that every ideal in $\mathbb{F}[x_1, x_2, \ldots, x_n]$ can be generated by finitely many polynomials.

(g) Prove Hilbert's basis theorem from Dickson's lemma.

In particular, Hilbert's basis theorem is equivalent to the "trivial combinatorial fact" given in Gordan's lemma.

It is obviously false that the Cartesian product of countably infinitely many copies of $\mathbb{N}$ is a well-quasi-order. However, an intermediate result holds. Let Q be a quasi-ordered set. The quasi-order Seq(Q) is the set of all *finite* sequences of Q, quasi-ordered by $(a_1, a_2, \ldots, a_l) \leq (b_1, b_2, \ldots, b_m)$ if there is an increasing injection $f : \{1, 2, \ldots, l\} \to \{1, 2, \ldots, m\}$ such that $a_i \leq b_{f(i)}$ for every $i \in \{1, 2, \ldots, l\}$. A sequence $(a_1, a_2, \ldots, a_m)$ is often regarded as a *word* $a_1 a_2 \cdots a_m$ and the sequence order is called the *divisibility* or *subword* order. For example, if $x_i \leq x_i'$ in Q, then

$$x_1 x_2 x_3 x_4 \leq y_1 y_2 x_4 x_1' y_3 y_2 y_2 x_2' x_3' x_1' y_4 y_5 x_4' x_1 y_6 y_7.$$

The quasi-order Set(Q) is the set of all finite subsets of Q, quasi-ordered by $A \leq B$ if there is an injection $f : A \to B$ such that $a \leq f(a)$ for every $a \in A$.
(h) Prove *Higman's lemma*. If Q is a well-quasi-order, then Seq(Q) and Set(Q) are well-quasi-orders.

Higman's lemma gives the most useful cases of a more general theorem, also due to Higman. One can regard Seq(Q) as a monoid, with concatenation as the binary operation. Instead of just concatenation, we can have a well-quasi-ordered set Ω of k-ary operations on Q, where $0 \leq k \leq k_0$ for some fixed positive integer k_0. Consider the set of all expressions formed from elements of Q using the k-ary operations, quasi-ordered by a generalization of the subword order. Intuitively, this means we consider finite words, with various kinds of brackets added. These more general quasi-orders constructed from a well-quasi-ordered algebra Q with well-quasi-ordered operations Ω are well-quasi-ordered (see Higman, 1952, for a formal statement).

An expression with k-ary operations can be represented as a labeled tree. Thus, a natural next step is to consider quasi-orders on finite trees. Recall that a *tree* is a connected graph without cycles. A *rooted* tree is a tree with a distinguished vertex x_0 called the *root*. There is exactly one path from the root x_0 to any other vertex. If the vertex x lies on the path from x_0 to y, then y is *above* x. If, in addition, $\{x, y\}$ is an edge, then y is a *successor* of x. Let T_1 and T_2 be rooted trees. An *admissible map* $f : T_1 \to T_2$ is a function from the vertex set $V(T_1)$ of T_1 into the vertex set $V(T_2)$ of T_2 such that for every vertex v in T_1, the images of the successors of u are equal to or above distinct successors of $f(u)$.

(i) Prove *Kruskal's theorem.* The set $\mathcal{T}$ of all finite rooted trees, quasi-ordered by $T_1 \leq T_2$ if there is an admissible function $f : T_1 \to T_2$, is a well-quasi-order.

A graph H is the *minor* of a graph G if H can be obtained from G by deleting or contracting edges. (Isolated vertices are ignored.) Kuratowski's theorem says that a graph is planar if and only if it does not contain the complete graph K_5 and the complete bipartite graph $K_{3,3}$ as minors. Being planar is a property *closed under minors,* in the sense that if G is planar, so are all its minors. Thus, the ultimate conceptual extension of Kuratowski's theorem is that if P is a property of graphs closed under minors, then there is a finite set $M_1, M_2, \ldots, M_r$ of graphs such that a graph has property P if and only if it does not contain any of the graphs $M_1, M_2, \ldots, M_r$ as minors. This very general theorem is in fact true, and follows from the fact that the set of finite graphs, ordered by being a minor, is a well-quasi-order.

(j) *The Robertson–Seymour graph minor theorem.* Show that the set of all finite graphs ordered under minors is a well-quasi-order.

(k) *The matroid minor project.* Show that the set $\mathcal{L}(q)$ of all matroids representable over the finite field GF(q) of order q, ordered under minors, is a well-quasi-order.

1.3 Lattices

Ideals give a representation of any lattice as a lattice of sets. This representation is given in the following folklore lemma.

1.3.1. Lemma. Let L be a lattice, 2^L be the Boolean algebra of subsets of all the lattice elements, and $I : L \to 2^L$ be the function sending an element a to the principal ideal $I(a)$ generated by a. Then I preserves arbitrary meets,

$$I\left(\bigwedge_{x:x \in A} x\right) = \bigcap_{x:x \in A} I(x),$$

and is "superadditive" on joins,

$$I(x \vee y) \supseteq I(x) \cup I(y).$$

A function $\varphi : L \to M$ between lattices is a *(lattice) homomorphism* if φ preserves both meets and joins. The function I in Lemma 1.3.1 is not a lattice homomorphism.

The representation using ideals can be strengthened for distributive lattices. The classical theorem of this type is the Birkhoff representation theorem for finite distributive lattices.[15]

An element j in a lattice L is a *join-irreducible* if it is not equal to the minimum (if one exists) and $j = a \vee b$ implies that $j = a$ or $j = b$. Equivalently, j is a join-irreducible if and only if j covers a unique element. Dually, an element m is a *meet-irreducible* if it is not equal to the maximum (if one exists) and $m = a \wedge b$ implies that $j = a$ or $j = b$, or equivalently, m is covered by a unique element. An element a is a *point* or an *atom* if it covers the minimum $\hat{0}$. Atoms are join-irreducibles. *Copoints* or *coatoms* are elements covered by the maximum $\hat{1}$. Denote by $J(L)$ the set of join-irreducibles of L and $M(L)$ the set of meet-irreducibles of L. The sets $J(L)$ and $M(L)$ are partially ordered by the order of L. If the lattice L is finite (or, more generally, have finite rank), then every element is a join of join-irreducibles and every element is the meet of meet-irreducibles. Indeed,

$$x = \bigvee\{j\colon j \in J(L) \text{ and } j \le x\} \tag{JR}$$

and

$$x = \bigwedge\{m\colon m \in M(L) \text{ and } m \ge x\}. \tag{MR}$$

A lattice is *atomic* if every element is a join of atoms. It is *coatomic* if every element is a meet of coatoms.

A lattice is *distributive* if it satisfies the distributive axioms (see Section 1.1). Concrete examples of distributive lattices are Boolean algebras, chains, and product of chains. The set of positive integers, ordered under divisibility, is a distributive lattice (by the distributive axioms of arithmetic). This lattice is in fact a product of infinitely many chains.

Distributive lattices satisfy two properties directly analogous to properties of primes and prime factorizations in arithmetic.

1.3.2. Lemma. Let j be a join-irreducible in a distributive lattice. Then $j \le x \vee y$ implies $j \le x$ or $j \le y$.

Proof. Since $j \le x \vee y$, we have $j = j \wedge (x \vee y)$. By the distributive axiom,

$$j = (j \wedge x) \vee (j \wedge y).$$

As j is a join-irreducible, $j = j \wedge x$ or $j = j \wedge y$. □

[15] Birkhoff (1933). Harper and Rota commented in *Matching theory* that this representation theorem "is far more important than the classical theorem of Boole, popularized in the current frenzy for the 'new math', stating that every finite Boolean algebra is isomorphic to the lattice of all subsets of a finite set."

1.3.3. Unique factorization lemma. Every element x of a finite distributive lattice is the join of a unique antichain $J(x)$ of join-irreducibles. The antichain $J(x)$ is the set of maximal elements in the order ideal $\{a \in J(L)\colon a \leq x\}$ in the partially ordered set $J(L)$ of join-irreducibles of L.

Proof. Suppose that

$$j_1 \vee j_2 \vee \cdots \vee j_k = x = h_1 \vee h_2 \vee \cdots \vee h_n,$$

where $n \geq k$ and $\{j_i\}$ and $\{h_i\}$ are two antichains of join-irreducibles. By the distributive axiom,

$$j_i = j_i \wedge x = (j_i \wedge h_1) \vee (j_i \wedge h_2) \vee \cdots \vee (j_i \wedge h_n).$$

Hence, by Lemma 1.3.2, for each index i, there exists an index $\pi(i)$ such that $j_i = j_i \wedge h_{\pi(i)}$. Reversing the argument, for each i, there exists an index $\sigma(i)$ such that $h_i = h_i \wedge j_{\sigma(i)}$. Combining this, we conclude that from each index i,

$$j_i \leq h_{\pi(i)} \leq j_{\sigma(\pi(i))}.$$

Since $\{j_i\}$ is an antichain, $j_{\sigma(\pi(i))} = j_i$ and hence $\sigma(\pi(i)) = i$. This implies that $n = k$ and π is a permutation.

Observing that the nonmaximal elements on the right of Equation (JR) can be removed, x is the join of the antichain of maximal elements in $\{a \in J(L)\colon a \leq x\}$. The second assertion now follows from uniqueness. □

Recall that a *ring* $\mathcal{D}$ of sets is a collection of sets closed under unions and intersections. Under set-containment, $\mathcal{D}$ forms a distributive lattice. The collection $\mathcal{D}(P)$ of order ideals of a partially ordered set P is a ring of subsets. The next theorem says that a finite distributive lattice is representable as a ring of order ideals of its partially ordered set of join-irreducibles.

1.3.4. The Birkhoff representation theorem. Let L be a finite distributive lattice. Then L is isomorphic to the lattice $\mathcal{D}(J(L))$ of order ideals of the partially ordered set $J(L)$ of join-irreducibles.

Proof. If x is in L, let $I(x) = \{j\colon j \in J(L), j \leq x\}$. Let $\varphi : L \to \mathcal{D}(J(L))$, $x \mapsto I(x)$. We will show that φ is a lattice homomorphism. Since

$$\begin{aligned}\varphi(x \wedge y) &= \{j\colon j \in J(L),\ j \leq x \text{ and } j \leq y\}\\ &= I(x) \cap I(y)\\ &= \varphi(x) \cap \varphi(y),\end{aligned}$$

φ preserves meets.

Next, observe that if $j \leq x$ or $j \leq y$, then $j \leq x \vee y$. Therefore, $I(x) \cup I(y) \subseteq I(x \vee y)$. Suppose that $j \in I(x \vee y)$; that is, $j \in J(L)$ and $j \leq x \vee y$. Then

$$j = j \wedge (x \vee y) = (j \wedge x) \vee (j \wedge y).$$

As j is a join-irreducible, $j = j \wedge x$ and $j \leq x$, or $j = j \wedge y$ and $j \leq y$. We conclude that $j \in I(x) \cup I(y)$. Hence, $I(x \vee y) = I(x) \cup I(y)$ and $\varphi(x \vee y) = \varphi(x) \cup \varphi(y)$.

By Lemma 1.3.3, the set of maximal elements of $I(x)$ equals $J(x)$ and x determines $J(x)$ uniquely. Hence, φ is injective. To show that φ is surjective, let I be an order ideal in $J(L)$ and x be the join of all the elements in I. Then $I \subseteq I(x)$. Suppose that $j \in I(x)$. Then

$$j = j \wedge x = j \wedge \bigvee_{y: y \in I} y = \bigvee_{y: y \in I} (j \wedge y).$$

Since j is a join-irreducible, $j = j \wedge y$, and thus, $j \leq y$, for some $y \in I$. Since I is an order ideal, $j \in I$. We conclude that $I = I(x) = \varphi(x)$.

An alternate way to show that φ is a lattice homomorphism is to show that φ is order-preserving and use Exercise 1.3.3. □

Inverting the order, Birkhoff's theorem says that a finite distributive lattice is isomorphic to the lattice of filters of the partially ordered set of meet-irreducibles.

1.3.5. Corollary. A finite lattice is distributive if and only if it can be represented as a ring of sets.

Much research has been done on infinite versions of Birkhoff's theorem.[16]

Birkhoff's theorem delineates the extent to which a finite lattice may be represented by sets so that meets and joins are intersections and unions. For lattices which are not distributive, one might ask for representations by subsets for which meets are intersections and there is a natural way to construct the joins. Such representations appeared in the paper[17] by Richard Dedekind. Dedekind viewed lattices as lattices of "subalgebras." Many of these lattices satisfy a weaker version of the distributive axiom. A lattice is *modular* if for all elements x, y, z such that $x \geq z$, the distributive axiom holds; that is, if $x \geq z$, then

$$x \wedge (y \vee z) = (x \wedge y) \vee (x \wedge z) = (x \wedge y) \vee z.$$

We shall study modular lattices in greater detail in Sections 3.4 and 3.5.

[16] See the survey in Grätzer (2003, appendix B). [17] Dedekind (1900).

Exercises

1.3.1. *The Knaster–Tarski fixed-point theorem.*[18]

A lattice L is *complete* if the supremum and infimum exist for any set (of any cardinality) of elements in L. A partially ordered set P has the *fixed-point property* if for every order-preserving function $f : P \to P$, there exists a point x in P such that $f(x) = x$.

(a) Show that a lattice L is complete if and only if L has the fixed-point property.

(b) Show that if a finite partially ordered set has a maximum, then it has the fixed-point property.

(c) (Research problem posed by Crawley and Dilworth[19]) Characterize the finite partially ordered sets with the fixed-point property.

1.3.2. Let P be a partially ordered set with a maximum $\hat{1}$ such that the infimum exists for every subset of elements. Then P is a lattice.

1.3.3. Let $\varphi : L \to M$ be a bijective function between lattices L and M. Show that φ is a lattice homomorphism if and only if both φ and its inverse φ^{-1} are order-preserving.

1.3.4. *Join- and meet-irreducibles in finite distributive lattices.*[20]

Let L be a finite distributive lattice.

(a) Show that a maximal chain C in L has length $|J(L)|$, the number of join-irreducibles in L.

(b) Show that if j is a join-irreducible, then there exists a unique meet-irreducible $m(j)$ such that $j \not\leq m(j)$. Show that the function $J(L) \to M(L)$, $j \mapsto m(j)$ is an injection.

(c) Conclude from duality and (a) or (b) that $|J(L)| = |M(L)|$.

1.3.5. Show that the group of lattice automorphisms of a finite distributive lattice is isomorphic to the group of order-preserving automorphisms of its partially ordered set of join-irreducibles.

[18] Davis (1955), Knaster (1928), and Tarski (1955). [19] Crawley and Dilworth (1973, p. 17).
[20] Part (a) is in Grätzer (2003, p. 44). Part (b) is due to Dilworth (unpublished).

1.3.6. Let L be a lattice. Show that the following conditions are equivalent:

Mod1. L is modular.

Mod2. L satisfies the *shearing identity*

$$x \wedge (y \vee z) = x \wedge ((y \wedge (x \vee z)) \vee z).$$

Mod3. L does not contain a pentagon as a sublattice (see Exercise 1.1.1).

1.3.7. *Consistency and the Kurosh–Ore property.*[21]

Let x be an element in a lattice L. A decomposition

$$x = j_1 \vee j_2 \vee \cdots \vee j_m$$

into join-irreducibles is *irredundant* if for every i,

$$x \neq j_1 \vee j_2 \vee \cdots \vee j_{i-1} \vee j_{i+1} \vee \cdots \vee j_m.$$

The lattice L satisfies the *Kurosh–Ore replacement property* if

$$j_1 \vee j_2 \vee \cdots \vee j_m = x = h_1 \vee h_2 \vee \cdots \vee h_l$$

are two irredundant decompositions of an element x into join-irreducibles; then there exists an index i, $1 \leq i \leq l$, so that

$$x = h_i \vee j_2 \vee j_3 \vee \cdots \vee j_m. \qquad (*)$$

In other words, for any join-irreducible, say j_1, in the first decomposition, there exists a join-irreducible h_i which can replace j_1 to obtain a decomposition of x.

(a) Show that if x is an element in L satisfying the Kurosh–Ore property, then every irredundant decomposition of x into join-irreducibles has the same number of join-irreducibles.

A join-irreducible j in a lattice L is *consistent* if for every element x in L, the join $x \vee j$ either equals x or is a join-irreducible in the lattice $\{y\colon y \geq x\}$. A lattice L is *consistent* if all join-irreducibles in L are consistent.

(b) Show that if $j_1 \vee j_2 \vee \cdots \vee j_m = x = h_1 \vee h_2 \vee \cdots \vee h_l$ are two irredundant decompositions into join-irreducibles for x and j_1 is a consistent join-irreducible, then there exists a join-irreducible h_i which can replace j_1 so that Equation $(*)$ holds. Conclude that if L is a lattice, then L satisfies the Kurosh–Ore property if and only if L is consistent. Conclude also that a modular lattice satisfies the Kurosh–Ore replacement property.

[21] The classic references are Kurosch (1935) and Ore (1936). Other references are Gragg and Kung (1992), Kung (1985), and Reuter (1989).

(c) (Research problem posed by Crawley and Dilworth[22]) A lattice L satisfies the *numerical Kurosh–Ore property* if for every element x in L, every irredundant decomposition of x into a join of join-irreducibles has the same number of join-irreducibles. Find a lattice condition equivalent to the numerical Kurosh–Ore property.

1.3.8. *Lattice polynomials, the word problem, varieties, free lattices, and free distributive lattices.*[23]

A lattice polynomial is an expression formed by combining variables with the operations of meet and joins. For example,

$$x_1, \quad (x_1 \vee x_2) \vee x_1, \quad ((x_1 \vee x_2) \wedge x_3) \wedge x_2$$

are lattice polynomials. Formally, the *set of lattice polynomials* $\mathcal{F}(n)$ on the variables $x_1, x_2, x_3, \ldots, x_n$ is defined recursively as the smallest set satisfying

LP1. $x_1, x_2, x_3, \ldots, x_n \in \mathcal{F}(n)$.

LP2. If $\beta(x_1, x_2, \ldots, x_n), \gamma(x_1, x_2, \ldots, x_n) \in \mathcal{F}(n)$, then

$$\beta(x_1, x_2, \ldots, x_n) \wedge \gamma(x_1, x_2, \ldots, x_n),$$
$$\beta(x_1, x_2, \ldots, x_n) \vee \gamma(x_1, x_2, \ldots, x_n) \in \mathcal{F}(n).$$

Every element in the lattice generated by the elements $x_1, x_2, \ldots, x_n$ can be expressed as a lattice polynomial in $x_1, x_2, \ldots, x_n$. However, the expression is not unique; for example, $x \wedge x = x$ and $x \vee y = y \vee x$. Indeed, each lattice axiom gives a way to obtain different expressions for the same element. The *word problem* for lattices is to find an algorithm or deduction system to decide whether two lattice polynomials are "equal" under the lattice axioms. Whitman gave such an algorithm to decide whether an inequality among lattice polynomials holds in every lattice. Since $\alpha = \beta$ if and only if $\alpha \leq \beta$ and $\alpha \geq \beta$, Whitman's algorithm gives a solution to the word problem for lattices.

Suppose that we wish to decide whether a given inequality holds in all lattices. Then suppose that the inequality is false and apply the following four *deduction rules:*

- From "$\alpha \vee \beta \leq \gamma$" is false, deduce "$\alpha \leq \gamma$" is false or "$\beta \leq \gamma$" is false.
- From "$\alpha \wedge \beta \leq \gamma$" is false, deduce "$\alpha \leq \gamma$" is false and "$\beta \leq \gamma$" is false.
- From "$\gamma \leq \alpha \vee \beta$" is false, deduce "$\gamma \leq \alpha$" is false and "$\gamma \leq \beta$" is false.
- From "$\gamma \leq \alpha \wedge \beta$" is false, deduce "$\gamma \leq \alpha$" is false or "$\gamma \leq \beta$" is false.

[22] Crawley and Dilworth (1973, p. 56).
[23] Birkhoff (1935), Freese et al. (1995), Grätzer (2003, chapter 1), and Whitman (1941, 1942).

Doing this repeatedly, a lattice inequality is broken up into a conjunction or disjunction of simple inequalities that are easily checked to be true or false. For example, to prove the distributive inequality (in Exercise 1.1.1), suppose that

$$(x \wedge y) \vee (x \wedge z) \leq x \wedge (y \vee z)$$

is false. Then at least one of the inequalities

$$x \wedge y \leq x \wedge (y \vee z) \text{ or } x \wedge z \geq x \wedge (y \vee z)$$

is false; continuing in the same way, at least one of the inequalities

$$x \wedge y \leq x \text{ or } x \wedge y \leq y \vee z \text{ or } x \wedge z \leq x \text{ or } x \wedge z \leq y \vee z$$

is false. However, all four inequalities are true by definition of meets and joins. (If this were not absolutely clear or one is writing a computer program, then one can reduce the inequalities further.) Hence, the original inequality cannot be false. If we wish, we can reverse the deduction to obtain a direct proof. Whitman's deduction system gives an automatic way to prove such inequalities.

Two lattice polynomials α and β are *equal* if $\alpha = \beta$ can be proved using Whitman's algorithm. The set $\mathcal{F}(n)$ (under the equivalence relation of equality) forms a lattice with meets and joins defined "formally": the meet $\beta \wedge \gamma$ in the lattice is the lattice polynomial $\beta \wedge \gamma$, and the join is defined similarly. This lattice is the *free lattice* $\mathcal{F}(n)$ on n generators.

(a) Show that the free lattice $\mathcal{F}(n)$ generated by $x_1, x_2, \ldots, x_n$ satisfies the *universal property:* if L is generated by the elements $a_1, a_2, \ldots, a_n$, then there is a unique lattice homomorphism ϕ from $\mathcal{F}(n)$ to L such that $\phi(x_i) = a_i$.

Note that if a lattice M satisfies the universal property, then it is isomorphic to $F(n)$. Thus, one may use the universal property to *define* free lattices.

If L is a lattice and $\beta(x_1, x_2, \ldots, x_n)$ is a lattice polynomial, then β defines a *polynomial function* from $L \times L \times \cdots \times L \to L$, $(a_1, a_2, \ldots, a_n) \mapsto \beta(a_1, a_2, \ldots, a_n)$.

(b) Show that a lattice polynomial function β is *monotone*; that is, if $a_1 \leq b_1$, $a_2 \leq b_2$, ..., and $a_n \leq b_n$, then $\beta(a_1, a_2, \ldots, a_n) \leq \beta(b_1, b_2, \ldots, b_n)$.

An *identity* on lattice polynomials is an expression of the form $\alpha(x_1, x_2, \ldots, x_n) = \beta(x_1, x_2, \ldots, x_n)$, where α and β are lattice polynomials. A lattice L *satisfies* the identity $\alpha(x_1, x_2, \ldots, x_n) = \beta(x_1, x_2, \ldots, x_n)$ if for every n-tuple $(a_1, a_2, \ldots, a_n)$ of elements of L, $\alpha(a_1, a_2, \ldots, a_n) = \beta(a_1, a_2, \ldots, a_n)$. The *variety* or *equational class* defined by a set $\{\alpha_i = \beta_i\}$

of identities is the collection of all lattices satisfying all the identities $\alpha_i = \beta_i$. Two sets of identities are *equivalent* if they define the same variety.

(c) Show that any finite set of identities is equivalent to a single identity (with possibly more variables).

Let $\mathbf{C}$ be a class of lattices. The *free lattice* $\mathcal{F}(\mathbf{C}, n)$ *in* $\mathbf{C}$ *on* n *generators* $x_1, x_2, \ldots, x_n$ is the lattice satisfying the *universal property:* if L is generated by the elements $a_1, a_2, \ldots, a_n$, then there is a unique lattice homomorphism ϕ from $\mathcal{F}(n)$ to L such that $\phi(x_i) = a_i$.

G. Birkhoff defined varieties and proved two fundamental theorems:[24]

- Let $\mathbf{C}$ be a family of lattices closed under taking sublattices and direct products. If $\mathbf{C}$ contains at least two (nonisomorphic) lattices, then for every cardinal n, $\mathbf{V}$ has a free lattice on n generators.
- A collection of lattices is a variety if and only if it is closed under taking homomorphic images, sublattices, and direct products.

Free lattices in a variety can be obtained as a quotient of a free lattice by the congruence generated by the identities. In general, this is no explicit description or construction for the free lattice of a variety and the word problem may not be decidable. An exception is the variety of distributive lattices.

Let $x_1, x_2, \ldots, x_n$ be variables. A *meet-monomial* or *conjunction* is a polynomial of the form $\bigwedge_{i:i \in E} x_i$, where $E \subseteq \{1, 2, \ldots, n\}$.

(d) Assuming the distributive axioms, show that every polynomial in the free distributive lattice can be written uniquely as a join of meet-monomials or a *disjunction of conjunctions*

$$\bigvee_{E:\, E \in \mathcal{A}} \left[\bigwedge_{i:\, i \in E} x_i \right],$$

where $\mathcal{A}$ is an antichain in the Boolean algebra $2^{\{1,2,\ldots,n\}}$. Conclude from this that there is a bijection between elements in the free distributive lattice on n generators and nonempty antichains not equal to $\{\emptyset\}$ in the Boolean algebra $2^{\{1,2,\ldots,n\}}$.

1.3.9. *Boolean polynomials and disjunctive normal form.*

A *Boolean function on* n *variables* $x_1, x_2, \ldots, x_n$ is a function on n variables from the Cartesian product $\underline{2}^n$ to $\underline{2}$. If x is a Boolean variable, define formally $x^1 = x$ and $x^0 = x^c$, the complement of x. A *Boolean polynomial* $\beta(x_1, x_2, \ldots, x_n)$ is an expression formed from the variables and the Boolean

[24] Birkhoff's theorems are results in universal algebra and apply to general "algebras." See, for example, Cohn (1981) and Grätzer (2002, chapter 5).

operations. For example,

$$((x_1 \wedge x_2^c) \vee x_3) \wedge (x_1 \wedge x_3) \vee x_4^c$$

is a Boolean polynomial. Usually, the symbol $\wedge$ for meet is suppressed, so that the polynomial just given is written as $(x_1x_2^c \vee x_3)(x_1x_3 \vee x_4^c)$. In addition, $\vee$ may be written as $+$; however, we shall not do this. A Boolean polynomial defines a Boolean function.

A *conjunctive* or *meet-monomial* is a Boolean polynomial of the form

$$x_1^{\epsilon_1} x_2^{\epsilon_2} \cdots x_n^{\epsilon_n},$$

where ϵ_i equals the formal exponents 0 or 1. There are 2^n conjunctions.

(a) Show using the distributive axioms and the De Morgan laws that every Boolean polynomial $\beta(x_1, x_2, \ldots, x_n)$ can be written as a disjunction or join of Boolean monomials.

(b) The *truth table* of a Boolean function $\beta(x_1, x_2, \ldots, x_n)$ is the table of the values of β at the 2^n possible inputs $(\epsilon_1, \epsilon_2, \ldots, \epsilon_n)$, where ϵ_i equals 0 or 1. Show that

$$\beta(x_1, x_2, \ldots, x_n) = \bigvee_{(\epsilon_i):\, \beta(\epsilon_i)=1} x_1^{\epsilon_1} x_2^{\epsilon_2} \cdots x_n^{\epsilon_n},$$

where the union ranges over all inputs $(\epsilon_1, \epsilon_2, \ldots, \epsilon_n)$ for which $\beta(\epsilon_1, \epsilon_2, \ldots, \epsilon_n) = 1$. In particular, this shows that every Boolean function is expressible as a Boolean polynomial.

A Boolean function is *monotone* if $\beta(\epsilon_1, \epsilon_2, \ldots, \epsilon_n) \le \beta(\tilde{\epsilon}_1, \tilde{\epsilon}_2, \ldots, \tilde{\epsilon}_n)$ whenever $\epsilon_i \le \tilde{\epsilon}_i$ for all i, $1 \le i \le n$.

(c) Show that β is monotone if and only if β is a disjunction of monomials not involving complements.

(d) Using complementation and the De Morgan laws, derive the conjunctive normal-form theorem: every Boolean polynomial can be expressed as a conjunction of "join-monomials."

(e) Define free Boolean algebras and show that the free Boolean algebra on n generators is isomorphic to

$$\underline{2}^{\underline{2}^{\{1,2,\ldots,n\}}}.$$

1.3.10. *Lattices with unique complements.*[25]

Let x be an element in a lattice L. An element y in L is a *complement* of x if $x \vee y = \hat{1}$ and $x \wedge y = \hat{0}$. A lattice is *complemented* if every element in L

[25] An excellent exposition is in Saliĭ (1988). For the latest developments, see Grätzer (2007).

has a complement. A lattice L *has unique complements* if every element x in L has exactly one complement.

(a) Show that in a lattice with unique complements, if $x \leq y$, x' is a complement of x, and y' is a complement of y, then $y' \leq x'$.

(b) Let x be an element in a distributive lattice. Show that x has at most one complement.

It follows from (b) that a Boolean algebra has unique complements. The subject of lattices with unique complements began with an error of C.S. Pierce.[26] As reported by G. Birkhoff in the first edition of *Lattice Theory* (1940, p. 74), "It is curious that C.S. Pierce should have thought that every lattice is distributive. He even said that [the distributive axioms] are 'easily proved, but the proof is too tedious to give'! His error was demonstrated by Schröder [and] A. Korselt. Pierce at first gave way before these authorities, but later boldly defended his original view."

Pierce's property for a lattice is the following condition: if $x \not\leq y'$ for every complement y' of y, then $x \wedge y \neq \hat{0}$.

(c) Show that a complemented lattice L satisfying Pierce's property has unique complements.

(d) Prove *Pierce's theorem.* A complemented lattice L satisfying Pierce's property is distributive, and hence a Boolean algebra.

Pierce's theorem led naturally to *Huntington's conjecture:* a lattice with unique complements is distributive, and hence a Boolean algebra. There are several partial results.

(e) Show that a modular lattice L with unique complements is distributive.[27]

(f) Prove the *Birkhoff–Ward theorem.* An atomic lattice L with unique complements is a Boolean algebra.[28]

However, Huntington's conjecture is false in the strongest possible sense. This is a consequence of a famous theorem of Dilworth.

(g) Prove *Dilworth's theorem on lattices with unique complements.* Every lattice is a sublattice of a lattice with unique complements.[29]

[26] For a carefully documented and nuanced account, see Saliĭ (1988, pp. 36–38).

[27] This theorem has been attributed to many mathematicians. The published source is Birkhoff (1948).

[28] See Birkhoff and Ward (1939), McLaughlin (1956), and Ogasawara and Sasaki (1949).

[29] Dilworth (1945). Dilworth gave an account of how he approached and proved this theorem in *The Dilworth Theorems*, Bogart et al. (1990, pp. 39–40).

(h) Prove that the free lattice with unique complements generated by two unordered elements contains a sublattice generated by countable number of incomparable elements.

1.3.11. *Heyting algebras.*[30]

Let a and b be elements in a lattice. If the set $\{x : a \wedge x \leq b\}$ has a maximum, then the *pseudocomplement of* a *relative to* b, denoted $[a \to b]$, exists and is defined to be that maximum.

(a) Show that if $[a \to b]$ exists, then $a \wedge x \leq b$ if and only if $x \leq [a \to b]$.

(b) Show that if L is a Boolean algebra, then $[a \to b] = a^c \vee b$.

Pseudocomplements are generalizations of complements. In a Boolean algebra, $(x^c)^c = x$. This corresponds to the law of excluded middle in classical logic: for a proposition P, $P = \text{not}\,(\text{not}\, P)$. In many natural situations (such as intuitionistic logic and the axiomatics of forcing), one has a weaker law: P implies not (not P). Pseudocomplements capture the weaker law.

A *Heyting algebra* H is a lattice with a minimum $\hat{0}$ in which $[a \to b]$ exists for every pair a and b. Heyting algebras are to intuitionistic logic what Boolean algebras are to classical logic.

(c) Show that a Heyting algebra has a maximum and is distributive.

(d) Let X be a topological space with closure operator $C \mapsto \overline{C}$. Show that the open sets of X form a Heyting algebra under union and intersection, with pseudocomplement given by

$$[A \to B] = \text{Int}(A^c \cup B), \tag{I}$$

where the *interior* $\text{Int}(C)$ of a subset C is defined to be the open set $(\overline{C^c})^c$.

A *closure algebra* P is a Boolean algebra with a closure operator (see Section 1.5) satisfying the extra properties: $\overline{x \vee y} = \overline{x} \vee \overline{y}$ and $\overline{\hat{0}} = \hat{0}$. An element x in P is *open* if its complement x^c is closed; that is, $x^c = \overline{x^c}$. The subset of open elements in P forms a Heyting algebra under the partial ordering inherited from P with pseudocomplement given by Equation (I).

(e) Prove the McKinsey–Tarski theorem.[31] Every Heyting algebra can be represented as the lattice of open elements of a closure algebra.

1.4 Functions, Partitions, and Entropy

The number of functions from a set S of size n to a set X of size x is x^n. Let $f : S \to X$ be a function. If $A \subseteq S$, the *image* $f(A)$ of A is the subset

[30] For an excellent exposition, Balbes and Dwinger (1974). [31] McKinsey and Tarski (1944).

$\{f(a)\colon a \in A\}$ of X. If $x \in X$, the *inverse image* or *fiber* $f^{-1}(x)$ is the subset $\{a\colon f(a) = x\}$ of S. If $A \subseteq X$, then $f^{-1}(A)$ is the union $\bigcup_{x: x \in A} f^{-1}(x)$. The following relations hold: For images,

$$f(A \cup B) = f(A) \cup f(B)$$
$$f(A \cap B) \subseteq f(A) \cap f(B).$$

For inverse images,

$$f^{-1}(A \cup B) = f^{-1}(A) \cup f^{-1}(B)$$
$$f^{-1}(A \cap B) = f^{-1}(A) \cap f^{-1}(B)$$
$$f^{-1}(A^c) = (f^{-1}(A))^c.$$

In general, $f(A^c) \neq f(A)^c$, and no containment relation, either way, holds.

Noting that $f^{-1}(\emptyset) = \emptyset$ and $f^{-1}(X) = S$, the inverse image f^{-1}, as a function on subsets, preserves all the Boolean operations. Hence, the function $f : S \to X$ gives a Boolean morphism $\phi : 2^X \to 2^S$ defined by $\phi(Y) = f^{-1}(Y)$ for $Y \in 2^X$.

1.4.1. Theorem. Let S and X be finite sets. Then there is a bijection between Boolean morphisms $2^X \to 2^S$ and functions $S \to X$.

Proof. The set X is the set of atoms or one-element subsets in 2^X. Thus, from a Boolean morphism $\phi : 2^X \to 2^S$, we can define a function $f : S \to X$ by $f(a) = x$ whenever $a \in \phi(\{x\})$. □

Theorem 1.4.1 says that the category of finite sets and functions is contravariant to the category of finite Boolean algebras and Boolean morphisms.

A *permutation* on the set S is a one-to-one function from S onto itself. If $|S| = n$, then there are $n!$ permutations on S. If f and g are functions from S to itself, then f is *conjugate* to g if there exists a permutation γ such that

$$g = \gamma^{-1} f \gamma.$$

Given a function $f : S \to S$, construct the directed graph Graph(f) on the vertex set S with directed edges $(a, f(a))$, $a \in S$. Two functions are conjugate if and only if their directed graphs are isomorphic. The graph of a permutation γ on S is a disjoint union of directed cycles. Let $c_i(\gamma)$ be the number of cycles with i vertices (or equivalently, i edges) in Graph(γ). The *cycle structure* of γ is the n-tuple

$$(c_1(\gamma), c_2(\gamma), \ldots, c_n(\gamma)).$$

This n-tuple may be regarded as an integer partition of n. Two permutations are conjugate if and only if they have the same cycle structure.

A *partition* σ of a finite set S is a collection $\{B_1, B_2, \ldots, B_m\}$ of nonempty subsets or *blocks* such that $B_1 \cup B_2 \cup \cdots \cup B_m = S$ and $B_i \cap B_j = \emptyset$ whenever $i \neq j$. A partition defines an equivalence relation and, conversely, an equivalence relation defines a partition: the blocks of the partition correspond to the equivalence classes of the relation. The *type* of the partition is the integer partition of $|S|$ obtained by writing down the sizes $|B_i|$ of the blocks in nondecreasing order. Two partitions σ and τ have the same type if and only if there is a permutation p of S such that B is a block in σ implies that $p(B)$ is a block in τ.

Partitions are partially ordered by *reverse refinement*: $\tau \leq \sigma$ or τ is *finer* than σ if every block of τ is contained in some block of σ, or equivalently, for each block C of σ, there are blocks $B_{i_1}, B_{i_2}, \ldots, B_{i_k}$ of τ such that $C = B_{i_1} \cup B_{i_2} \cup \cdots \cup B_{i_k}$. In particular, τ covers σ if τ is obtained from σ by merging two blocks or σ is obtained from τ by splitting a block into two. If $E(\tau)$ is the equivalence relation defined by the partition τ, then $\tau \leq \sigma$ if and only if for all x and y in S, $x \sim_{E(\tau)} y$ implies $x \sim_{E(\sigma)} y$, that is, if $E(\tau) \subseteq E(\sigma)$ (as subsets of $S \times S$).

This partial order of reverse refinement is a lattice. The minimum is the partition in which every block is a one-element subset. The maximum is the partition with one single block, the entire set. The meet $\tau \wedge \sigma$ is the partition whose blocks are nonempty subsets of the form $B \cap C$, where B is a block of τ and C is a block of σ. The join $\tau \vee \sigma$ is the partition whose blocks are the equivalence classes of the transitive closure of the union of the equivalence relations defined by τ and σ.

If $f : S \to X$ is a function, then the relation $s \sim t$ if $f(s) = f(t)$ is an equivalence relation and those inverse images $f^{-1}(x)$ that are nonempty form a partition $\pi(f)$ of the set S. The partition $\pi(f)$ is the *coimage* or *kernel* of f. We can also think of a function $f : S \to X$ as a random variable on a sample space $(S, \Pr)$ with probability measure Pr, so that the blocks of the coimage $\pi(f)$ are the nonempty events defined by f.

The lattice of partitions plays for information the role that the Boolean algebra of subsets plays for size or probability. Indeed, the coimage $\pi(f)$ contains the "information" provided by f. To see this, consider the *Goblin coin-weighing problem.* We are given four labeled coins, some of which may be counterfeit. The counterfeit coins are lighter. There are 16 possible events, denoted by

$$\emptyset, 1, 2, 3, 4, 12, 13, 14, 23, 24, 34, 123, 124, 134, 234, 1234,$$

where 13, say, is the event that coins 1 and 3 are genuine. This gives a set S with 16 elements. Now suppose that we are also given a balance that will compare two coins and tell us if they have the same weight, the left coin is heavier, or the right coin is heavier.

Suppose that we compare coins 1 and 2 on the balance. If coins 1 and 2 weigh the same, then either they are both genuine or they are both counterfeit. In this case, the events ∅, 12, 123, 124, 1234, 3, 4, 34 remain possible. If coin 1 weighs more, then 1, 13, 14, 134 remain possible. If coin 2 weighs more, then 2, 23, 24, 234 remain possible. Thus, a comparison of coins 1 and 2 yields the following partition σ:

$$\{\emptyset, 12, 123, 124, 1234, 3, 4, 34\}, \{1, 13, 14, 134\}, \{2, 23, 24, 234\}.$$

The partition σ encodes the information given by comparing coins 1 and 2. We can also think of σ as the coimage of a random variable w_{12} on S taking three values. Let w_{ij} be the random variable defined by comparing coins i and j on the balance.

Next, suppose that we compare coins 2 and 3. Then each block in σ is further partitioned into smaller blocks. The block {∅, 12, 123, 124, 1234, 3, 4, 34} splits into

$$\{3, 34\}, \{\emptyset, 123, 1234, 4\}, \{12, 124\},$$

the block {1, 13, 14, 134} splits into

$$\{1, 14\}, \{13, 134\},$$

and the block {2, 23, 24, 234} splits into

$$\{2, 24\}, \{23, 234\},$$

yielding a partition τ finer than σ. The partition τ can also be obtained by taking the meet of the coimages $\pi(w_{12})$ and $\pi(w_{23})$. Thus, the result of any sequence of coin weighings is given by taking all possible meets of the coimages $\pi(w_{ij})$, $1 \le i, j \le 4$. The minimum partition obtainable is the partition with 1 two-element block {∅, 1234} and 14 one-element blocks. This corresponds to the fact that we cannot distinguish the case where all coins are genuine or counterfeit with a balance.

The coin-weighing example shows why Kolmogorov, one of the founders of modern probability theory, suggested calling the partition $\pi(f)$ of a random variable f the "experiment" of f. If $\pi(g)$ is finer than $\pi(f)$, then g contains all the information given by f. In statistical terminology, g is a *sufficient statistic* for f if $\pi(g) \le \pi(f)$. In addition, the meet $\pi(f) \wedge \pi(g)$ gives all "correlations" between f and g.

The information contained in a partition can be measured by entropy, as formulated by Claude Shannon.[32] The *discrete Shannon entropy* H is defined on partitions σ of finite sample spaces $(S, \Pr)$ by

$$H(\sigma) = \sum_{B:\, B \in \sigma} -\Pr(B) \log \Pr(B),$$

where log is logarithm to base 2 and $0 \log 0$ is interpreted as the right-hand limit $\lim_{x \to 0^+} x \log x$ and is defined to be 0. Since $0 \leq \Pr(B) \leq 1$, the entropy function is nonnegative. We choose binary logarithms (so that information is measured in bits). Other bases, such as e, can *naturally* be chosen. The entropy functions thus defined differ from each other by a nonzero constant factor. The entropy function depends only on the multiset $\{\Pr(B) \colon B \in \sigma\}$ of nonnegative real numbers. Thus, we can also think of the entropy function H as the real-valued function defined on finite multisets $\{p_1, p_2, \ldots, p_n\}$ of nonnegative numbers such that $p_1 + p_2 + \cdots + p_n = 1$ by

$$H(p_1, p_2, \ldots, p_n) = \sum_{i=1}^{n} -p_i \log p_i,$$

For example, consider tossing a biased coin, with probability p of a head. Then a toss partitions a sample space into two events B_0, the event that a head occurs, and B_1, the event that a tail occurs. Then

$$H(B_0, B_1) = -p \log p - (1 - p) \log(1 - p).$$

If $p = 1$, then a head always occurs and no information is obtained by tossing the coin; as expected, $H(B_0, B_1) = 0$. If $p = 0.9$, say, then the toss yields somewhat more information, quantified by $H(B_0, B_1) \approx 0.47$. If $p = 0.5$, the case of a fair coin, then $H(B_0, B_1) = 1$. That is, one bit of information is obtained by tossing a fair coin. This is the maximum information obtainable by a coin toss, since we know which of the two equally likely events actually happened. The amount of information then decreases as p decreases to 0. In particular, if one were to obtain information by asking a question with a yes-or-no answer, then the maximum information is obtained when one asks a question to which it is equally likely that the answer is yes or no. This is a special case of the following "equipartition-maximizes-information" lemma.

1.4.2. Lemma. Suppose that the partition π has n blocks. Then

$$H(\pi) \leq \log n,$$

with equality if and only if the blocks in π have the same probability $\frac{1}{n}$.

[32] Shannon (1948).

Proof. Changing the base of the logarithm changes H and log by the same nonzero constant. Hence, we may use natural logarithms to base e. We shall use an elementary inequality from calculus: if $x > 0$, $\log x \leq x - 1$, with equality if and only if $x = 1$. (Indeed, the function $\log x - x + 1$ has a unique maximum at $x = 1$ on the positive real axis.)

Let $B_1, B_2, \ldots, B_n$ be the blocks of π. Then,

$$\begin{aligned} H(\pi) - \log n &= \sum_{i=1}^{n} -\Pr(B_i)[\log \Pr(B_i) + \log n] \\ &= \sum_{i=1}^{n} -\Pr(B_i) \log(\Pr(B_i)n) \\ &= \sum_{i=1}^{n} \Pr(B_i) \log\left(\frac{1}{\Pr(B_i)n}\right) \\ &\leq \sum_{i=1}^{n} \Pr(B_i)\left(\frac{1}{(\Pr(B_i)n} - 1\right) \\ &= \sum_{i=1}^{n} \frac{1}{n} - \sum_{i=1}^{n} \Pr(B_i) \\ &= 0, \end{aligned}$$

where on the third line, we use the inequality $\log x \leq x - 1$. Equality holds when for all i, $\Pr(B_i)n = 1$; that is, $\Pr(B_i) = \frac{1}{n}$. □

The partitions σ and τ are (*stochastically*) *independent* if

$$\Pr(B \cap C) = \Pr(B)\Pr(C)$$

for any two blocks $B \in \sigma$ and $C \in \tau$. As a typical example, let $p_1, p_2, \ldots, p_m$ and $q_1, q_2, \ldots, q_n$ be nonnegative real numbers such that $p_1 + p_2 + \cdots + p_m = 1$ and $q_1 + q_2 + \cdots + q_n = 1$. Let $S = \{1, 2, \ldots, m\} \times \{1, 2, \ldots, n\}$ be the sample space in which element (i, j) has probability $p_i q_j$. Let σ be the partition into "rows,"

$$\begin{gathered} \{(1,1), (1,2), \ldots, (1,n)\}, \\ \{(2,1), (2,2), \ldots, (2,n)\}, \\ \vdots \\ \{(m,1), (m,2), \ldots, (m,n)\}, \end{gathered}$$

and τ be the partition into "columns,"

$$\begin{gathered}\{(1,1),(2,1),\ldots,(m,1)\},\\ \{(1,2),(2,2),\ldots,(m,2)\},\\ \vdots\\ \{(1,n),(2,n),\ldots,(m,n)\}.\end{gathered}$$

Then σ and τ are independent. Conversely, given two independent partitions σ and τ, we can display S as a rectangular "chessboard," with rows the blocks of σ, columns the blocks of τ, and squares the blocks of the meet $\sigma \wedge \tau$. Put another way, independent partitions arise from the rows and columns of a Cartesian product.

1.4.3. The chessboard construction. Let $p_1, p_2, \ldots, p_m$ and $q_1, q_2, \ldots, q_n$ be nonnegative real numbers such that $p_1 + p_2 + \cdots + p_m = 1$ and $q_1 + q_2 + \cdots + q_n = 1$. Then there exists a finite sample space $(S, \Pr)$ and partitions σ and τ such that as multisets $\{\Pr(B)\colon B \in \sigma\} = \{p_1, p_2, \ldots, p_m\}$, $\{\Pr(C)\colon C \in \tau\} = \{q_1, q_2, \ldots, q_n\}$, and σ and τ are independent.

The next result says that the information given by independent partitions is additive.

1.4.4. Theorem. Let σ and τ be independent partitions. Then

$$H(\tau \wedge \sigma) = H(\tau) + H(\sigma).$$

Proof. This follows from the following computation:

$$\begin{aligned} H(\sigma \wedge \tau) &= -\sum_{B,C\colon B\in\sigma,\, C\in\tau} \Pr(B\cap C)\log\Pr(B\cap C)\\ &= -\sum_{B,C\colon B\in\sigma,\, C\in\tau} \Pr(B)\Pr(C)[\log\Pr(B)+\log\Pr(C)]\\ &= -\left[\sum_{B\colon B\in\sigma}\Pr(B)\right]\left[\sum_{C\colon C\in\tau}\Pr(C)\log\Pr(C)\right]\\ &\quad -\left[\sum_{C\colon C\in\tau}\Pr(C)\right]\left[\sum_{B\colon B\in\sigma}\Pr(B)\log\Pr(B)\right]\\ &= H(\sigma)+H(\tau), \end{aligned}$$

using the fact that $\sum_{B\colon B\in\sigma}\Pr(B) = \sum_{C\colon C\in\tau}\Pr(C) = 1$. □

A consequence of Theorem 1.4.4 and the chessboard construction is the following proposition.

1.4.5. Proposition. Let $p_1, p_2, \ldots, p_m$ and $q_1, q_2, \ldots, q_n$ be nonnegative real numbers such that $p_1 + p_2 + \cdots + p_m = 1$ and $q_1 + q_2 + \cdots + q_n = 1$. Then

$$\begin{aligned} &H(p_1q_1, p_1q_2, \ldots, p_iq_j, \ldots, p_mq_n) \\ &\quad = H(p_1, p_2, \ldots, p_m) + H(q_1, q_2, \ldots, q_n). \end{aligned}$$

The notion of conditional entropy allows us to generalize Lemma 1.4.4 to pairs of partitions that are not independent. If τ is finer than σ and B is a block of σ, let $H(\tau|B)$ be the entropy of the partition on the sample space B (with normalized probability measure $\frac{1}{\Pr(B)}\Pr$) given by the blocks of τ contained in B. Explicitly,

$$H(\tau|B) = \sum_{C:\, C\in\tau, C\subseteq B} -\frac{\Pr(C)}{\Pr(B)} \log \frac{\Pr(C)}{\Pr(B)}.$$

The *conditional entropy* $H(\tau|\sigma)$ *of the partition* τ *given the partition* σ is defined by

$$H(\tau|\sigma) = \sum_{B:\, B\in\sigma} \Pr(B)H(\tau|B).$$

In other words, $H(\tau|\sigma)$ is the expected increase of entropy when one refines σ into τ. There is a simpler formula for the expected increase in entropy: $H(\tau) - H(\sigma)$. These two quantities are equal.

1.4.6. Lemma. If τ is finer than σ,

$$H(\tau|\sigma) = H(\tau) - H(\sigma).$$

In particular, entropy increases under refinement.

Proof. The formula follows from a simple calculation. The second assertion follows from the fact that $H(\tau|\sigma) \geq 0$. □

1.4.7. Corollary. If τ and σ are partitions, then

$$H(\tau \wedge \sigma) = H(\tau) + H(\tau \wedge \sigma|\tau) = H(\sigma) + H(\tau \wedge \sigma|\sigma).$$

If τ and σ are independent, then

$$H(\tau \wedge \sigma|\sigma) = H(\tau).$$

1.4.8. Theorem. If τ and σ are partitions,

$$H(\tau \wedge \sigma|\sigma) \leq H(\tau).$$

Proof. We shall need Jensen's inequality: let $f(x)$ be concave on the interval $[a, b]$, $a \leq x_1 \leq x_2 \leq \cdots \leq x_n \leq b$, and $\lambda_1, \lambda_2, \ldots, \lambda_n$ be nonnegative real numbers satisfying $\lambda_1 + \lambda_2 + \cdots + \lambda_n = 1$. Then

$$\sum_{i=1}^{n} \lambda_i f(x_i) \le f\left(\sum_{i=1}^{n} \lambda_i x_i\right).$$

We begin by changing the order of summation in the definition of conditional entropy:

$$\begin{aligned} H(\tau \wedge \sigma|\sigma) &= -\sum_{B:\,B\in\sigma} \Pr(B)\left[\sum_{C:\,C\in\tau} \frac{\Pr(B\cap C)}{\Pr(B)} \log \frac{\Pr(B\cap C)}{\Pr(B)}\right] \\ &= -\sum_{C:\,C\in\tau}\left[\sum_{B:\,B\in\sigma} \Pr(B)\frac{\Pr(B\cap C)}{\Pr(B)} \log \frac{\Pr(B\cap C)}{\Pr(B)}\right]. \end{aligned}$$

Since $f(x) = -x\log x$ is concave when $x \ge 0$, we can apply Jensen's inequality to the inner sums, with $\lambda_B = \Pr(B)$, $x_B = \Pr(B\cap C)/\Pr(B)$, so that $\sum_B \lambda_B x_B = \sum_B \Pr(B\cap C) = \Pr(C)$. We obtain, for a block C in σ,

$$-\sum_{B:\,B\in\sigma} \Pr(B)\frac{\Pr(B\cap C)}{\Pr(B)} \log \frac{\Pr(B\cap C)}{\Pr(B)} \le -\Pr(C)\log\Pr(C).$$

From this, we conclude that

$$H(\tau\wedge\sigma|\sigma) \le -\sum_{C:\,C\in\tau} \Pr(C)\log\Pr(C) = H(\tau). \qquad \square$$

Theorem 1.4.8 and Corollary 1.4.7 imply

$$H(\tau\wedge\sigma) \le H(\sigma) + H(\tau).$$

Thus, we have the following inequality.

1.4.9. Corollary. If $\tau_1, \tau_2, \ldots, \tau_n$ are partitions of a finite sample space, then

$$H\left(\bigwedge_{i=1}^{n} \tau_i\right) \le \sum_{i=1}^{n} H(\tau_i).$$

When $\bigwedge_{i=1}^{n} \tau_i$ is the minimum partition $\hat{0}$ into single-element blocks, then Corollary 1.4.9 yields the "information-theoretic lower bound." For example, if one wishes to locate a point in a finite sample space of size n using a set of random variables $w_1, w_2, \ldots, w_r$, then it is necessary that

$$H(\pi(w_1)) + H(\pi(w_2)) + \cdots + H(\pi(w_r)) \ge H(\hat{0}) = \log n.$$

In particular, if each random variable w_i takes two possible values, then at least $\log n$ random variables are needed.

The entropy function can be characterized by axioms. One set of axioms, in which conditional entropy plays the starring role, is the following:

En1. H is continuous and is not identically zero.
En2. $H(p_1, p_2, \ldots, p_n, 0) = H(p_1, p_2, \ldots, p_n)$, that is, H depends only on the nonzero numbers in the multiset.
En3. $H(p_1, p_2, \ldots, p_n) \le H(\frac{1}{n}, \frac{1}{n}, \ldots, \frac{1}{n})$.
En4. $H(\tau|\sigma) = H(\tau) - H(\sigma)$.

1.4.10. Theorem. If H is a function satisfying axioms En1, En2, En3, and En4, then there is a positive constant C such that

$$H(p_1, p_2, \ldots, p_n) = -C \sum_{i=1}^{n} p_i \log p_i.$$

The axioms are chosen so that the proof requires a minimum of analysis. We begin with the following lemma.

1.4.11. Lemma. If n is a positive integer, let $f(n) = H(\frac{1}{n}, \frac{1}{n}, \ldots, \frac{1}{n})$. Then

$$f(n) = -C \log(1/n),$$

where C is a positive constant.

Proof. By En4,

$$H(\{S\}) = H(\{S\}|\{S\}) = H(\{S\}) - H(\{S\}) = 0,$$

and hence, $f(1) = H(\{S\}) = 0$.

By En2 and En3,

$$\begin{aligned} f(n) = H\left(\frac{1}{n}, \frac{1}{n}, \ldots, \frac{1}{n}\right) &= H\left(\frac{1}{n}, \frac{1}{n}, \ldots, \frac{1}{n}, 0\right) \\ \le H\left(\frac{1}{n+1}, \frac{1}{n+1}, \ldots, \frac{1}{n+1}\right) &= f(n+1). \end{aligned}$$

Hence, $f(n)$ is increasing. In particular, since $f(1) = 0$, $f(2) \ge 0$.

Next, let σ be a partition of a sufficiently large sample space into n^{k-1} blocks, each having the same probability $1/n^{k-1}$. Let τ be the refinement obtained by partitioning each block of σ into n blocks, each with probability $1/n^k$. If B is a block of σ, $H(\tau|B) = f(n)$ and hence, $H(\tau|\sigma) = f(n)$. By En4,

$$f(n) = H(\tau|\sigma) = H(\tau) - H(\sigma) = f(n^k) - f(n^{k-1}).$$

Applying this k times, we obtain the functional equation

$$f(n^k) = kf(n).$$

When $n = 2$, we have $f(2^k) = kf(2)$. If $f(2) = 0$, then $f(2^k) = 0$ for all k. Since f is increasing, this would imply that $f(n)$ is identically zero. We conclude that $f(2) > 0$.

Fix two positive integers n and k. As the function 2^x increases to infinity, there is an integer b such that $2^b \leq n^k \leq 2^{b+1}$. Since $f(n)$ is increasing,

$$f(2^b) \leq f(n^k) \leq f(2^{b+1}).$$

Applying the functional equation, we obtain

$$bf(2) \leq kf(n) \leq (b+1)f(2).$$

Dividing by $kf(2)$, this yields

$$\frac{b}{k} \leq \frac{f(n)}{f(2)} \leq \frac{b+1}{k}.$$

The function $\log n$ satisfies a similar inequality. As log is increasing,

$$\log 2^b \leq \log n^k \leq \log 2^{b+1}.$$

Hence,

$$\frac{b}{k} \leq \log n \leq \frac{b+1}{k}.$$

From the two inequalities, both $f(n)/f(2)$ and $\log n$ are in the interval $[\frac{b}{k}, \frac{b+1}{k}]$, that is to say,

$$|f(n)/f(2) - \log n| \leq \frac{1}{k}$$

for any positive integer k. By En1, f is continuous, and hence,

$$f(n) = f(2)\log(n) = -f(2)\log(1/n),$$

where, as observed earlier, $f(2) > 0$. □

Consider next a multiset $\{p_1, p_2, \ldots, p_n\}$ of positive rational numbers such that $p_1 + p_2 + \cdots + p_n = 1$. Let N be the least common multiple of their denominators and write $p_i = a_i/N$, where a_i are positive integers and $a_1 + a_2 + \cdots + a_n = N$. Let σ be a partition of a sample space S into n blocks $B_1, B_2, \ldots, B_n$ so that $\Pr(B_i) = p_i$. Let τ be the refinement of σ obtained by dividing B_i into a_i blocks, each with the same probability $1/N$. In particular, τ partitions S into N blocks, each with probability $1/N$. Observing that $H(\tau|B_i) = H(1/a_i, 1/a_i, \ldots, 1/a_i) = f(a_i)$,

$$H(\tau|\sigma) = \sum_{i=1}^{n} p_i H(\tau|B_i) = \sum_{i=1}^{n} p_i f(a_i) = -C \sum_{i=1}^{n} p_i \log(a_i).$$

On the other hand,

$$H(\tau) - H(\sigma) = f(N) - H(\sigma) = -C \log(N) - H(\sigma).$$

We conclude, by En4, that

$$\begin{aligned} H(\sigma) &= -C \log(N) + C \sum_{i=1}^{n} p_i \log(a_i) \\ &= C \sum_{i=1}^{n} p_i [\log(a_i) - \log(N)] \\ &= C \sum_{i=1}^{n} p_i \log(a_i/N) \\ &= C \sum_{i=1}^{n} p_i \log(p_i). \end{aligned}$$

Since the rational numbers are dense, the theorem now follows from En1, that H is continuous. □

We end with a combinatorial interpretation of the entropy function. This interpretation was the motivation for Boltzmann's definition of entropy.[33]

1.4.12. Proposition. Let $n, n_1, n_2, \ldots, n_r$ be nonnegative integers such that $n_1 + n_2 + \cdots + n_r = n$. Then if n is sufficiently large,

$$nH\left(\frac{n_1}{n}, \frac{n_2}{n}, \ldots, \frac{n_r}{n}\right) \approx \log \binom{n}{n_1, n_2, \ldots, n_r},$$

where

$$\binom{n}{n_1, n_2, \ldots, n_r} = \frac{n!}{n_1! n_2! \cdots n_r!},$$

the number of ways to distribute n distinguished balls into r distinguished boxes so that the i box has n_i balls.

Proof. Use Stirling's approximation: for large n, $\log n! \approx n \log n - n$. □

[33] See Cercignani (1998, p. 121).

Our brief account of Shannon entropy is intended to illustrate how partitions, information, and coimages of functions are "dual notions" to subsets, size, and images of functions. To a very limited extent, Shannon entropy has been extended to σ-subalgebras, the continuous analogs of partitions. There is another concept of entropy, Boltzmann entropy, which is based on probability densities (see Exercise 1.4.6 for a glimpse). However, "an abyss of ill-understood ideas separates Shannon entropy from Boltzmann entropy. No one to this day has seen through the dilemma of choosing between entropy as a measure of randomness of probability densities of random variables and entropy as a measure of randomness of σ-subalgebras." There is much rethinking to be done here.[34]

Exercises

1.4.1. *Functions and finiteness.*

Assuming that S is finite, show that a function $f : S \to S$ is injective (or one to one) if and only if it is surjective (or onto).

1.4.2. Show that a function preserving $\vee$ and $\cdot^c$ is a Boolean morphism.

1.4.3 *Theorems of Whitman and Pudlák–Tùma.*[35]

Whitman showed in 1946 that every lattice is isomorphic to a sublattice of the lattice of partitions of an infinite set. Jónsson refined this theorem in 1953.

Let L be a lattice. Then an element a in L can be represented as the principal ideal $I(a)$, where $I(a) = \{x\colon x \leq a\}$. Thinking of the ideal $I(a)$ as the partition of the set L into the blocks $I(a)$ and $\{y\}$, $y \not\leq a$, we have a meet-preserving representation of elements of L as partitions.

It is much harder to construct an injection preserving both meets and joins. One way to do this is to consider equivalence relations defined by distances. If A is a set and L is a lattice with a minimum $\hat{0}$, then an *L-valued distance* on A is a function $\delta : A \times A \to L$ satisfying three properties:

Symmetry: $\delta(x, y) = \delta(y, x)$.

Normalization: $\delta(x, x) = \hat{0}$ for all $x \in L$.

The triangle inequality: $\delta(x, y) \vee \delta(y, z) \geq \delta(x, z)$.

If δ is a distance and a is a fixed element of L, then the relation $E(\delta, a)$

$$x \sim y \text{ whenever } \delta(x, y) \leq a$$

[34] See problem 4 in *Twelve problems*. Among the many books on information theory, we single out Rényi (1984).

[35] Jónsson (1953), Pudlák and Tùma (1980), and Whitman (1946). See Grätzer (2002, pp. 250–257) for an exposition.

is an equivalence relation. If $A = L$, then $\delta(x, y) = x \vee y$ if $x \neq y$ and $\delta(x, x) = \hat{0}$ is a distance. In this case, $E(\delta, a)$ defines the partition with blocks $I(a)$ and $\{y\}$, $y \not\leq a$, given earlier.

Using cleverer distances, Jónsson constructed a representation of a lattice by partitions preserving both meets and joins, giving a natural proof of Whitman's theorem. (The join in a partition lattice is calculated by taking the transitive closure of a union of equivalence relations; thus, a countable number of unions may be necessary to calculate the join and the partition lattice may not be finite when L is finite.)

Whitman's theorem. Every lattice is a sublattice of a partition lattice.

Whitman's theorem also implies that if an identity holds for all partition lattices, then it holds for all lattices. Thus, one cannot characterize partition lattices by identities.

Whitman asked whether every finite lattice is a sublattice of a finite partition lattice. Whitman's conjecture was proved by Pudlák and Tůma.

The Pudlák–Tůma theorem. Every finite lattice is a sublattice of a finite partition lattice.

1.4.5 *Hartmanis n-partitions.*[36]

Let S be a finite set. Partially order the antichains in 2^S by $\mathcal{A}_1 \leq \mathcal{A}_2$ if for each subset A_1 in $\mathcal{A}_1$, there exists a subset A_2 in $\mathcal{A}_2$ such that $A_1 \subseteq A_2$. Let $\mathcal{B}$ be an antichain in 2^S. An antichain $\mathcal{A}$ *partitions* $\mathcal{B}$ if $\mathcal{A} \geq \mathcal{B}$ and every subset B in $\mathcal{B}$ is contained in exactly one subset of $\mathcal{A}$. The subsets $\{B \in \mathcal{B}: B \subseteq A\}$, where A ranges over all $A \in \mathcal{A}$, are blocks of a partition of $\mathcal{B}$. Let $\Pi(\mathcal{B}, S)$ be the set of antichains partitioning $\mathcal{B}$ partially ordered as earlier. The partial order $\Pi(\mathcal{B}, S)$ has minimum $\mathcal{B}$ and maximum the one-element antichain $\{S\}$.

(a) Show that the partial order $\Pi(\mathcal{B}, S)$ is a lattice.

An *n-partition* σ of S is an antichain partitioning the collection of all size-n subsets of S; that is, σ is a collection of subsets of S of size at least n such that every n-subset of S is contained in exactly one subset in σ. Let $\Pi(n; S)$ be the lattice of all n-partitions of S.

(b) Show that $\Pi(n; S)$ is isomorphic to a sublattice of $\Pi(n+1; S \cup \{a\})$, where a is a new element not in S.

(c) Show that $\Pi(n; S)$ is complemented.

[36] Crapo (1970) and Hartmanis (1956, 1957).

Consider a fixed 2-partition on a finite set S. We can think of the elements of S as *points* and the blocks of the 2-partition as *lines* of a *line geometry.* The defining property of a 2-partition translates into the axiom: *two points determine a unique line.* A subset of points A is *line-closed* if whenever two points a and b of a line ℓ are in A, then the line ℓ is contained in A. The line-closed sets are closed under intersection and they form a lattice LG(S). (A rank-3 simple matroid is a line geometry. However, line-closed subsets are not necessarily closed under the matroid closure.)

Motivated by Whitman's problem, Hartmanis showed two sublattice-embedding theorems for lattices of line-closed subsets and 2-partitions.

(d) Show that every finite atomic lattice L is isomorphic to the lattice of line-closed sets of a finite line geometry.

(e) Show that every lattice of line-closed sets is isomorphic to a sublattice of $\Pi_2(T)$ for some (larger) set T. In particular, every finite atomic lattice is isomorphic to a sublattice of a lattice of 2-partitions of a finite set.

(f) (Research problem) If one can construct a sublattice-embedding of a lattice of 2-partitions of a finite set into a lattice of partitions of a finite set, then one has another proof of the Pudlák–Tůma theorem. Find such a construction.

1.4.6. (Research problem) There is no information or entropy interpretation of the join of two partitions. Find an interesting one.

1.4.7. Let G be a finite group and $L(G)$ the lattice of subgroups of G.

(a) Show that the function defined from $L(G)$ to the lattice of partitions of the (underlying) set G sending a subgroup H to the partition $\pi(H)$ of G by the cosets of H is a lattice homomorphism: that is, if H and K are subgroups of G, then $\pi(H \cap K) = \pi(H) \wedge \pi(K)$ and $\pi(H \vee K) = \pi(H) \vee \pi(K)$.

(b) Let $HK = \{hk\colon h \in H, k \in K\}$. Show that the partitions $\pi(H)$ and $\pi(K)$ are independent if $HK = KH$ and $|HK| = |G|$.

1.4.8. *Boltzmann entropy.*

If X is a continuous random variable with density f, the *Boltzmann entropy* $H(X)$ is defined by

$$H(X) = \int_{-\infty}^{\infty} -f(x)\log f(x)\,dx,$$

where log is natural logarithm. Boltzmann entropy has many applications. We will give only one: the characterization of probability distributions by a maximum entropy condition.

(a) Let $X : S \to [a, b]$ be a continuous random variable. Then $H(X)$ attains its maximum value when X has the uniform distribution $[a, b]$. The maximum value is $\log(b - a)$.

(b) Let $X : S \to [0, \infty)$ be a continuous random variable with mean $1/\lambda$. Then $H(X)$ attains its maximum value when X is exponentially distributed. The maximum value is $1 - \log \lambda$.

(c) Let $X : S \to \mathbb{R}$ be a continuous random variable with variance s^2. Then $H(X)$ attains its maximum value when X is normally distributed. The maximum value is $\frac{1}{2} + \log s\sqrt{2\pi}$.

(d) Find a random variable X so that $H(X) < 0$. To a philosophical mind, the existence of such random variables causes some uneasiness.

1.4.9. *Lattices of Boolean σ-subalgebras.*

Let S be a finite set. Show that there is a bijection between the Boolean subalgebras of 2^S and partitions of S.

This easy exercise indicates that a continuous analog of the partition lattice is the lattice Σ of all σ-subalgebras of a given σ-algebra Π. The partial order is given by $\tau \le \tau'$ if $\tau \supseteq \tau'$ as families of sets. The join is the set-theoretic intersection of the two families and the meet is the σ-subalgebra generated by τ and τ'. (See Rota's *Twelve problems* for research problems about lattices of σ-subalgebras.)

1.4.10. *A universal entropy group.*

In *Twelve problems,* Rota constructed a universal entropy group. Using Proposition 1.4.5, we can define a finite analog of this group. Let F be the free Abelian group generated by all finite multisets $\{r_1, r_2, \ldots, r_k\}$ of nonnegative real numbers such that $r_1 + r_2 + \cdots + r_k = 1$, and let R be the subgroup generated by elements of the form

$$\{p_1q_1, p_1q_2, \ldots, p_iq_j, \ldots, p_mq_n\} - \{p_1, p_2, \ldots, p_m\} - \{q_1, q_2, \ldots, q_n\}.$$

The quotient group F/R satisfies the following *universal property:* let $\mathcal{P}$ be the collection of pairs $(\sigma, (S, \Pr))$, where σ is a partition of a finite probability space $(S, \Pr)$. Let $H : \mathcal{P} \to \mathbb{A}$ be a function taking values in an additive Abelian group $\mathbb{A}$ satisfying two conditions:

(a) The value $H(\{B_1, B_2, \ldots, B_k\})$ depends only on the multiset $\{\Pr(B_1), \Pr(B_2), \ldots, \Pr(B_k)\}$.

(b) If σ and τ are independent partitions, then $H(\sigma \wedge \tau) = H(\sigma) + H(\tau)$.

Then H factors through F/R, that is, there is an Abelian group homomorphism $h : F/R \to \mathbb{A}$ such that $H = h \circ \iota$, where $\iota : \mathcal{P} \to F/R$ is the function sending $(\{B_1, B_2, \ldots, B_k\}, (S, \Pr))$ to $\{\Pr(B_1), \Pr(B_2), \ldots, \Pr(B_k)\}$. Put another way,

F/R is a "Grothendieck group" containing all possible functions H satisfying the two conditions. Study the group F/R.

1.4.11. (Research problem) Proposition 1.4.12 suggests several questions: Is there an entropy interpretation of Sperner's theorem (or Mešalkin's extension)? More generally, is there a role for entropy in extremal set theory? Are there q-analogs of entropy?

1.5 Relations

A *relation* R from the set S and X is a subset of the Cartesian product $S \times X$. As with functions, S is the *domain* and X the *codomain* of R. Relations were defined by Augustus De Morgan.[37]

There are many ways to think about relations. One way is as a bipartite graph. For example, consider the relation

$$R = \{(a,t),(a,u),(a,v),(b,s),(b,t),(c,s),(c,u),(c,v)\}$$

on $\{a,b,c\}$ and $\{s,t,u,v\}$. Then R defines a (directed) bipartite graph on the vertex set $\{a,b,c\} \cup \{s,t,u,v\}$ with edge set equal to R. In particular, we will often refer to an ordered pair in R as an *edge* of R.

Another way is to *picture* the Cartesian product $S \times T$ as a rectangle and mark, in some distinctive way, those ordered pairs in the relation R. For example, the relation R_1 has the picture

$$\begin{array}{ccccc} & s & t & u & v \\ a & & * & * & * \\ b & * & * & & \\ c & * & & * & * \end{array}$$

Replacing asterisks by a nonzero number and blank spaces by 0, we obtain a matrix *supported by* the relation R.

Let $R : S \to X$ be a relation. If $a \in S$, let

$$R(a) = \{x\colon (a,x) \in R\}.$$

In analogy, if $A \subseteq S$, then let

$$R(A) = \bigcup_{a:a \in A} R(a) = \{x\colon (a,x) \in R \text{ for some } a \in A\}.$$

[37] De Morgan (1864). It was further developed by C.S. Pierce, E. Schröder, and others. A succinct history can be found in the book by Maddux (2006).

The labeled list $R(a)$, $a \in S$, determines the relation R. If we forget the labels, then R is determined up to a permutation of S. For a similar result, see Lemma 2.5.2. Two relations $R : S \to X$ and $Q : T \to Y$ are *isomorphic* if there exist bijections $\sigma : S \to T$ and $\tau : X \to Y$ such that

$$(a, x) \in R \text{ if and only if } (\sigma(a), \tau(x)) \in Q.$$

For subsets A and B of S,

$$R(A \cup B) = R(A) \cup R(B),$$
$$R(A \cap B) \subseteq R(A) \cap R(B).$$

A function $h : P \to Q$ from the Boolean algebra P to the Boolean algebra Q is a *hemimorphism* if $h(\emptyset) = \emptyset$ and

$$h(A \cup B) = h(A) \cup h(B).$$

A relation R defines a hemimorphism $h_R : 2^S \to 2^X$ in the following way:

$$h_R(A) = R(A).$$

1.5.1. Theorem. Let $R : S \to X$ be a relation between finite sets S or X. The construction h_R gives a bijection between the set of relations $S \to X$ and the set of hemimorphisms $2^S \to 2^X$.

Proof. Let $h : 2^S \to 2^X$ be a given hemimorphism. Define the relation $R : S \to X$ by $R(a) = h(a)$. Then as h is a hemimorphism,

$$h_R(A) = R(A) = \bigcup_{a: a \in A} h(a) = h(A)$$

for all $A \subseteq S$; that is, $h = h_R$. Thus, the construction $h \mapsto R$ is the inverse of the construction $R \mapsto h_R$. □

We shall discuss next Galois connections and closure operators.[38] Let P be a partially ordered set. A function $P \to P, x \mapsto \overline{x}$ is a *closure operator* if

CL1. $x \le \overline{x}$.
CL2. $x \le y$ implies $\overline{x} \le \overline{y}$.
CL3. $\overline{x} = \overline{\overline{x}}$.

[38] Everett (1944) and Ore (1944).

A *coclosure operator* is a closure operator on the order-dual $P^{\downarrow}$. An element x is *closed* if $x = \overline{x}$. The *quotient*, denoted $P/\text{-}$ or $\mathrm{Cl}(P)$, is the subset of closed elements, with the partial order induced from P.

The next result characterizes closure operations in finite lattices.

1.5.2. Proposition. A nonempty subset C in a finite lattice L is the set of closed elements of a closure operator on L if and only if C is closed under meets.

Proof. Let $x \mapsto \overline{x}$ be a closure operator on L. Suppose x and y are closed elements. Since $\overline{x \wedge y} \leq \overline{x} = x$, $\overline{x \wedge y} \leq \overline{y} = y$, and $x \wedge y \leq \overline{x \wedge y}$, it follows that $x \wedge y = \overline{x \wedge y}$. Thus, the set of closed elements is closed under meets. Conversely, given a set C closed under meets, define

$$\overline{x} = \bigwedge \{c\colon c \in C,\ c \geq a\}.$$

It is routine to show that this defines a closure operator. □

The set of all closure operators on a partially ordered set P can be partially ordered in the following way: if $x \mapsto \eta(x)$ and $x \mapsto \nu(x)$ are two closure operators on P, then $\eta \leq \nu$ if for all $x \in P$, $\eta(x) \leq \nu(x)$. Not much is known about partial orders of closure operators.

One source of closure operators is Galois connections. Galois connections were motivated by the Galois correspondence between the partially ordered sets of normal extensions of a field and their automorphism groups. This correspondence is order-reversing: the larger the extension, the smaller the automorphism group fixing it.

Let P and Q be partially ordered sets. A *Galois connection* between P and Q is a pair of functions $\varphi : P \to Q$ and $\psi : Q \to P$ satisfying the following axioms:

GC1. Both φ and ψ are order-reversing.
GC2. For $x \in P$, $\psi(\varphi(x)) \geq x$, and for $y \in Q$, $\varphi(\psi(y)) \geq y$.

A Galois coconnection between P and Q is a Galois connection between P and the order-dual $Q^{\downarrow}$. Explicitly, a *Galois coconnection* between P and Q is a pair of functions $\varphi : P \to Q$ and $\psi : Q \to P$ satisfying the following axioms:

GCC1. Both φ and ψ are order-preserving.
GCC2. For $x \in P$, $\psi(\varphi(x)) \geq x$, and for $y \in Q$, $\varphi(\psi(y)) \leq y$.

When φ and ψ is a Galois coconnection, the function ψ is sometimes called the *residual* of φ.

1.5.3. Proposition. Let $\varphi : P \to Q$ and $\psi : Q \to P$ be a Galois connection. Then

(a) $\varphi\psi\varphi = \varphi$ and $\psi\varphi\psi = \psi$.

(b) $\psi\varphi : P \to P$ is a closure operator on P and $\varphi\psi : Q \to Q$ is a closure operator on the order dual $Q^{\downarrow}$.

(c) The quotients $P/\psi\varphi$ and $Q/\varphi\psi$ of closed elements are anti-isomorphic; indeed, ϕ and ψ, restricted to $P/\psi\varphi$ and $Q/\varphi\psi$, are inverses of each other, and in particular they are order-reversing bijections.

Proof. Let $x \in P$. Then, using GC2, $\psi\varphi(x) \geq x$, and using GC1, $\varphi\psi\varphi(x) \leq \varphi(x)$. On the other hand, using GC2 applied to $\varphi(x)$, $\varphi\psi\varphi(x) \geq \varphi(x)$. We conclude that $\varphi\psi\varphi(x) = \varphi(x)$. Next, we show that $\psi\varphi$ is a closure operator on P. CL1 follows from GC2. Next, if $x \leq y$, then $\varphi(x) \geq \varphi(y)$ and, in turn, $\psi\varphi(x) \leq \psi\varphi(y)$, Finally, CL3 follows from (a). The other assertions in (a) and (b) can be proved in the same way.

If x is closed in P, then $x = \psi\varphi(x)$. Hence, when restricted to the quotients $P/\psi\varphi$ and $Q/\varphi\psi$, φ and ψ are inverses of each other. □

Galois connections may be regarded as abstractions of relations. Let $R : S \to X$ be a relation. If $A \subseteq S$, let

$$A^* = \bigcap_{a:\, a \in A} R(a) = \{x \in T\colon (a, x) \in R \text{ for all } a \text{ in } A\}.$$

If $Y \subseteq X$, the set Y^* is defined similarly. The functions

$$A \mapsto A^*,\ 2^S \to 2^X \quad \text{and} \quad Y \mapsto Y^*,\ 2^X \to 2^S$$

form a Galois connection. The *lattice of the relation* R is the lattice of closed sets in 2^S defined by the Galois relation. Every finite lattice L is the lattice of some relation. Just take the relation $\leq : L \to L$.

Taking the complementary relation, we obtain an orthogonality. If $R : S \to X$ is a relation, let $\tilde{R} = (S \times X) \setminus R$,

$$A^{\perp} = \bigcap_{a:\, a \in A} \tilde{R}(a), \text{ and } Y^{\perp} = \bigcap_{x:\, x \in Y} \tilde{R}^{-1}(x).$$

Then $A \mapsto A^{\perp}$ and $X \mapsto X^{\perp}$ form a Galois connection. Such Galois connections are called *orthogonalities*. For example, let U and V be vector spaces (over the same field) paired by a bilinear form $\langle u, v\rangle$. Then the functions between 2^U and 2^V given by $A \mapsto A^{\perp}$ and $Y \mapsto Y^{\perp}$, where

$$A^{\perp} = \{v\colon \langle u, v\rangle = 0 \text{ for all } u \text{ in } A\}, \quad \text{and}$$
$$Y^{\perp} = \{u\colon \langle u, v\rangle = 0 \text{ for all } v \text{ in } Y\},$$

form a Galois connection. If the bilinear form is nondegenerate, then the lattice of closed sets in 2^U is the lattice of subspaces of U. In a very loose sense, a bilinear form is a vector space analog of a relation.

We end this section with a brief account of relation algebras. Relations, rather than functions, are central to logic. For example, a model can be defined as a set with a collection of relations satisfying certain axioms.

Relations have several natural operations defined on them. Let $R : S \to X$ be a relation. The *converse* R^{-1} is the relation $X \to S$ defined by

$$(x, a) \in R^{-1} \text{ if and only if } (a, x) \in R.$$

The *contrary* or *complement* $\tilde{R}$ is the relation $S \to X$ defined by

$$(a, x) \in \tilde{R} \text{ if and only if } (a, x) \notin R.$$

Now let $Q : X \to Z$ be a relation. Then the *composition* $Q \circ R$ or *relative product* $R \mid Q$ is the relation $S \to Z$ defined by $(a, z) \in Q \circ R$ if and only if there exists an element $x \in X$, $(a, x) \in R$ and $(x, z) \in Q$. (The reversal of order is intended in the relative product. Pierce said "relatives" rather than "relations." Thus, relative product just means products of relations.)

From these natural operations, one can define a fourth operation. The *relative sum* of the relations $R : S \to X$ and $Q : X \to Z$ is the relation $R \dagger Q$ defined by

$$R \dagger Q = \widetilde{\tilde{R} \mid \tilde{Q}}.$$

Besides these four operations, others have been proposed. Indeed, Pierce studied all 64 binary operations obtainable from converse, complementation, and composition.

The idea behind a relation algebra is to axiomatize the natural operations on sets and relations. For sets, it is generally accepted that the Boolean algebra operations and axioms capture the essence of set operations. For relations, many operations and axiom systems have been proposed. Thus, there are definitions rather than a definition of a relation algebra. A popular definition is an equational definition (with ten axioms) given by Tarski and Givant.[39] A survey of relation algebras is given in the 2006 book of Maddux. Whether relation algebras have any applications in combinatorics is unclear.

[39] Tarski and Givant (1987).

Exercises

1.5.1. *Equivalence and difunctional relations.*[40]

A relation $R: S \to S$ is an *equivalence relation* if it is *reflexive,* that is, $\{(a, a): a \in S\} \subseteq R$; *symmetric,* that is, $R^{-1} = R$; and *transitive,* that is, $R \circ R \subseteq R$.

(a) Show that an equivalence relation $R: S \to S$ defines a partition on S and, conversely, a partition on S defines an equivalence relation on S.

Two relations $R: S \to X$ and $Q: X \to S$ are *mutually transitive* if

$$R \circ Q \circ R \subseteq R \quad \text{and} \quad Q \circ R \circ Q \subseteq Q.$$

The first condition, say, of mutual transitivity can be visualized as a square as follows: let a, a', b, and b' be placed cyclically at the corners of a square. Then if $(a, a') \in R$, $(b, b') \in R$, and $(a', b) \in S$, then $(a, b') \in R$.

(b) Let R and Q be mutually transitive. Show that $R \circ S$ and $S \circ R$ are transitive. Show that the three pairs Q and R, R^{-1} and Q^{-1}, Q^{-1} and R^{-1} are mutually transitive.

A relation $R: S \to S$ is *self-transitive* or *difunctional* (as opposed to dysfunctional) if R and R^{-1} are mutually transitive.

(c) Prove that a relation is difunctional if one can rearrange S and X so that the picture of R is a union of squares with disjoint domains and codomains; that is, there are subsets $S' \subseteq S$ and $X' \subseteq X$ and partitions $S_1, S_2, \ldots, S_k$ of $S \backslash S'$ and $X_1, X_2, \ldots, X_k$ of $X \backslash X'$ such that $|S_i| = |X_i|$ and

$$R = (S_1 \times X_1) \cup (S_2 \times X_2) \cup \cdots \cup (S_k \times X_k).$$

1.5.2. Two relations $R, Q: S \to S$ *commute* if $R \circ Q = Q \circ R$.

(a) Given a relation R, describe all the relations commuting with it.
(b) Study the hexagonal condition

$$R \circ Q \circ R = Q \circ R \circ Q.$$

The motivation for this exercise is a theorem of B. Jónsson (see Exercise 3.5.10).

1.5.3. *Ferrers relations.*[41] A relation $R: S \to X$ is a *Ferrers* relations if there exists an ordering $s_1, s_2, \ldots, s_n$ of S such that $R(s_i) \supseteq R(s_{i+1})$. Informally, the

[40] Jacotin-Dubreil (1950), Ore (1962, section 11.4), and Riguet (1950). [41] Riguet (1951).

picture of a Ferrers relation is the Ferrers diagram of an integer partition. Note that if any one of the four relations R, $\tilde{R}$, R^{-1}, $\tilde{R}^{-1}$ is Ferrers, then all are Ferrers.

(a) Show that R is Ferrers if and only if R and the converse of its complement $\tilde{R}^{-1}$ are mutually transitive.

(b) Show that if R is Ferrers, then $R \circ \tilde{R}^{-1} \circ R$ is also Ferrers.

(c) Show that R is Ferrers if and only if the lattice of closed sets of the Galois connection formed from R is a chain.

1.5.4. *Counting relations.*[42]

The total number of binary relations on a set S of size n is 2^{n^2}. A general problem is to enumerate relations with given properties:

(a) Show that the number of reflexive antisymmetric binary relations on a set of size n is $3^{n(n-1)/2}$.

(b) Part (a) gives an upper bound on the number $q(n)$ of partial orders on a set of size n. Improve the upper bound to

$$q(n) \leq n! 2^{n(n-1)/2}.$$

Show also the lower bound

$$q(n) \geq 2^{\lfloor n/2 \rfloor \lceil n/2 \rceil}.$$

1.5.5. *Completions of partially ordered sets.*[43]

Let P be a partially ordered set. If $A \subseteq P$, let

$$A_\flat = \bigcap_{a:\,a \in A} I(a)$$
$$A^\# = \bigcap_{a:\,a \in A} F(a),$$

where $I(a)$ is the principal ideal generated by a and $F(a)$ is the principal filter generated by a.

(a) Show that $A \mapsto A_\flat$ and $A \mapsto A^\#$ form a Galois connection between 2^P and 2^P, and hence, $\nu : A \mapsto (A^\#)_\flat$ is a closure operator on 2^P.

Note that

$$\nu(A) = \bigcap_{u:\,A \subseteq I(u)} I(u),$$

where the intersection is over all *upper bounds u of A*, that is, elements u such that $u \geq a$ for all $a \in A$ or, equivalently, $A \subseteq I(u)$. The ν-closed subsets

[42] Kleitman and Rothschild (1970, 1975). [43] MacNeille (1937) and Robison and Wolk (1957).

in 2^P are closed under intersections and have a maximum, the set P. Hence they form a (complete) lattice $\overline{P}$. The function $\nu : P \to \overline{P}, a \mapsto \nu(\{a\})$ is an order-preserving injection and $\overline{P}$ "extends" P to a complete lattice. The lattice $\overline{P}$ is usually called the *normal completion* of P.

The closure operator ν satisfies the additional property:

Em. For every element $a \in P$, $\nu(\{a\}) = I(a)$.

An *embedding operator* is a closure operator on 2^P satisfying the property (Em). Embedding operators are partially ordered as a suborder of the partially order of all closure operators on 2^P.

(b) Show that ν is the unique maximal element in the set of embedding operators of P under the partial order restricted from the partial order on all closure operators. Let

$$\eta(A) = \bigcup_{a:\, a \in A} I(a).$$

Show that η is the unique minimal element in the partial order of embedding operators.

(c) Show that a finite partially ordered set P and its normal completion have the same order dimension.

1.5.6. *Ideals and varieties.*

Let $R = \mathbb{F}[x_1, x_2, \ldots, x_n]$, the ring of polynomials in n variables over the field $\mathbb{F}$, and $A = \mathbb{F}^n$.

(a) Show that the functions

$V : 2^R \to 2^A$,

$J \mapsto \{(a_1, a_2, \ldots, a_n)\colon p(a_1, a_2, \ldots, a_n) = 0 \text{ for all } p(x_1, x_2, \ldots, x_n) \text{ in } J\}$,

$I : 2^A \to 2^R$,

$B \mapsto \{p(x_1, x_2, \ldots, x_n)\colon p(a_1, a_2, \ldots, a_n) = 0 \text{ for all } (a_1, a_2, \ldots, a_n) \text{ in } B\}$

form a Galois connection between 2^R and 2^A.

(b) Assuming that $\mathbb{F}$ is algebraically closed, show that $I(V(J))$ is the radical ideal generated by J, the ideal of all polynomials p such that for some positive integer n, p^n is in the ideal generated by J. (This is Hilbert's *Nullstellensatz*.)

1.5.7. *Markowsky's representation of lattices.*[44]

Let L be a finite lattice, $J(L)$ be the set of join-irreducibles, $M(L)$ be the set of meet-irreducibles, and $U : J(L) \to M(L)$ be the relation $j \leq m$. Show that L is isomorphic to the lattice of the relation U.

[44] This is the finite case of a theorem of Markowsky. See Markowsky (1975) and Crapo (1982).

1.5.8. *De Morgan's rule for changing places.*[45]

Let R, S, and T be relations with suitable domains and codomains. Then the following conditions are equivalent:

(a) $R \mid S \subseteq T$.

(b) $R^{-1} \mid \tilde{T} \subseteq \tilde{S}$.

(c) $\tilde{T} \mid S^{-1} \subseteq \tilde{R}$.

1.5.9. *Relative sums of partial functions.*

Show that if R and Q are partial functions with suitable domains and codomains, then $R \dagger Q$ is a partial function.

1.6 Further Reading

The founding paper in lattice theory is the 1900 paper by Dedekind. Among other ideas, this paper contains the isomorphism theorems of "modern" algebra, treated at the right level of generality. Emmy Noether is reported to have said that "everything is already in Dedekind." It is historically interesting that in 1942, Dilworth learned lattice theory from Dedekind's paper. This changed with the first edition of *Lattice Theory* by Garrett Birkhoff. The three editions show how the book and the subject responded to and changed each other. Three other classics are the books of Balbes and Dwinger, Crawley and Dilworth, and Grätzer. The third book is comprehensively updated in a new edition in 2003. The first two books are currently out of print. A short elementary introduction is the book of Davey and Priestley. An excellent account of partially ordered sets, with emphasis on the order dimension, can be found in the book of Trotter.

R. Balbes and P. Dwinger, *Distributive Lattices*, University of Missouri Press, Columbia, MI, 1974.

G. Birkhoff, *Lattice Theory*, 1st edition, American Mathematical Society, New York, 1940; 2nd edition, Providence, RI, 1948; 3rd edition, Providence, RI, 1967.

P. Crawley and R.P. Dilworth, *Algebraic Theory of Lattices*. Prentice-Hall, Englewood Cliffs, NJ, 1973.

B.A. Davey and H.A. Priestley, *Introduction to Lattices and Order*, 2nd edition, Cambridge University Press, New York, 2002.

R. Dedekind, Über die drei Moduln erzengte Dualgruppe, *Math. Ann.* 53 (1900) 317–403.

G. Grätzer, *General Lattice Theory*, 2nd edition, Birkhäuser, Basel, 2003.

W.T. Trotter, *Combinatorics and Partially Ordered Sets. Dimension Theory,* Johns Hopkins University Press, Baltimore, MD, 1992.

[45] De Morgan (1864).

2
Matching Theory

2.1 What Is Matching Theory?

An answer to this question can be found in the survey paper *Matching theory* of L.H. Harper and G.-C. Rota:

> Roughly speaking, matching theory is concerned with the possibility and the number of ways of covering a large, irregularly shaped combinatorial object with several replicas of a given small regularly shaped object, subject usually to the requirement that the small objects shall not overlap and that the small objects be "lined up" in some sense or other.

An elementary example of matching theory is the following puzzle (popularized by R. Gomory): given a standard 8×8 chessboard, can it always be covered by dominoes (a piece consisting of two squares) if one arbitrary black square and one arbitrary white square are deleted? Whether this is a problem in matching theory depends on one approaches it (see Exercises 2.1.1).

Another puzzle which involves the idea of a matching is the following 1979 Putnam problem:[1] let n *red* points $r_1, r_2, \ldots, r_n$ and n *blue* points $b_1, b_2, \ldots, b_n$ be given in the Euclidean plane. Show that there exists a permutation π of $\{1, 2, \ldots, n\}$, matching the red points with the blue points, so that no pair of finite line segments $\overline{r_i\, b_{\pi(i)}}$ and $\overline{r_j\, b_{\pi(j)}}$ intersects. Although this puzzle is about matchings, it is perhaps not matching theory in our sense. The reason – and a hint for its solution – is that it is really a *geometry* problem.

Our third example is indisputably a theorem in matching theory.[2] This theorem settles the question, often asked when one is first told that left and

[1] The problem can be found in Winkler (2004).

[2] König (1916). This theorem was cited by Philip Hall in 1935, as a motivation for the marriage theorem, in spite of the fact that in this paper, König has also proved the "König–Egerváry theorem," a more general theorem than the marriage theorem. Such anomalies occur rather often in matching theory.

right cosets of a subgroup may not be the same, whether there exists a set of elements acting simultaneously as left and right coset representatives.

2.1.1. König's theorem. Let S be a set of size mn. Suppose that S is partitioned into m subsets, all having size n, in two ways: $A_1, A_2, \ldots, A_m$ and $B_1, B_2, \ldots, B_m$. Then there exist m distinct elements $a_1, a_2, \ldots, a_m$ and a permutation π of $\{1, 2, \ldots, m\}$ such that for all i, $a_i \in A_i \cap B_{\pi(i)}$.

Exercises

2.1.1. (a) The answer to Gomory's puzzle is "yes." Find a proof.

(b) Can an 8×8 chessboard be covered by dominoes if two arbitrary black squares and two arbitrary white squares are removed?

(c) Study Gomory's puzzle for $m \times n$ chessboards.

(d) Develop a theory for domino-coverings of $m \times n$ chessboards with some squares removed.[3]

2.1.2. Do the Putnam problem.

2.1.3. Give an elementary proof (that is, a proof not using the marriage theorem) of König's theorem.

2.2 The Marriage Theorem

A subrelation M of a relation $R : S \to X$ is a *partial matching* of R if no two edges (or ordered pairs) in M have an element in common. A partial matching M is a *matching* if $|M| = |S|$, or equivalently, the subrelation $M : S \to X$ is a one-to-one (but not necessarily onto) function.

Observe that a relation $R : S \to X$ is determined, up to a permutation of S, by the list of subsets $R(a)$, where a ranges over the set S. This gives another way of looking at matchings. A *transversal* or *system of distinct representatives* of a list $X_1, X_2, \ldots, X_n$ of subsets of a set X is a labeled set of n (distinct) elements $\{a_1, a_2, \ldots, a_n\}$ of X such that for each i, $a_i \in X_i$. Thus, transversals are, more or less, ranges of matchings. A transversal exists for the list $X_1, X_2, \ldots, X_n$ of subsets if and only if the relation $R : \{1, 2, \ldots, n\} \to X$ with $R(i) = X_i$ has a matching.

Suppose a matching M exists in $R : S \to X$. If $A \subseteq S$, then $M(A) \subseteq R(A)$. Since M is a one-to-one function, $|M(A)| = |A|$, and hence, $|R(A)| \geq |A|$. Thus, the condition

$$\text{for all subsets } A \text{ in } S,\ |R(A)| \geq |A|$$

[3] Such a theory is developed in Brualdi (1975).

is a necessary condition for the existence of a matching. This condition is called the *(Philip) Hall condition*. The fundamental result in matching theory is that the Hall condition is also sufficient.

2.2.1. The marriage theorem. A relation $R : S \to X$ has a matching if and only if for every subset A in S, $|R(A)| \geq |A|$.

An equivalent way to state the marriage theorem is in terms of transversals of subsets.

2.2.2. Theorem. Let $X_1, X_2, \ldots, X_n$ be a list of subsets of a set X. Then there exists a transversal if and only if for all subsets $A \subseteq \{1, 2, \ldots, n\}$,

$$\left| \bigcup_{i:i \in A} X_i \right| \geq |A|.$$

We shall give several proofs, each with a different idea. Since necessity has already been proved, we need only prove sufficiency.

First proof of the marriage theorem.[4] This proof is due to Easterfield, Halmos, and Vaughan. We proceed by induction, observing that the case $|S| = 1$ obviously holds. We distinguish two cases:

Case 1. For every nonempty subset A strictly contained in S,

$$|R(A)| > |A|.$$

Choose any edge (a, x) in the relation R and consider the relation R' obtained by restricting R to the sets $S \backslash \{a\}$ and $X \backslash \{x\}$. If $A \subseteq S \backslash \{a\}$, $|R'(A)|$ equals $|R(A)|$ or $R(A) - 1$, depending on whether x is in $R(A)$. Thus, as $|R(A)| > |A|$, the smaller relation R' satisfies the Hall condition and has a matching by induction. Adding the ordered pair (a, x) to the matching for R', we obtain a matching for R.

Case 2. There is a nonempty subset A strictly contained in S such that $|R(A)| = |A|$. Then the restriction $R|_A : A \to R(A)$ of R to the subsets A and $R(A)$ has a matching by induction.

Let $R'' : S \backslash A \to X \backslash R(A)$ be the relation obtained by restricting R to $S \backslash A$ and $X \backslash R(A)$. Consider a subset B in the complement $S \backslash A$. Then $R(A \cup B)$ is the union of the disjoint sets $R(A)$ and $R''(B)$. Since A and B are disjoint, R

[4] Easterfield (1946) and Halmos and Vaughan (1950). The paper of Halmos and Vaughan popularized the matrimonial interpretation of Theorem 2.2.1.

satisfies the Hall condition, and $|R(A)| = |A|$,

$$\begin{aligned}|R''(B)| &= |R(A \cup B)| - |R(A)| \\ &\geq |A \cup B| - |A| \\ &= |B|.\end{aligned}$$

Hence, R'' satisfies the Hall condition and has a matching by induction. The matchings for $R|_A$ and R'' are disjoint. Taking their union, we obtain a matching for R. □

The argument in the proof of Easterfield, Halmos, and Vaughan can be modified to prove the following theorem of Marshall Hall.[5]

2.2.3. Marshall Hall's theorem. Let $R : S \to X$ be a relation satisfying the Hall condition. Suppose that, in addition, every element in S is related to at least k elements in X. Then R has at least $k!$ matchings.

Proof. We proceed by induction on S and k, the case $|S| = 1$ being obvious. As in the previous proof, we distinguish two cases:

Case 1. For all nonempty proper subsets A in S, $|R(A)| > |A|$. Choose an element a in S and let $R(a) = \{x_1, x_2, \ldots, x_m\}$. Consider the relations R_i obtained by restricting R to $S\backslash\{a\}$ and $X\backslash\{x_i\}$. In R_i, every element in $S\backslash\{a\}$ is related to at least $k - 1$ elements in $X\backslash\{x_i\}$. By induction, the relation R_i has at least $(k-1)!$ matchings. Adding (a, x_i) to any matching of R_i yields a matching of R. Doing this for all the relations R_i, we obtain $m(k-1)!$ matchings. Since $m \geq k$, there are at least $k(k-1)!$ matchings in R.

Case 2. There exists a nonempty proper subset A in S so that $|A| = |R(A)|$. As in the earlier proof, we can break up R into two smaller matchings $R|_A$ and R'', both satisfying the Hall condition. In the relation $R|_A : A \to R(A)$, every element in A is related to at least k elements in $R(A)$. Hence, by induction, $R|_A$ has at least $k!$ matchings. Taking the union of a fixed matching of R'' and a matching in $R|_A$, we obtain at least $k!$ matchings for R. □

2.2.4. Corollary. Let $R : S \to X$ be a relation and suppose that every element in S is related to at least k elements in X. Then R has a matching implies that R has at least $k!$ matchings.

[5] Hall (1948). Hall's argument is similar to the one given here. See also Exercise 2.2.3.

Note that an easy argument shows that the number of matchings in $R : S \to X$ is bounded above by $\prod_{i=1}^{n} |R(i)|$, with equality if and only if the sets $R(i)$ are pairwise disjoint.

Second proof of the marriage theorem.[6] This proof is based on the following lemma, which allows edges to be removed while preserving the Hall condition.

2.2.5. Rado's lemma. Let $R : S \to X$ be a relation satisfying the Hall condition. If (c, x) and (c, y) are two edges in R, then at least one of the relations $R\backslash\{(c, x)\}$ and $R\backslash\{(c, y)\}$ satisfies the Hall condition.

Proof. Suppose the lemma is false. Let $R' = R\backslash\{(c, x)\}$ and $R'' = R\backslash\{(c, y)\}$. Then we can find subsets A and B in S such that $c \notin A$, $c \notin B$,

$$|R'(A \cup \{c\})| < |A \cup \{c\}|,$$

and

$$|R''(B \cup \{c\})| < |B \cup \{c\}|.$$

Since $R'(A) = R(A)$ and $R''(B) = R(B)$, it follows that $R'(c) \subseteq R(A)$ and $R''(c) \subseteq R(B)$. We conclude that

$$|R'(A \cup \{c\})| = |R(A)| = |A|$$

and

$$|R''(B \cup \{c\})| = |R(B)| = |B|.$$

Using these two equalities, we have

$$\begin{aligned}
|A| + |B| &= |R'(A \cup \{c\})| + |R''(B \cup \{c\})| \\
&= |R'(A \cup \{c\}) \cup R''(B \cup \{c\})| + |R'(A \cup \{c\}) \cap R''(B \cup \{c\})| \\
&= |R(A \cup B \cup \{c\})| + |R(A) \cap R(B)| \\
&\geq |R(A \cup B \cup \{c\})| + |R(A \cap B)| \\
&\geq |A \cup B \cup \{c\}| + |A \cap B| \\
&\geq |A \cup B| + 1 + |A \cap B| \\
&= |A| + |B| + 1,
\end{aligned}$$

a contradiction. □

We can now prove the marriage theorem by removing edges until we reach a matching.

[6] Rado (1967b).

Third proof of the marriage theorem.[7] The third proof combines ideas from the first and second proof. This proof begins with an easy lemma, which we state without proof.

2.2.6. Lemma. Let $R : S \to X$ be a relation satisfying the Hall condition. Suppose that A and B are subsets of S such that $|A| = |R(A)|$ and $|B| = |R(B)|$. Then $|A \cup B| = |R(A \cup B)|$ and $|A \cap B| = |R(A \cap B)|$.

Choose an element a in S. Suppose that for some x in $R(a)$, the relation $R_x : S\backslash\{a\} \to X\backslash\{x\}$ obtained by restricting R to $S\backslash\{a\}$ and $X\backslash\{x\}$ satisfies the Hall condition. Then by induction, R_x has a matching, and adding (a, x) to that matching, we obtain a matching for R.

Thus, we need only deal with the case $(*)$: for all x in $R(a)$, the relation R_x does not satisfy the Hall condition; that is to say, for each x in $R(a)$, there exists a subset B_x of $S\backslash\{a\}$ such that $|B_x| > |R_x(B_x)|$. Since R satisfies the Hall condition, $|R(B_x)| \geq |B_x|$. In addition,

$$R_x(B_x) \subseteq R(B_x) \subseteq R_x(B_x) \cup \{x\}.$$

Thus, $R(B_x)$ equals $R_x(B_x) \cup \{x\}$, x is in $R(B_x)$, and $|R(B_x)| = |R_x(B_x)| + 1$. Since R satisfies the Hall condition, it follows that for all x in $R(a)$,

$$|R(B_x)| = |B_x|.$$

Let

$$B = \bigcup_{x:x \in R(a)} B_x.$$

By Lemma 2.2.6, $|B| = |R(B)|$. However, $R(a) \subseteq R(B)$. Together, this implies that

$$|B \cup \{a\}| > |R(B)|,$$

contradicting the assumption that R satisfies the Hall condition. Thus, the case $(*)$ is impossible and the proof is complete. □

Fourth proof of the marriage theorem.[8] This is Philip Hall's proof. Let $R : S \to X$ be a relation with at least one matching. Let

$$H(R) = \bigcap_M M(S),$$

[7] Everett and Whaples (1949).

[8] Hall (1935). Although citing the paper of Hall is obligatory, Hall's proof seems not to have appeared in any textbook.

where the intersection ranges over all matchings M in R; in other words, $H(R)$ is the subset of elements in X that must occur in the range of any matching of R. The set $H(R)$ may be empty.

2.2.7. Lemma. Assume that $R : S \to X$ has a matching. If $A \subseteq S$ and $|A| = |R(A)|$, then $R(A) \subseteq H(R)$.

Proof. Suppose $|A| = |R(A)|$. Then for any matching M in R, $M(A) = R(A)$. Hence, $R(A)$ occurs as a subset in the range of every matching, and assuming R has a matching, $R(A) \subseteq H(R)$. □

2.2.8. Lemma. Let R have a matching, M be a matching of R, and

$$I = \{a\colon M(a) \in H(R)\}.$$

Then $R(I) = H(R)$. In particular, $|H(R)| = |I|$.

Proof. We will use an alternating path argument.[9] An *alternating path* (*relative to the matching* M) is a sequence $x, a_1, x_1, a_2, \ldots, x_{l-1}, a_l, x_l$ such that the edges

$$(x, a_1), (a_1, x_1), (x_1, a_2), (a_2, x_2), \ldots, (a_{l-1}, x_{l-1}), (x_{l-1}, a_l), (a_l, x_l)$$

are all in R, the edges (a_i, x_i) are in the matching M, and except for the head x and the tail x_l, which may be equal, all the elements in the path are distinct. If x is not in the range $M(S)$, then it is easy to check that the matching M^Δ defined by

$$\begin{aligned} M^\Delta = {} & (M \backslash \{(a_1, x_1), (a_2, x_2), \ldots, (a_l, x_l)\}) \\ & \cup \{(x, a_1), (x_1, a_2), (x_2, a_3), \ldots, (x_{l-1}, a_l)\} \end{aligned}$$

is a matching of R and x_l is not in the range $M^\Delta(S)$.

Let $H'(M)$ be the subset of elements x in X connected by an alternating path to an element y in $H(R)$. Then $H(R) \subseteq H'(M)$. We shall show that, in fact,

$$H(R) = H'(M).$$

To do this, we first show that $H'(M) \subseteq M(S)$. Suppose that $x \in H'(M)$ but $x \notin M(S)$. Then y is not in the range of the new matching M^Δ defined by the alternating path from x to y, contradicting the assumption that $y \in H(R)$.

Next, let $J = \{a\colon M(a) \subseteq H'(M)\}$. If $a \in J$ and $M(a) = \{x\}$, then there is an alternating path P starting with an edge (x, b), where $b \in S$ and $b \neq a$, and ending at a vertex in $H(R)$. If $y \in R(a)$ and $y \neq x$, then (y, a) is an edge

[9] Alternating paths first appeared in König (1916).

in R, (a, x) is an edge in M, and y, a, P is an alternating path ending at a vertex in $H(R)$. We conclude that if $a \in J$, then $R(a) \subseteq H'(M)$. In particular, $R(J) \subseteq H'(M)$. Thus, we have

$$M(J) \subseteq R(J) \subseteq H'(M) = M(J).$$

From this, it follows that $|J| = |R(J)|$. By Lemma 2.2.7, $H'(M) \subseteq H(R)$. We conclude that $H(R) = H'(M)$. In particular, $I = J$ and $H(R) = R(I)$. □

We now proceed by induction on n. Let $R : S \to X$ be a relation satisfying the Hall condition and let $b \in S$. Since the Hall condition holds for the restriction $R|_{S\backslash\{b\}} : S\backslash\{b\} \to X$, the smaller relation $R|_{S\backslash\{b\}}$ has a matching M by induction. Let $I = \{a\colon M(a) \subseteq H(R|_{S\backslash\{b\}})\}$. If $R(b) \subseteq H(R|_{S\backslash\{b\}})$, then

$$R(I \cup \{b\}) = H(R|_{S\backslash\{b\}})$$

and $|R(I \cup \{b\})| = |I| < |I| + 1$, contradicting the assumption that the Hall condition holds. Thus it must be the case that $R(b)$ is not contained in $H(R|_{S\backslash\{b\}})$. For any element x in $R(b)$ not in $H(R|_{S\backslash\{b\}})$, the union $M \cup \{(b, x)\}$ is a matching for R. □

We end this section with two easy but useful extensions of the marriage theorem.

2.2.9. Theorem.[10] A relation $R : S \to X$ has a partial matching of size $|S| - d$ if and only if R satisfies the *Ore condition*:

$$\text{for every subset } A \text{ in } S,\ |R(A)| \geq |A| - d.$$

Proof. Let D be a set of size d disjoint from X and $R' : S \to X \cup D$ be the relation $R \cup (S \times D)$. Then it is easy to check that the following statements hold:

- R' satisfies the Hall condition if and only if R satisfies the Ore condition.
- R' has a matching if and only if R has a partial matching of size $|S| - d$.

We can now complete the proof using the marriage theorem. □

Restating Theorem 2.2.9, we obtain another result of Ore.

2.2.10. Defect form of the marriage theorem. Let $R : S \to X$ be a relation and τ be the maximum size of a partial matching in R. Then

$$\tau = |S| + \min\{|R(A)| - |A|\colon A \subseteq S\}.$$

[10] Ore (1955); see Exercise 2.4.7 for Ore's proof.

Exercises

2.2.1. *The Hungarian method.*[11]

Use alternating paths (defined in the fourth proof) to obtain a proof of the marriage theorem. (This has become the standard proof in introductory books. Among its many advantages, alternating paths give a polynomial-time algorithm for finding a maximum-size partial matching; see Lovász and Plummer, 1986.)

2.2.2. *Cosets of groups and König's theorem.*[12]

(a) Let H be a subgroup of a group G such that the index $|G|/|H|$ is finite. Then there are coset representatives that are simultaneously right and left coset representatives.

(b) Use the method in (a) to prove König's theorem in Section 2.1.

(c) Conclude that if H and K are subgroups with the same finite index, then there exists one system that is a system of coset representives for both H and K.

(d) Let $S_1, S_2, \ldots, S_m$ and $T_1, T_2, \ldots, T_m$ be two partitions of a set S. Find nice conditions for the conclusion in König's theorem to hold.

2.2.3. *Regular relations.*

Let $R : S \to X$ be a relation in which every element a in S is related to the same number k of elements in X and every element in X is related to the same number l in S. Show that the Hall condition holds, and hence R contains a matching.

2.2.4. *Extending Latin rectangles.*[13]

Let $\mathcal{A}$ be an alphabet with n letters. If $1 \leq r, s \leq n$, an $r \times s$ array with entries from $\mathcal{A}$ is a *Latin rectangle* if each letter occurs at most once in any row or column. An $n \times n$ Latin rectangle is called a *Latin square.*

(a) Show that if $m \leq n - 1$, every $m \times n$ Latin rectangle can be extended to an $(m+1) \times n$ Latin rectangle. In particular, every $m \times n$ Latin rectangle can be extended to a Latin square.

(b) Use Theorem 2.2.2 to show that there exist at least

$$n!(n-1)!(n-2)! \cdots (n-m+1)!$$

$m \times n$ Latin rectangles.

[11] Kuhn (1955). The name is chosen in honor of D. König. [12] Ore (1958).

[13] Hall (1948) and Ryser (1951). The bound given in (b) for the number of $m \times n$ Latin rectangles is far from sharp. For better bounds, see Murty and Liu (2005), and the references cited in that paper.

(c) Let $1 \leq r, s \leq n$. Show that an $r \times s$ Latin rectangle L can be extended to a Latin square if and only if every letter in $\mathcal{A}$ occurs at least $r + s - n$ times in L.

2.2.5. *The number of matchings.*[14]

Let $R : \{1, 2, \ldots, n\} \to X$ be a relation such that $|R(i)| \leq |R(j)|$ whenever $i < j$. (This condition can always be achieved by rearranging $\{1, 2, \ldots, n\}$.) Show that if R has at least one matching, then the number of matchings in R is at least

$$\prod_{i=1}^{n} \max(1, |R(i)| - i + 1),$$

with equality holding if $R(i) \subseteq R(j)$ whenever $i < j$.

2.2.6. *Easterfield's demobilization problem.*[15]

Find all possible ways of obtaining the minimum sum of n entries of a given $n \times n$ matrix of positive integers, with no two entries from the same row or column. (*Comment.* Easterfield gives an algorithm for doing this. In the discussion of this algorithm, he discovered the marriage theorem. There may be a more efficient algorithm than Easterfield's.)

2.3 Free and Incidence Matrices

Let $R : S \to X$ be a relation. A matrix (c_{ax}) with row set S and column set X is *supported by* the relation R if the entry c_{ax} is nonzero if and only if $(a, x) \in R$.

There are several ways to choose the nonzero entries. One way is to create independent variables (or elements transcendental over some ground field) $X_{a,x}$, one for each edge (a, x) in the relation R, and set $c_{a,x} = X_{a,x}$. This choice of nonzero entries gives the *free* or *generic matrix* of the relation R. At the other extreme, one can set $c_{a,x} = 1$. This gives the $(0, 1)$-*incidence matrix* of R.

We begin by considering determinants of free matrices.

2.3.1. Lemma. Let $R : S \to X$ be a relation with $|S| = |X|$.

(a) If C is a matrix supported by R, then $\det C \neq 0$ implies that R has a matching.

(b) If C is the free matrix of R, then the number of matchings in R equals the number of nonzero monomials in the expansion of the determinant of the free matrix C of R. In particular, $\det C \neq 0$ if and only if R has a matching.

[14] Ostrand (1970) and Rado (1967a). [15] Easterfield (1946).

Proof. Expanding the determinant of C, we have

$$\det C = \sum_{\sigma} \prod_{a: a \in S} c_{a,\sigma(a)},$$

where the sum ranges over all bijections $\sigma : S \to X$. If $\det C \neq 0$, then there is a nonzero product in the expansion. The bijection σ giving that product is a matching of R. This proves Part (a). Part (b) follows since independent variables have no algebraic relations. □

Since the determinant of a square matrix is nonzero if and only if it has full rank, we have the following corollary.

2.3.2. Corollary. Let $R : S \to X$ be a relation and C be its free matrix. Then the following are equivalent:

(a) R has a matching of size $|S| - d$.

(b) The free matrix C has rank $|S| - d$.

(c) There are subsets $S' \subseteq S$ and $X' \subseteq X$, both having size $|S| - d$, such that the $(|S| - d) \times (|S| - d)$ square submatrix $C[S'|X']$ formed by restricting C to the rows in S' and the columns in X' has nonzero determinant.

Using Lemma 2.3.1 and its corollary, we can reformulate the marriage theorem as a result about the rank of free matrices.

2.3.3. Edmonds' theorem.[16] Let $m \leq n$ and C be a free $m \times n$ matrix of a relation R. Then C has rank strictly less than m if and only if there exists a set H of h rows such that the submatrix of C consisting of the rows in H has $h - 1$ or fewer nonzero columns.

Proof. Suppose that C contains an $h \times (n - h + 1)$ submatrix with all entries zero. Then every square $m \times m$ submatrix C' of C contains an $h \times (m - h + 1)$ zero submatrix. In the expansion of the determinant of C', every product $\prod c_{a,\sigma(a)}$ has at least one term $c_{b,\sigma(b)}$ in the $h \times (m - h + 1)$ zero submatrix. Thus, the determinant of every square $m \times m$ submatrix is zero, and C has rank strictly less than m.

Now suppose that C has rank strictly less than m. Then its rows are linearly dependent. Let H be a minimal linearly dependent set of rows. Relabeling, we may assume that $H = \{1, 2, \ldots, h\}$. Let $C[H]$ be the $h \times n$ submatrix consisting of the rows in H. Since H is a minimal linearly dependent set, the submatrix $C[H]$ has rank $h - 1$ and there exists a set K of $h - 1$ columns such that the $h \times (h - 1)$ submatrix $C[H|K]$, obtained by restricting C to the

[16] Edmonds (1967).

rows in H and columns in K, has rank $h-1$. Relabeling, we may assume that $K = \{1, 2, \ldots, h-1\}$.

We shall show that if a column in $C[H]$ is not in K, then every entry in it is zero. Consider the $h \times h$ square matrix obtained by adding a column l in $C[H]$, not in C, to $C[H|K]$. Since $C[H]$ has rank $h-1$, the square matrix is singular and its determinant equals zero. Taking the Laplace expansion along the added column, we obtain

$$\sum_{i=1}^{h}(-1)^i \det C[H\backslash\{i\}|K]c_{i,l} = 0. \qquad \text{(Det)}$$

As H is a minimal linearly dependent set of rows, *every* subdeterminant in the left-hand sum in the equation is nonzero. Hence, each subdeterminant is a nonzero polynomial in the variables X_{ij} with $1 \leq j \leq h-1$. If $l \geq r$ and $c_{i,l}$ is nonzero for some i, then Equation (Det) gives a nontrivial polynomial relation among the variables X_{ij}, $1 \leq j \leq h-1$, and the variables X_{il}, $(i,l) \in R$, contradicting the declaration that the variables X_{ij} are independent. We conclude that every entry in every column in $C[H]$ but not in K is zero. □

Note that by Corollary 2.3.2, Edmonds' theorem is equivalent to the marriage theorem.

Let C be a matrix on row set S and column set X. A pair (A, Y), where $A \subseteq S$ and $Y \subseteq X$, *covers* C if for every nonzero entry $c_{a,x}$ in C, $a \in A$ or $x \in Y$.

2.3.4. The König–Egerváry theorem. Let C be a matrix supported by the relation R and let τ be the maximum size of a partial matching in R. Then

$$\tau = \min\{|A| + |Y|: (A, Y) \text{ covers } C\}.$$

In particular, if C is the free matrix of R, then

$$\text{rk}(C) = \min\{|A| + |Y|: (A, Y) \text{ covers } C\}.$$

Proof. Let (A, Y) be a cover of C. If (a, x) is an edge in a partial matching, then $a \in A$, $x \in Y$, or both. Hence, $|A| + |Y| \geq \tau$.

On the other hand, $(S\backslash A, R(A))$ is a cover of C. Hence, by the defect form of the marriage theorem (2.2.10),

$$\begin{aligned} \tau &= \min\{|R(A)| + (|S| - |A|): A \subseteq S\}. \\ &\geq \min\{|A| + |Y|: (A, Y) \text{ covers } C\}. \end{aligned}$$

□

Other proofs are given in the exercises. The König–Egerváry theorem says for a rank-m free matrix C, we can permute the rows and columns so that M has the following form:

$$\begin{pmatrix}
? & ? & ? & ? & ? & \bullet & ? & \dots & ? \\
? & & & ? & \bullet & ? & & & ? \\
? & & ? & \bullet & 0 & 0 & 0 & \dots & 0 \\
? & ? & \bullet & ? & 0 & & & & 0 \\
? & \bullet & ? & ? & 0 & & & & 0 \\
\bullet & ? & & ? & 0 & & & & 0 \\
? & & & ? & 0 & & & & 0 \\
\vdots & & & \vdots & \vdots & & & & \vdots \\
? & ? & ? & ? & 0 & 0 & 0 & \dots & 0
\end{pmatrix},$$

where the entries $c_{m,1}, c_{m-1,2}, \dots, c_{1,m}$ in the sub-antidiagonal are nonzero and there are nonnegative integers a and b, $a + b = m$, such that all the nonzero entries in C lie in the union of the two rectangles $\{1, 2, \dots, a\} \times \{1, 2, \dots, n\}$ and $\{1, 2, \dots, m\} \times \{1, 2, \dots, b\}$. In the preceding example, $m = 6$, $a = 2$, $b = 4$, a "$\bullet$" is a nonzero entry (in a maximum-size partial matching), a "?" is an entry which may be zero or nonzero, and a "0" is, naturally, a zero entry.

Exercises

2.3.1. *Graph-theoretic form of the König–Egerváry theorem.*

Let G be a graph. A *partial matching* M in G is a subset of edges, with no two edges in M sharing a vertex. A *vertex cover* C is a subset of vertices such that every edge is incident on at least one vertex in C. Show that the König–Egerváry theorem is equivalent to *König's minimax theorem*: in a bipartite graph G, the maximum size of a partial matching equals the minimum size of a vertex cover. Find graph-theoretic proofs of König's minimax theorem.

2.3.2. *Bapat's extension of the König–Egerváry theorem.*[17]

Let M be a matrix with rows labeled by S and columns labeled by T. If $A \subseteq S$ and $B \subseteq T$, then let $M[B|A]$ be the submatrix labeled by rows in A

[17] Bapat (1994).

and columns in A. A submatrix $M[B|A]$ has *zero type* if for every $b \in B$ and $a \in A$,

$$\operatorname{rank} M[B|A] = \operatorname{rank} M[B\backslash\{b\}|A\backslash\{a\}].$$

A submatrix with all entries zero has zero type. Another example is a $2 \times n$ matrix (with $n > 1$) where the second row is a nonzero multiple of the first.

(a) Show that if $\operatorname{rank} M[T|S] < \min\{|T|, |S|\}$, then there exists a submatrix $M[B|A]$ of zero type such that

$$|T| + |S| - \operatorname{rank} M[T|S] = |B| + |A| - \operatorname{rank} M[B|A].$$

(b) Show that if M is the free matrix of a relation, then a submatrix N has zero type if and only if every entry in N is zero. Hence, deduce the König–Egerváry theorem from Bapat's theorem.

2.3.3. *Frobenius' irreducibility theorem.*[18]

Let M be the $n \times n$ square free matrix of the complete relation $\{1, 2, \ldots, n\} \times \{1, 2, \ldots, n\}$. A *general matrix function* $G(X_{ij})$ over the field $\mathbb{F}$ is a polynomial of the form

$$\sum_{\pi} c_\pi X_{1,\pi(1)} X_{2,\pi(2)} \cdots X_{n,\pi(n)},$$

the sum ranging over all permutations π of $\{1, 2, \ldots, n\}$ and c_π is in $\mathbb{F}$.

(a) Show that a general matrix function factors in the form

$$G = PQ,$$

where P and Q have positive degrees m and $n - m$, if and only if there are square submatrices M_1 and M_2 of size m and $n - m$ and P and Q are general matrix functions of M_1 and M_2.

A matrix M is said to be *decomposable* or *reducible* if there exist permutations of the row and column sets so that

$$M = \begin{pmatrix} M_1 & U \\ 0 & M_2 \end{pmatrix},$$

where the matrix in the lower left corner is a zero matrix with at least one row and one column.

(b) Let M be an $n \times n$ square free matrix (of an arbitrary relation). Then the determinant of M factors nontrivially if and only if M is decomposable.

[18] Frobenius (1917). Frobenius proved his theorem for determinants. The more general form stated here is from Brualdi and Ryser (1991, p. 295). A good account of the Frobenius' work on combinatorial matrix theory can be found in Schneider (1977).

2.3.4. *Converting determinants to permanents.*[19]

Let (a_{ij}) be an $n \times n$ (0, 1)-matrix such that $\mathrm{per}(a_{ij}) > 0$. If there exists a matrix (b_{ij}) such that $b_{ij} = \epsilon_{ij} a_{ij}$, where ϵ_{ij} equals -1 or 1, and $\mathrm{per}\, A = \det B$, then (a_{ij}) has at most $(n^2 + 3n - 2)/2$ nonzero entries.

2.3.5. *The bipartite graph case of Kasteleyn's theorem.*

If the bipartite graph associated with the relation is planar, then there exists an assignment of signs to the incidence matrix so that the determinant of the incidence matrix equals the number of matchings.[20]

2.4 Submodular Functions and Independent Matchings

Let S be a set and $\rho : 2^S \to \mathbb{R}$ be a function from the subsets of S to the real numbers. The function ρ is *submodular* if for all subsets A and B of S,

$$\rho(A) + \rho(B) \geq \rho(A \cup B) + \rho(A \cap B).$$

Note that if $A \subseteq B$ or $B \subseteq A$, then the inequality holds trivially as an equality. The function ρ is *increasing* if $\rho(B) \leq \rho(A)$ whenever $B \subseteq A$. Submodular functions need not be increasing. (There is an example on a set of size 2.)

Submodular functions occur naturally from valuations when the intersection is smaller than expected. For example, let $R : S \to X$ be a relation. Then,

$$R(A \cap B) \subseteq R(A) \cap R(B)$$

and it is possible that strict containment occurs. Thus, the function

$$\rho : 2^S \to \{0, 1, 2, \ldots\}, \quad A \mapsto |R(A)|$$

is not a valuation in general. However, it is a submodular function.

Another example arises from matrices. Let M be a matrix with column set S. The *(column) rank function* rk is the function defined on S sending a subset A in S to the rank of the submatrix of M formed by the columns in A.

2.4.1. Lemma. Let M be a matrix with column set S. Then the column rank function rk is a submodular function.

Proof. We shall use *Grassmann's identity* from linear algebra. If U and V are subspaces of a vector space, then

$$\dim(U) + \dim(V) = \dim(U \vee V) + \dim(U \cap V),$$

[19] Gibson (1971). [20] Kasteleyn (1963).

where $U \vee V$ is the subspace spanned by the vectors in the union $U \cup V$. Since $\text{rk}(A)$ equals the dimension of the subspace spanned by the column vectors in A, we have

$$\text{rk}(A) + \text{rk}(B) = \text{rk}(A \cup B) + \dim(\overline{A} \cap \overline{B}),$$

where $\overline{A}$ and $\overline{B}$ are the subspaces spanned by A and B. If a spanning set of the subspace intersection $\overline{A} \cap \overline{B}$ happens to be in the set intersection $A \cap B$, then

$$\text{rk}(A) + \text{rk}(B) = \text{rk}(A \cup B) + \text{rk}(A \cap B).$$

However, in general, we only have the submodular inequality

$$\text{rk}(A) + \text{rk}(B) \geq \text{rk}(A \cup B) + \text{rk}(A \cap B). \qquad \square$$

A matrix rank function rk satisfies two additional conditions:

Normalization: $\text{rk}(\emptyset) = 0$.

Unit increase: For a subset A and an element a in X,

$$\text{rk}(A) \leq \text{rk}(A \cup \{a\}) \leq \text{rk}(A) + 1.$$

In his 1935 paper "On the abstract properties of linear dependence,"[21] H. Whitney defined a *(matroid) rank function* to be an integer-valued normalized submodular function satisfying the unit-increase condition. A matroid rank function defines a *matroid* on the set X. Whitney's intuition was that the three axioms for a matroid rank function capture all the combinatorics or "abstract properties" of rank functions of matrices. His intuition was confirmed by several independent rediscoveries of matroid axioms. Many concepts and results in elementary linear algebra extend to matroids. For example, one can extend the notion of linear independence by defining a subset I to be *independent* (relative to the rank function rk) if $|I| = \text{rk}(I)$.

Let $R : S \to X$ be a relation, C be its free matrix, and rk be the matrix rank function on the columns in C. Then by Corollary 2.3.2, if $Y \subseteq X$, $\text{rk}(Y)$ is the maximum size of a partial matching in the relation $R_Y : S \to Y$ obtained by restricting R to Y, so that $R_Y(a) = R(a) \cap Y$. Thus, by the defect form of the marriage theorem (2.2.10),

$$\text{rk}(Y) = |S| + \min\{|R(A) \cap Y| - |A| \colon A \subseteq S\}. \qquad \text{(TM)}$$

[21] Whitney (1935). There are many ways to do matroid theory. Three different accounts of matroids can be found in Crapo and Rota (1970), Kung (1996a), and Oxley (1992). There are strong connections between matching theory and matroids. See, for example, chapter 12 of Oxley's book.

The matroid defined in this way is the *transversal matroid defined by R on X*.

There are two operations on matroids, restriction and contraction. Let $\mathrm{rk} : 2^X \to \{0, 1, 2, \ldots\}$ be a matroid rank function on X. If $Y \subseteq X$, then the *restriction* of rk to Y is the restriction $\mathrm{rk} : 2^Y \to \{0, 1, 2, \ldots\}$ of rk to the subsets in Y. If $Z \subseteq X$, then the rank function rk_Z obtained by *contracting* Z is the function $2^{X \backslash Z} \to \{0, 1, 2, \ldots\}$ defined as follows: for $B \subseteq X \backslash Z$,

$$\mathrm{rk}_Z(B) = \mathrm{rk}(B \cup Z) - \mathrm{rk}(Z).$$

It is routine to show that rk_Z is a matroid rank function.

A partial matching M of $R : S \to X$ is *independent* if its range $M(S)$ is an independent set. An independent matching is an independent partial matching of size $|S|$.

2.4.2. The marriage theorem for matroids.[22] Let $R : S \to X$ be a relation and X be equipped with a matroid rank function rk. Then R has an independent matching if and only if R satisfies the Rado–Hall condition:

$$\text{for all subsets } A \text{ in } S,\ \mathrm{rk}(R(A)) \geq |A|.$$

Proof. As in the marriage theorem, necessity is clear. To prove sufficiency, we use a modification of the argument of Easterfield, Halmos, and Vaughan (see Section 2.2). We proceed by induction, observing that the theorem holds for $|S| = 1$.

Case 1. For every nonempty proper subset A of S,

$$\mathrm{rk}(R(A)) > |A|.$$

Choose an element a in S. Since $\mathrm{rk}(R(a)) > 1$, there exists an element x in $R(a)$ such that $\mathrm{rk}(\{x\}) = 1$. Consider the relation R' obtained by restricting R to $S \backslash \{a\} \to X \backslash \{x\}$ and let $X \backslash \{x\}$ be equipped with the contraction rank function $\mathrm{rk}_{\{x\}}$. By induction, the relation R' has an independent matching M' relative to the contraction rank function $\mathrm{rk}_{\{x\}}$. Adding the edge (a, x) to M' yields a matching M of R. The matching M is independent because

$$\begin{aligned} \mathrm{rk}(M(S)) &= \mathrm{rk}(M(S \backslash \{a\}) \cup \{x\}) \\ &= \mathrm{rk}_{\{x\}}(M(S \backslash \{a\})) + 1 \\ &= (|S| - 1) + 1. \end{aligned}$$

22 Rado (1942).

Case 2. There is a proper nonempty subset A such that

$$\mathrm{rk}(R(A)) = |A|.$$

Consider the restriction $R|_A : A \to R(A)$ equipped with the restriction of rk to the subset $R(A)$ of X. This restriction satisfies the Rado–Hall condition and, by induction, $R|_A$ has an independent matching $M|_A$ of size $|A|$. Its range is $R(A)$.

Let $R'' : S\backslash A \to X\backslash R(A)$ be the restriction of R to the sets indicated, equipped with the contraction rank function $\mathrm{rk}_{R(A)}$ defined by

$$\mathrm{rk}_{R(A)}(Y) = \mathrm{rk}(Y \cup R(A)) - \mathrm{rk}(R(A)),$$

when Y is a subset of $X\backslash R(A)$. Consider a subset B in $S\backslash A$. Then, as in the Easterfield–Halmos–Vaughan proof,

$$R(A \cup B) = R''(B) \cup R(A).$$

Since R satisfies the Hall–Rado condition and $\mathrm{rk}(R(A)) = |A|$, we have

$$\begin{aligned} \mathrm{rk}_{R(A)}(R''(B)) &= \mathrm{rk}(R''(B) \cup R(A)) - \mathrm{rk}(R(A)) \\ &= \mathrm{rk}(R(A \cup B)) - \mathrm{rk}(R(A)) \\ &\geq |A \cup B| - |A| \\ &= |B|. \end{aligned}$$

Hence, R'' satisfies the Hall–Rado condition and has a matching M'' by induction. Taking the union of the matchings $M|_A$ and M'', we obtain a matching for R. This is an independent matching because

$$\begin{aligned} |S\backslash A| &= \mathrm{rk}_{R(A)}(M''(S\backslash A)) \\ &= \mathrm{rk}(M''(S\backslash A) \cup R(A)) - |A|, \end{aligned}$$

and hence,

$$\mathrm{rk}(M''(S\backslash A) \cup M|_A(A)) = |S\backslash A| + |A| = |S|. \qquad \square$$

2.4.3. Corollary. Let $R : S \to X$ be a relation and X be equipped with a matroid rank function rk. Then R has an independent partial matching of size $|S| - d$ if and only if for all subsets A in S, $\mathrm{rk}(R(A)) \geq |A| - d$. In particular, the maximum size of an independent partial matching equals

$$|S| + \min\{\mathrm{rk}(R(A)) - |A| \colon A \subseteq S\}.$$

Sketch of proof. Apply Theorem 2.4.3 to the relation $R' : S \to X \cup D$, where D is a set of size d disjoint from S, $P' = R \cup (S \times D)$, and $X \cup D$ is equipped with the rank function: if $A \subseteq X \cup D$, then

$$\mathrm{rk}(A) = \mathrm{rk}(A \cap X) + |A \cap D|.$$

(For matroid theorists, the elements of D are added as isthmuses or coloops.) □

Two relations $P : S \to X$ and $Q : T \to X$ have a *common partial transversal of size* τ if there is a subset Y of size τ in X such that both the relations $P' : S \to Y$ and $Q' : T \to Y$ (obtained by restricting the relations P and Q to Y) have a partial matching of size τ. If $|S| = |T|$ and P and Q have a common partial transversal of size $|S|$, then P and Q have a *common transversal.*

2.4.4. The Ford–Fulkerson common transversal theorem.[23] The relations $P : S \to X$ and $Q : T \to X$ have a common partial transversal of size τ if and only if for all pairs $A \subseteq S$ and $B \subseteq T$,

$$|P(A) \cap Q(B)| \geq |A| + |B| + \tau - |S| - |T|.$$

In particular,

$$\tau = |S| + |T| + \min\{|P(A) \cap Q(B)| - |A| - |B|\}. \qquad \text{(FF)}$$

Proof. It suffices to prove (FF). Let rk be the rank function of the transversal matroid on X defined by the relation Q. Then a common partial transversal of size τ exists if and only if an independent partial matching of size τ exists for P relative to rk. By Rado's theorem, this occurs if and only if for all subsets $A \subseteq S$,

$$\mathrm{rk}(P(A)) \geq |A| - (|S| - \tau).$$

To obtain Equation (FF), use the fact that, by (TM),

$$\mathrm{rk}(P(A)) = |T| + \min\{|Q(B) \cap P(A)| - |B|\colon B \subseteq T\}.$$ □

Exercises

2.4.1. *Submodular functions and matroids.*

Prove the following results:

(a) If ρ and σ are submodular functions and α and β are nonnegative real numbers, then $\alpha\rho + \beta\sigma$ is a submodular function.

[23] Ford and Fulkerson (1958). Ford and Fulkerson give a proof using the maximum-flow minimum-cut theorem. Our proof is from Mirsky and Perfect (1967).

(b) Let $\rho : 2^X \to \mathbb{Z}$ be an integer-valued, increasing, submodular function. Then the subsets I such that for all nonempty subsets $J \subseteq I$, $|J| \leq \rho(J)$ are the independent sets of a matroid $M(\rho)$ on X.

(c) If ρ is assumed to be normalized, that is, $\rho(\emptyset) = 0$, then the matroid rank function rk of $M(\rho)$ is given by

$$\mathrm{rk}(A) = \min\{\rho(B) + |A \backslash B|: B \subseteq A\}.$$

(d) Let $R : S \to X$ be a relation and ρ the submodular function on S defined by $A \mapsto |R(A)|$. Characterize such submodular functions ρ and the matroids $M(\rho)$ constructed from them.

2.4.2. *Welsh's version of the matroid marriage theorem.*[24]

Adapt Rado's proof (the second proof given in Section 2.2) of the marriage theorem to the matroid case. This will prove the following more general form of the matroid marriage theorem: let $R : S \to X$ and let ρ be a nonnegative, integer-valued, increasing, submodular function defined on X. Then there exists a matching M in R with $\rho(M(T)) \geq |T|$ for all subsets T in S if and only if $\rho(R(T)) \geq |T|$ for all subsets T in S.

2.4.3. Adapt other proofs of the marriage theorem to the matroid case. Is there a matroid analog of Theorem 2.2.2?

2.4.4. *Contractions and Gaussian elimination.*

Suppose that rk is the rank function of a matrix M with column set X and a a nonzero column of M. Construct the matrix M' as follows: choose a row index i such that the ia-entry is nonzero. Subtracting a suitable multiple of row i from the other rows, reduce the matrix M so that the only nonzero entry in column a is the ia-entry. Let M' be the matrix obtained from the reduced matrix by deleting row i and column a. Show that the rank function $\mathrm{rk}_{\{a\}}$ obtained by contracting $\{a\}$ is the rank function on the matrix M'.

2.4.5. Prove the matroid marriage theorem with rk a matrix rank function using the Binet–Cauchy formula for determinants and the idea in Edmonds' proof.

2.4.6. *Common transversals, compositions of relations, and matrix products.*

(a) Let $P : S \to X$ and $Q : T \to X$ be relations. Show that P and Q have a common partial transversal of size τ if and only if the composition $PQ^{-1} : T \to S$ has a partial matching of size τ.

[24] Welsh (1971).

(b) Let C and D be the free matrices of P and Q. Then P and Q have a common partial transversal of size τ if and only if the matrix product CD^T has rank τ.

(c) Find an analog of the König–Egerváry theorem for the product of two free matrices.

2.4.7. *Ore's excess function.*[25]

Let $R : S \to X$ be a relation. The *excess function* η is defined by

$$\eta(A) = |R(A)| - |A|.$$

(a) Prove that η is a submodular function.

(b) Observe that the Hall condition is equivalent to $\eta(A) \geq 0$ for every subset A in S.

(c) Let $\eta_0(R) = \min\{\eta(A)\colon A \subseteq S\}$. Note that $\eta_0(R)$ may be negative. Define a subset A of S to be *critical* if $\eta(A) = \eta_0$. Show that if A and B are critical, then $A \cup B$ and $A \cap B$ are also critical.

(d) Define the *core* of the relation R to be the intersection of all the critical subsets of S or, equivalently, the minimum critical subset. Let C be the core of R. Show (without the assumption that R satisfies the Hall condition) that the restriction $R : S \backslash C \to X$ of R to the complement of C always satisfies the Hall condition.

(e) Every element x in $R(C)$ is related to at least two elements of C.

(f) Let $a \in S$ and let R' be the relation obtained from R by removing a and all ordered pairs (a, x) in R containing a. Show that

$$\eta_0(R) \leq \eta_0(R') \leq \eta_0(R) + 1.$$

Indeed, show that if $a \notin C$, then $\eta_0(R') = \eta_0(R)$ and if $a \in C$, then $\eta_0(R') = \eta_0(R) + 1$.

(g) Show that if $\eta_0(R) < 0$, then there exist $-\eta_0(R)$ elements in S such that if we remove them and all ordered pairs containing them, we obtain a relation satisfying the Hall condition. Using the marriage theorem, deduce Ore's theorem (Theorem 2.2.8): let $R : S \to X$ be a relation. Then the maximum size of a partial matching is $|S| + \eta_0(R)$. (Ore gave an independent proof of the marriage theorem using the excess function η. However, this proof is quite similar to the Easterfield–Halmos–Vaughan proof.)

(h) Let $R^{-1} : X \to S$ be the "reverse" relation. Show that

$$\text{core}\, R \cap R^{-1}(\text{core}\, R^{-1}) = \emptyset.$$

[25] Ore (1955, 1962, chapter 10).

2.4.8. *Rota's basis conjecture.*[26]

Let $B_1, B_2, \ldots, B_n$ be bases of an n-dimensional vector space (or more generally, a rank-n matroid). Then the n^2 vectors occurring in the bases can be arranged in an $n \times n$ square so that each row and each column is a basis. Such a square can be considered a vector-analog of a Latin square.

2.5 Rado's Theorem on Subrelations

The marriage theorem is about when a relation contains a one-to-one function. In this form, it seems quite specialized. However, when it is applied to suitably constructed relations, it generates results that are, or appear to be, more general.

We begin with a simple example due to Halmos and Vaughan.[27] Let $R : S \to X$ be a relation and $m : S \to \{0, 1, 2, \ldots\}$ be a *multiplicity function* that assigns a nonnegative integer to each element of S. A relation $H : S \to X$ is an *m-matching* of R if $H \subseteq R$, $|H(a)| = m(a)$, and $H(a) \cap H(b) = \emptyset$ whenever $a \neq b$.

2.5.1. Theorem. Let $R : S \to X$ be a relation and $m : S \to \{0, 1, 2, \ldots\}$ be a multiplicity function. Then there exists an m-matching of R if and only if

$$\text{for all subsets } A \text{ in } S, \quad |R(A)| \geq \sum_{a \in A} m(a). \qquad \text{(HV)}$$

Proof. As in the marriage theorem, necessity is clear. To prove sufficiency, let S^* be the set

$$\bigcup_{a \in S} \{a_1, a_2, \ldots, a_{m(a)}\}$$

obtained from S by replacing each element a with $m(a)$ copies $a_1, a_2, \ldots, a_{m(a)}$. Consider the relation $R^* : S^* \to X$ defined by $(a_i, x) \in R^*$ if and only if $(a, x) \in R$. Then it is easy to show that

- condition (HV) holds for R if and only if the Hall condition holds for R^*, and
- an m-matching exists in R if and only if a matching exists in R^*.

We can now complete the proof using the marriage theorem. □

Theorem 2.5.1 can be extended to the case when the sets $H(a)$ have specified overlaps. As usual, we need several definitions. Two relations $T : S \to X$ and

[26] Huang and Rota (1994).

[27] Halmos and Vaughan (1950). This theorem is sometimes called the harem theorem. Halmos and Vaughan cite a "conte drolâtique" of Balzac as motivation.

$T' : S \to Y$ are *combinatorially equivalent* if there exists a bijection $\sigma : X \to Y$ such that

$$(a, x) \in T \text{ if and only if } (a, \sigma(x)) \in T',$$

or, equivalently, for all a in S, $T'(a) = \sigma(T(a))$. Note that the set S is fixed in a combinatorial equivalence, and hence an isomorphism of relations is not necessarily a combinatorial equivalence.

We will modify Theorem 2.5.1 into a criterion for deciding when a relation $R : S \to X$ contains a subrelation equivalent to a given *template relation* $T : S \to Y$. To do so, we first construct from T a new relation $T^\flat$, and from it, a multiplicity function m_T. The idea is to partition Y into (disjoint) *blocks* so that the sets $T(a)$, $a \in S$, are unions of blocks. Let $T^\flat : 2^S \to Y$ be the relation defined as follows: if A is a subset of S, then

$$T^\flat(A) = \left(\bigcap_{a: a \in A} T(a) \right) \cap \left(\bigcap_{a: a \notin A} T(a)^c \right),$$

where $T(a)^c$ is the complement $Y \backslash T(a)$, that is, $T^\flat(A)$ is the subset of those elements in X that are in all of the subsets $T(a)$, $a \in A$, and none of the subsets $T(a), a \notin A$. It follows from the definition that $T^\flat(\emptyset) = \emptyset$, the sets $T^\flat(A)$ are pairwise disjoint, and their union is the range $T(S)$. The *multiplicity function* m_T defined by T is the function $2^S \to \{0, 1, 2, \ldots\}$ defined by

$$m_T(A) = |T^\flat(A)|.$$

Note that $T^\flat$ is an m_T-matching.

For example, suppose $T : \{1, 2, 3\} \to \{a, b, c, d, e, f\}$ is the relation defined by

$$T(1) = \{a, b, c\},$$
$$T(2) = \{b, c, d, e\},$$
$$T(3) = \{c, f\}.$$

Then

$$T^\flat(\emptyset) = \emptyset,$$
$$T^\flat(\{1\}) = \{a\},\ T^\flat(\{2\}) = \{d, e\},\ T^\flat(\{3\}) = \{f\},$$
$$T^\flat(\{1, 2\}) = \{b\}, T^\flat(\{1, 3\}) = T^\flat(\{2, 3\}) = \emptyset,$$
$$T^\flat(\{1, 2, 3\}) = \{c\}.$$

The next lemma gives a connection between multiplicity functions and combinatorial equivalence. This lemma uses results about Boolean polynomials presented in Exercise 1.3.7 in Section 1.3.

2.5.2. Lemma. Let $T : \{1, 2, \ldots, n\} \to X$ and $T' : \{1, 2, \ldots, n\} \to Y$ be relations such that $T(\{1, 2, \ldots, n\}) = X$ and $T'(\{1, 2, \ldots, n\}) = Y$. Then, the following are equivalent:

(a) The relations $T : \{1, 2, \ldots, n\} \to X$ and $T' : \{1, 2, \ldots, n\} \to Y$ are combinatorially equivalent.

(b) For every monotone Boolean polynomial $\beta(X_1, X_2, \ldots, X_n)$ in n variables,

$$|\beta(T(1), T(2), \ldots, T(n))| = |\beta(T'(1), T'(2), \ldots, T'(n))|.$$

(c) The multiplicity functions m_T and $m_{T'}$ are equal.

Proof. Let $\sigma : X \to Y$ be a bijection giving an equivalence of T and T'. Then σ preserves all the Boolean operations. Hence, for any Boolean polynomial β,

$$\begin{aligned}\beta(T'(1), T'(2), \ldots, T'(n)) &= \beta(\sigma(T(1)), \sigma(T(2)), \ldots, \sigma(T(n))) \\ &= \sigma(\beta(T(1), T(2), \ldots, T(n))).\end{aligned}$$

In particular,

$$|\beta(T(1), T(2), \ldots, T(n))| = |\beta(T'(1), T'(2), \ldots, T'(n))|.$$

This proves that (a) implies (b).

To show that (b) implies (c), observe that

$$T^\flat(A) = \left[\bigcap_{a:a\in A} T(a)\right] \setminus \left[\left(\bigcup_{a:a\notin A} T(a)\right) \cap \left(\bigcap_{a:a\in A} T(a)\right)\right]. \qquad \text{(B)}$$

Hence,

$$m_T(A) = |T^\flat(A)| = |\beta_1(T(i))| - |\beta_2(T(i))|,$$

where β_1 and β_2 are the monotone Boolean polynomials occurring in the left side of Equation (B). Since (b) holds, we have

$$|\beta_1(T(i))| - |\beta_2(T(i))| = |\beta_1(T'(i))| - |\beta_2(T'(i))|.$$

We conclude that for all $A \subseteq S$, $m_T(A) = m_{T'}(A)$.

Finally, assume that (c) holds. Since the sets $T^\flat(A)$ are pairwise disjoint, we can build a bijection $X \to Y$ by choosing bijections from $T^\flat(A)$ to $(T')^\flat(A)$ for each $A \subseteq S$ and putting them together into one bijection. □

Next, we construct a new relation from R. If $R : S \to X$ is a relation, then its *Boolean expansion* $R^\# : 2^S \to X$ is the relation defined by

$$R^\#(A) = \bigcap_{a:a\in A} R(a).$$

For example, suppose $R : \{1, 2, 3\} \to \{a, b, c, d, e, f\}$ is the relation defined by

$$\begin{aligned} R(1) &= \{a, b, c, d\}, \\ R(2) &= \{b, c, d, e\}, \\ R(3) &= \{a, c, f\}. \end{aligned}$$

Then $R^{\#}(\emptyset) = \emptyset$, $R^{\#}(\{1\}) = R(1)$, $R^{\#}(\{2\}) = R(2)$, $R^{\#}(\{3\}) = R(3)$, and

$$R^{\#}(\{1, 2\}) = \{b, c, d\},\ R^{\#}(\{1, 3\}) = \{a, c\},\ R^{\#}(\{2, 3\}) = \{c\},$$
$$R^{\#}(\{1, 2, 3\}) = \{c\}.$$

2.5.3. Rado's theorem on subrelations.[28] Let $T : \{1, 2, \ldots, n\} \to Y$ be a template relation such that $T(\{1, 2, \ldots, n\}) = Y$ and $m_T : 2^{\{1,2,\ldots,n\}} \to \{0, 1, 2, \ldots\}$ be its multiplicity function. Then the following are equivalent:

(a) The relation $R : \{1, 2, \ldots, n\} \to X$ has a subrelation combinatorially equivalent to T.

(b) The Boolean expansion $R^{\#} : 2^{\{1,2,\ldots,n\}} \to X$ has an m_T-matching.

(c) For every monotone Boolean polynomial $\beta(X_1, X_2, \ldots, X_n)$,

$$|\beta(R(1), R(2), \ldots, R(n))| \geq |\beta(T(1), T(2), \ldots, T(n))|.$$

Proof. To see that (a) implies (b), let $\hat{T}$ be a subrelation of R combinatorially equivalent to T. Then the Boolean expansion $R^{\#}$ contains a subrelation combinatorially equivalent to m_T-matching $\hat{T}^{\flat}$. Conversely, if $R^{\#}$ has an m_T-matching $H^{\flat}$, then let $H : \{1, 2, \ldots, n\} \to X$ be the relation defined by

$$H(a) = \bigcup_{A:\, a \in A} H^{\flat}(A).$$

Since the multiplicity function of H equals m_T, H is combinatorially equivalent to T by Lemma 2.5.2. We conclude that (b) implies (a).

Next, suppose (a) holds. Since T is combinatorially equivalent to a subrelation of R, there is an injection $\iota : Y \to X$ such that $\iota(T(i)) \subseteq R(i)$ for $1 \leq i \leq n$. Hence, for all monotone Boolean polynomials β,

$$\beta(\iota(T(1)), \iota(T(2)), \ldots, \iota(T(n))) \subseteq \beta(R(1), R(2), \ldots, R(n)).$$

The inequalities in (c) follow.

To finish the proof, we show that (c) implies (b). By Theorem 2.5.1, it suffices to prove that (HV) holds in $R^{\#}$. Let C be a collection of subsets in

[28] Rado (1938). We present Rado's theorem for finite sets. Our proof is based on the exposition in chapter 5 of Mirsky (1971b).

$\{1, 2, \ldots, n\}$. Then

$$\bigcup_{A:\, A \in C} R^{\#}(A)$$

is a monotone Boolean polynomial in $R(i)$. Hence, by (c),

$$\left| \bigcup_{A:\, A \in C} R^{\#}(A) \right| \geq \left| \bigcup_{A:\, A \in C} T^{\#}(A) \right| \geq \left| \bigcup_{A:\, A \in C} T^{\flat}(A) \right|.$$

The last derivation follows from $T^{\#}(i) \supseteq T^{\flat}(i)$. Since the sets $T^{\flat}(i)$ are pairwise disjoint,

$$\left| \bigcup_{A:\, A \in C} T^{\flat}(A) \right| = \sum_{A:\, A \in C} |T^{\flat}(A)| = \sum_{A:\, A \in C} m_T(A).$$

Hence, condition (HV) holds and by Theorem 2.5.1, $R^{\#}$ has an m_T-matching; that is, (c) implies (b). □

Exercises

2.5.1. Let $C : \{1, 2, \ldots, n\} \to \{1, 2, \ldots, n\}$ be the "cycle" relation with edges (i, i) and $(i, i + 1)$, $i = 1, 2, \ldots, n$, regarded as integers modulo n. Find a condition for a relation $R : \{1, 2, \ldots, n\} \to X$ to contain a subrelation combinatorially equivalent to C.

2.5.2. (Research problem) Is there a version of Rado's theorem where combinatorial equivalence is replaced by isomorphisms of relations?

2.6 Doubly Stochastic Matrices

A matrix D is *doubly stochastic* if it has nonnegative real entries and the sum of all the entries on each row or each column equals 1. In other words, D is doubly stochastic if $\underline{e}D = \underline{e}$ and $D\underline{e}^T = \underline{e}^T$, where $\underline{e}$ is the row vector having all coordinates equal to 1. Since the sum of all the entries in a doubly stochastic matrix equals both the number of rows and the number of columns, such a matrix must be a square matrix.

Doubly stochastic matrices are special cases of transition matrices of Markov chains. (Transition matrices are only required to have column sums equal to 1.) Also, it is easy to construct examples. If π is a permutation of $\{1, 2, \ldots, n\}$, then we can associate with it the matrix P_π with entries $p_{\pi(i),i}$ equal to 1 and all other entries equal to 0. Such matrices, called *permutation matrices*, are doubly stochastic. Also, if $c_1, c_2, \ldots, c_n$ are real numbers in the unit interval $[0, 1]$ and $c_1 + c_2 + \cdots + c_n = 1$, then the circulant matrix

$$\begin{pmatrix} c_1 & c_2 & c_3 & \cdots & c_{n-1} & c_n \\ c_n & c_1 & c_2 & \cdots & c_{n-2} & c_{n-1} \\ c_{n-1} & c_n & c_1 & \cdots & c_{n-3} & c_{n-2} \\ \vdots & \vdots & & & & \vdots \\ c_2 & c_3 & c_4 & \cdots & c_n & c_1 \end{pmatrix}$$

is one possible doubly stochastic matrix one can construct.

Several facts follow easily from the definition. If D and E are doubly stochastic, then the product DE and the transpose D^T are doubly stochastic. Since $\underline{e}D = \underline{e}$, 1 is always an eigenvalue of a doubly stochastic matrix. If $\lambda, \mu \in [0, 1]$, and $\lambda + \mu = 1$, then the convex combination $\lambda D + \mu E$ is doubly stochastic. The set of $n \times n$ matrices with real entries forms a n^2-dimensional real vector space. The set of $n \times n$ doubly stochastic matrices is a closed bounded convex subset of this vector space.

The fundamental theorem in the combinatorics of doubly stochastic matrices is the following theorem.

2.6.1. Birkhoff's Theorem.[29] Every doubly stochastic matrix is a convex combination of permutation matrices.

We give two proofs in the text and sketch three others as exercises. The first uses the marriage theorem and is based on Birkhoff's original proof.

First proof of Birkhoff's theorem. We proceed by induction on the number of nonzero entries. If D is an $n \times n$ doubly stochastic matrix, then D must have at least n nonzero entries. If D has exactly n nonzero entries, then D is a permutation matrix and the theorem holds.

We may now assume that D has more than n nonzero entries. Let (d_{ij}) be such a matrix. Consider the relation $B : \{1, 2, \ldots, n\} \to \{1, 2, \ldots, n\}$ defined by $(i, j) \in B$ whenever $d_{ij} \neq 0$. We will check that the relation B satisfies the Hall condition. Let H be a subset of rows and consider the submatrix

[29] Birkhoff (1946).

$D[H|B(H)]$ obtained by restricting D to the rows H and columns $B(H)$. Observe that the sum of all the entries in the submatrix $D[H|B(H)]$ equals $|H|$. Since the sum of all the entries in any column in $D[H|B(H)]$ is at most 1, there must be at least $|H|$ columns in that submatrix. We conclude that $|B(H)| \geq |H|$.

By the marriage theorem, B has a matching M. Let P be the permutation matrix with 1s exactly at the entries (i, j), where (i, j) is an edge in the matching M. Let ϵ be the minimum $\min\{d_{ij}: (i, j) \in M\}$. Then the matrix

$$\frac{1}{1-\epsilon}[D - \epsilon P]$$

is a doubly stochastic matrix with at least one fewer nonzero entry. By induction, it is a convex combination of permutation matrices. We conclude that D itself is a convex combination of permutation matrices. □

We remark that the main step in Birkhoff's proof is to prove that there is at least one nonzero term in the determinant expansion of a doubly stochastic matrix (even if the determinant is zero). This was stated by D. König in 1916 and he proved it in the following way:[30] by the König–Egerváry theorem, if all the terms in the determinant expansion of an $n \times n$ doubly stochastic matrix D are zero, then D contains an $h \times k$ zero submatrix, where $h + k > n$; that is, there exists a set H of h rows such that the submatrix $D[H]$ has at most $h - 1$ nonzero columns. As in the first proof of Birkhoff's theorem, the argument using the sum of entries now yields a contradiction.

In contrast to Birkhoff's combinatorial proof, the second proof uses some convexity theory. A point a in a convex set C is *extreme* if a cannot be a proper convex combination of two points in C; that is, $a = \lambda b + \mu c$, $b, c \in C$, $\lambda + \mu = 1$, and $\lambda, \mu \in [0, 1]$ imply that $a = b$ or $a = c$. The Krein–Milman theorem says that every closed bounded convex subset is the convex closure of its extreme points.[31]

Second proof of Birkhoff's theorem. As observed earlier, the set of doubly stochastic matrices is a closed bounded convex set in real n^2-dimensional space. Thus, by the Krein–Milman theorem, it suffices to prove the following lemma.

2.6.2. Lemma.[32] A doubly stochastic matrix is extreme if and only if it is a permutation matrix.

[30] König (1916). [31] See, for example, Webster (1994).
[32] J. von Neumann, unpublished; but see Exercise 2.6.7.

Proof. If a permutation matrix P is a convex combination $\lambda A + \mu B$ of two matrices with real nonnegative entries, then the ijth entries of A and B are zero whenever the ijth entry of P is zero. Since A and B are doubly stochastic, A and B equal P or the zero matrix. We conclude that P is extreme.

Let E be an extreme doubly stochastic matrix. If every nonzero entry of E equals 1, then E is a permutation matrix and the theorem holds. Thus, we may assume that there are nonzero entries in E which are strictly less than 1. Consider the undirected graph on the vertex set $\{(i, j)\colon 0 < m_{ij} < 1\}$, with adjacencies given by (i, j) is adjacent to (i', j') whenever $i = i'$ or $j = j'$. In other words, two ordered pairs on the same row or the same column are adjacent. Since the row and column sums all equal 1, every vertex is adjacent to at least two other vertices. Hence, this graph contains a cycle.

Starting at any vertex in the cycle, label the vertices in the cycle consecutively by $1, 2, \ldots$. Choose a positive real number ϵ sufficiently small so that when it is added to or subtracted from an entry occurring in the cycle, the entry remains in the unit interval $[0, 1]$. Let E_{even} be the matrix obtained from E by adding ϵ to all the even-labeled entries and subtracting ϵ from all the odd-labeled entries ϵ. The matrix E_{odd} is constructed similarly, with $-\epsilon$ replacing ϵ, so that ϵ is subtracted from the even vertices and added to the odd vertices. By construction,

$$E = \frac{E_{\text{even}} + E_{\text{odd}}}{2}.$$

We conclude that if E is not a permutation matrix, then E is not an extreme point. □

For example, in the matrix

$$\begin{pmatrix} .6 & 0 & .4 \\ .3 & .5 & .2 \\ .1 & .5 & .4 \end{pmatrix},$$

$(1, 1), (2, 1), (2, 2), (3, 2), (3, 3), (1, 3)$ is a cycle with $(1, 1), (2, 2), (3, 3)$ odd and $(2, 1), (3, 2), (1, 3)$ even and we have

$$\begin{pmatrix} .6 & 0 & .4 \\ .3 & .5 & .2 \\ .1 & .5 & .4 \end{pmatrix} = .5 \begin{pmatrix} .6-\epsilon & 0 & .4+\epsilon \\ .3+\epsilon & .5-\epsilon & .2 \\ .1 & .5+\epsilon & .4-\epsilon \end{pmatrix}$$
$$+ .5 \begin{pmatrix} .6+\epsilon & 0 & .4-\epsilon \\ .3-\epsilon & .5+\epsilon & .2 \\ .1 & .5-\epsilon & .4+\epsilon \end{pmatrix}.$$

We can choose ϵ to be any real number in the interval $(0, .3]$.

Doubly stochastic matrices play a central role in (discrete) majorization theory. Let $(r_1, r_2, \dots, r_n)$ and $(s_1, s_2, \dots, s_n)$ be two row vectors with nonnegative real coordinates and $(r'_1, r'_2, \dots, r'_n)$ and $(s'_1, s'_2, \dots, s'_n)$ be the vectors obtained by rearranging their coordinates in nonincreasing order. The *majorization order* $\preceq$ is the quasi-order defined by

$$(r_1, r_2, \dots, r_n) \preceq (s_1, s_2, \dots, s_n)$$

whenever

$$r_1 + r_2 + \cdots + r_n = s_1 + s_2 + \cdots + s_n,$$

and

$$\begin{aligned} r'_1 &\leq s'_1, \\ r'_1 + r'_2 &\leq s'_1 + s'_2, \\ &\vdots \\ r'_1 + r'_2 + \cdots + r'_{n-1} &\leq s'_1 + s'_2 + \cdots + s'_{n-1}. \end{aligned}$$

There are two basic results about the majorization order.

2.6.3. Theorem.[33] $(r_1, r_2, \dots, r_n) \preceq (s_1, s_2, \dots, s_n)$ if and only if $(r_1, r_2, \dots, r_n)$ is a convex combination of the $n!$ vectors

$$(s_{\pi(1)}, s_{\pi(2)}, \dots, s_{\pi(n)})$$

obtained by permuting the coordinates of $(s_1, s_2, \dots, s_n)$.

2.6.4. The Hardy–Littlewood–Pólya majorization theorem.[34] Let $\underline{r}$ and $\underline{s}$ be vectors with nonnegative real coordinates. Then $\underline{r} \preceq \underline{s}$ if and only if there exists a doubly stochastic matrix D such that $\underline{r} = \underline{s}D$.

Theorem 2.6.3 implies Theorem 2.6.4, and assuming Birkhoff's theorem, Theorem 2.6.4 implies Theorem 2.6.3. Thus, it suffices to prove Theorem 2.6.4.

We will now sketch a proof of Theorem 2.6.4.[35] A *transfer* T is a matrix of the form

$$\lambda I + (1 - \lambda)Q,$$

[33] This seems to be a folklore theorem. An equivalent form (stated for symmetric means) appeared in Rado (1952).

[34] This is Theorem 43 in the famous book by Hardy et al. (1952). We give the usual attribution, although the theorem has appeared earlier in Muirhead (1901).

[35] This proof is based on proofs in Muirhead (1901) and Hardy et al. (1952).

where $\lambda \in [0, 1]$, I is the identity matrix, and Q is a permutation with exactly two off-diagonal matrices (that is, Q is the matrix of a transposition). If Q is the matrix of the transposition switching j and k (keeping all the other indices fixed), then

$$\begin{aligned}&(s_1, s_2, \ldots, s_j, \ldots, s_k, \ldots, s_n)(\lambda I + (1-\lambda)Q)\\ &\quad= (s_1, s_2, \ldots, \lambda s_j + (1-\lambda)s_k, \ldots, \lambda s_k + (1-\lambda)s_j, \ldots, s_n).\end{aligned}$$

In particular, if $0 \leq \delta \leq s_j - s_k$, and we choose λ to be $1 - \delta/(s_j - s_k)$, then the $\lambda I + (1-\lambda)Q$ transforms $\underline{s}$ to

$$(s_1, s_2, \ldots, s_j - \delta, \ldots, s_k + \delta, \ldots, s_n).$$

Since transfers are doubly stochastic, Theorem 2.6.4 follows from the next lemma.

2.6.5. Lemma. If $\underline{r} \preceq \underline{s}$, then there exists a finite sequence $T_1, T_2, \ldots, T_m$ of transfers such that $\underline{r} = \underline{s}T_1T_2\cdots T_m$.

Sketch of proof. Assume that $\underline{r} \prec \underline{s}$. Let j and k be indices such that $j < k$, $r_j < s_j$, $r_k > s_k$, and $r_i = s_i$ for all indices $j < i < k$. Let $\delta = \min\{s_j - r_j, r_k - s_k\}$. Then we can subtract δ from s_j and add it to s_k so that the resulting vector $\underline{s}'$ satisfies $\underline{r} \preceq \underline{s}' \prec \underline{s}$ and at least one of $r_j = s'_j$ and $r_k = s'_k$ holds. This movement of δ can be effected by the transfer

$$T_1 = \left(1 - \frac{\delta}{s_j - s_k}\right) I + \frac{\delta}{s_j - s_k} Q,$$

where Q is the matrix of the transposition switching j and k. Since $\underline{s}' = \underline{s}Q$, Theorem 2.6.4 implies that $\underline{s}' \prec \underline{s}$. It is less immediate that $\underline{r} \preceq \underline{s}'$, but this can be checked by an easy argument. Now repeat the argument on $\underline{s}T_1$ and continue until $\underline{r} = \underline{s}'$. □

For example, let $\underline{r} = (20, 16, 16, 14, 8, 3, 3, 3)$ and $\underline{s} = (20, 20, 16, 11, 9, 3, 3, 1)$. Then $j = 2$, $k = 4$, and $\delta = 3$. We can subtract 3 from s_2 and add it to s_4 to obtain

$$(20, 17, 16, 14, 9, 3, 3, 1),$$

a vector between $\underline{r}$ and $\underline{s}$ in the majorization order. The new vector can be obtained from $\underline{s}$ by the transfer

$$\frac{2}{3}I + \frac{1}{3}Q,$$

where Q is the matrix of the transposition switching 2 and 4.

We end this section with Muirhead's theorem on symmetric means. Let $(a_1, a_2, \ldots, a_n)$ be a vector with nonnegative real values. Then its *symmetric mean* $[a_1, a_2, \ldots, a_n]$ is the function on the variables $x_1, x_2, \ldots, x_n$, defined by the formula

$$[a_1, a_2, \ldots, a_n] = \frac{1}{n!} \sum_{\pi} x_{\pi(1)}^{a_1} x_{\pi(2)}^{a_2} \cdots x_{\pi(n)}^{a_n},$$

where the sum ranges over all permutations of $\{1, 2, \ldots, n\}$. Note that the sum on the right can be further symmetrized. For example, because

$$[a_1, a_2, \ldots, a_n] = \frac{1}{n!} \sum_{\pi} x_{\pi(2)}^{a_1} x_{\pi(1)}^{a_2} x_{\pi(3)}^{a_3} \cdots x_{\pi(n)}^{a_n},$$

we have

$$[a_1, a_2, \ldots, a_n] = \frac{1}{2n!} \sum_{\pi} (x_{\pi(1)}^{a_1} x_{\pi(2)}^{a_2} + x_{\pi(2)}^{a_1} x_{\pi(1)}^{a_2}) x_{\pi(3)}^{a_3} \cdots x_{\pi(n)}^{a_n}.$$

Symmetric means include as special cases many of averages and symmetric functions. For example, $[1, 0, 0, \ldots, 0]$ is the *arithmetic mean*

$$\frac{x_1 + x_2 + \cdots + x_n}{n},$$

and $[1/n, 1/n, \ldots, 1/n]$ is the *geometric mean*

$$(x_1 x_2 \cdots x_n)^{1/n}.$$

Further, if there are m 1s, then

$$[1, 1, \ldots, 1, 0, 0, \ldots, 0] = \frac{m!(n-m)!}{n!} \sum_{\{i, j, \ldots, k\}} x_i x_j \cdots x_k,$$

where the sum ranges over all m-element subsets $\{i, j, \ldots, k\}$ of $\{1, 2, \ldots, n\}$. Thus, $[1, 1, \ldots, 1, 0, 0, \ldots, 0]$ is a constant multiple of the *mth elementary symmetric function*.

2.6.6. Muirhead's inequality.[36] Let $\underline{a}$ and $\underline{b}$ be two vectors with nonnegative real coordinates. Then

$$[\underline{a}] \leq [\underline{b}]$$

for all nonnegative real values $x_1, x_2, \ldots, x_n$ if and only if $\underline{a} \preceq \underline{b}$.

[36] Muirhead (1901).

Muirhead's inequality is a generalization of the *arithmetic–geometric mean inequality:*

$$(x_1x_2\cdots x_n)^{1/n} \le \frac{x_1+x_2+\cdots+x_n}{n}.$$

One way to prove this inequality (due to A. Hurwitz[37]) is to write the difference of the two means as a telescoping sum:

$$\begin{aligned}
&[1,0,0,\ldots,0]-\left[\frac{1}{n},\frac{1}{n},\frac{1}{n},\ldots,\frac{1}{n}\right]\\
&=\sum_{i=0}^{n-2}\left(\left[\frac{n-i}{n},\frac{1}{n},\ldots,\frac{1}{n},0,0,\ldots,0\right]\right.\\
&\quad\left.-\left[\frac{n-(i+1)}{n},\frac{1}{n},\ldots,\frac{1}{n},\frac{1}{n},0,\ldots,0\right]\right).
\end{aligned}$$

A summand in the right-hand sum can be rewritten as follows:

$$\begin{aligned}
&\left[\frac{n-i}{n},\frac{1}{n},\ldots,\frac{1}{n},0,0,\ldots,0\right]-\left[\frac{n-(i+1)}{n},\frac{1}{n},\ldots,\frac{1}{n},\frac{1}{n},0,\ldots,0\right]\\
&=\frac{1}{2n!}\sum_{\pi}\left(x_{\pi(1)}^{(n-i)/n}+x_{\pi(i+2)}^{(n-i)/n}-x_1^{(n-(i+1))/n}x_{\pi(i+2)}-x_{\pi(i+2)}^{(n-(i+1))/n}x_{\pi(1)}\right)\\
&\quad\cdot\left(x_{\pi(2)}^{1/n}x_{\pi(3)}^{1/n}\cdots x_{\pi(i+1)}^{1/n}\right)\\
&=\frac{1}{2n!}\sum_{\pi}\left(x_{\pi(1)}^{(n-(i+1))/n}-x_{\pi(i+2)}^{(n-i-1)/n}\right)\left(x_{\pi(1)}^{1/n}-x_{\pi(i+2)}^{1/n}\right)\left(x_{\pi(2)}^{1/n}x_{\pi(3)}^{1/n}\cdots x_{\pi(i+1)}^{1/n}\right).
\end{aligned}$$

Since $x_{\pi(1)}^{(n-i-1)/n}-x_{\pi(i+2)}^{(n-i-1)/n}$ and $x_{\pi(1)}^{1/n}-x_{\pi(i+2)}^{1/n}$ have the same sign, we conclude that each summand, and hence the sum itself, on the right is nonnegative.

Hurwitz's argument can be generalized to a proof of sufficiency in Muirhead's theorem. The telescoping sum is replaced by Lemma 2.6.5 and the nonnegativity of each difference is replaced by the following lemma.

2.6.7. Lemma. For any transfer T,

$$[\underline{a}]-[\underline{a}T]\ge 0$$

for all nonnegative real values $x_1,x_2,\ldots,x_n$.

Proof. We may assume that $\underline{a}$ is in nonincreasing order. Suppose that the transfer T sends $(a_1,\ldots,a_j,\ldots,a_k,\ldots,a_n)$ to $(a_1,\ldots,a_j-\delta,\ldots,a_k+\delta,\ldots,a_n)$,

[37] Hurwitz (1891).

where $\delta \le a_j - a_k$. Then

$$\begin{aligned}
&[\underline{a}] - [\underline{a}T] \\
&= \frac{1}{2n!}\sum_{\pi}(x_{\pi(j)}^{a_j}x_{\pi(k)}^{a_k} + x_{\pi(k)}^{a_j}x_{\pi(j)}^{a_k} - x_{\pi(j)}^{a_j-\delta}x_{\pi(k)}^{a_k+\delta} - x_{\pi(k)}^{a_j-\delta}x_{\pi(j)}^{a_k+\delta}) \prod_{i:i\neq j\, i\neq k} x_{\pi(i)}^{a_i} \\
&= \frac{1}{2n!}\sum_{\pi} x_{\pi(j)}^{a_k}x_{\pi(k)}^{a_k}(x_{\pi(j)}^{a_j-a_k-\delta} - x_{\pi(k)}^{a_j-a_k-\delta})(x_{\pi(j)}^{\delta} - x_{\pi(k)}^{\delta}) \prod_{i:i\neq j\, i\neq k} x_{\pi(i)}^{a_i}.
\end{aligned}$$

Since $x_{\pi(j)}^{a_j-a_k-\delta} - x_{\pi(k)}^{a_j-a_k-\delta}$ and $x_{\pi(j)}^{\delta} - x_{\pi(k)}^{\delta}$ have the same sign and $x_1, x_2, \ldots, x_n$ are nonnegative, the difference $[\underline{a}] - [\underline{a}T]$ is nonnegative. □

We can also prove sufficiency using elementary analysis. Suppose $\underline{a} \preceq \underline{b}$. Then by the Theorems 2.6.1 and 2.6.4,

$$\underline{a} = \sum h_Q \underline{b} Q,$$

where $\sum h_Q Q$ is a convex combination of permutation matrices.

By continuity, it suffices to prove that $[\underline{a}] \le [\underline{b}]$ for positive real values $x_1, x_2, \ldots, x_n$. Let $y_i = \log x_i$. If $\underline{c}$ is the vector $(c_1, c_2, \ldots, c_n)$, let

$$(\underline{c}|\underline{y}) = c_1y_1 + c_2y_2 + \cdots + c_ny_n.$$

If P is the matrix of the permutation π^{-1} of $\{1, 2, \ldots, n\}$, then

$$(\underline{c}P|\underline{y}) = c_1y_{\pi(1)} + c_2y_{\pi(2)} + \cdots + c_ny_{\pi(n)}.$$

Then

$$\begin{aligned}
n![\underline{a}] &= \sum_P \exp((\underline{a}P|\underline{y})) \\
&= \sum_P \exp((\sum h_Q \underline{b}QP|\underline{y})) \\
&\le \sum_P \sum h_Q \exp((\underline{b}QP|\underline{y})) \\
&= \sum h_Q \sum_P \exp((\underline{b}QP|\underline{y})) \\
&= \sum h_Q \sum_P \exp((\underline{b}P|\underline{y})) \\
&= \sum_P \exp((\underline{b}P|\underline{y})) \\
&= n![\underline{b}].
\end{aligned}$$

The inequality in this derivation follows the theorem from calculus that the exponential function is convex; that is,

$$\exp(\lambda y + \mu z) \leq \lambda \exp(y) + \mu \exp(z),$$

where y and z are real numbers, $\lambda, \mu \in [0, 1]$, and $\lambda + \mu = 1$. This completes the analysis-based proof of sufficiency.

Surprisingly, it is not hard to prove necessity. Since symmetric means are invariant under permutation of variables, we may assume that $\underline{a}$ and $\underline{b}$ are in nondecreasing order. We may also assume the hypothesis that $[\underline{a}] \leq [\underline{b}]$ for all nonnegative real values x_i.

Setting all the variables x_i equal to x and using the hypothesis, we conclude that

$$x^{a_n + a_{n-1} + \cdots + a_n} \leq x^{b_n + b_{n-1} + \cdots + b_n}$$

for all nonnegative real numbers x. This can happen for both large and small values of x only when

$$a_n + a_{n-1} + \cdots + a_1 = b_n + b_{n-1} + \cdots + b_1.$$

Next, let $1 \leq k \leq n - 1$. If we set $x_n = x_{n-1} = \cdots = x_k = x$ and $x_{k-1} = x_{k-2} = \cdots = x_1 = 1$, then the highest power of x occurring in $[\underline{a}]$ has exponent $a_n + a_{n-1} + \cdots + a_k$ and the highest power of x in $[\underline{b}]$ has exponent $b_n + b_{n-1} + \cdots + b_k$. By the hypothesis, $[\underline{a}] \leq [\underline{b}]$ for all sufficiently large x, when the highest power of x dominates. Hence, we conclude that

$$a_n + a_{n-1} + \cdots + a_k \leq b_n + b_{n-1} + \cdots + b_k.$$

All together, we have proved that $\underline{a} \preceq \underline{b}$.

This completes the proof of Muirhead's inequality.

2.6.8. Corollary. Let $(a_1, a_2, \ldots, a_n)$ be a vector with nonnegative real coordinates such that $a_1 + a_2 + \cdots + a_n = 1$. Then

$$(x_1 x_2 \cdots x_n)^{1/n} \leq [a_1, a_2, \ldots, a_n] \leq \frac{x_1 + x_2 + \cdots + x_n}{n}$$

for all nonnegative real values $x_1, x_2, \ldots, x_n$.

Exercises

2.6.1. Find ways of obtaining doubly stochastic matrices from Latin squares.

2.6.2. Show that a matrix D is doubly stochastic if and only if for every row vector $\underline{r}$ with nonnegative real coordinates, $\underline{r}D \preceq \underline{r}$.

2.6.3. (a) Show that $x \preceq xP \preceq x$ for all vectors x if and only if P is a permutation matrix.

(b) Suppose that P is a nonsingular doubly stochastic matrix. Show that the inverse matrix P^{-1} is doubly stochastic if and only if P is a permutation matrix.

2.6.4. *The assignment polytope or the polytope of doubly stochastic matrices.*
I. The linear inequalities.

Show that the convex set of doubly stochastic matrices (d_{ij}) in the vector space of real $n \times n$ matrices is the convex polyhedron defined by $n^2 + 4n - 2$ linear inequalities: the *nonnegativity constraints*

$$d_{ij} \geq 0, \quad 1 \leq i, j \leq n;$$

and the *row and column sum constraints*

$$\sum_{i=1}^{n} d_{ij} \leq 1, \quad \sum_{i=1}^{n} d_{ij} \geq 1, \quad 1 \leq j \leq n;$$
$$\sum_{j=1}^{n} d_{ij} \leq 1, \quad \sum_{j=1}^{n} d_{ij} \geq 1, \quad 1 \leq i \leq n-1.$$

Note that the two constraints $\sum_{i=1}^{n} d_{in} \leq 1$ and $\sum_{i=1}^{n} d_{in} \geq 1$ are implied by the others, so that we can reduce the number of constraints to $4n - 2$.

II. Another proof of Birkhoff's theorem.[38]

(a) Let E be an extreme point of the polyhedron of doubly stochastic matrices. Then at least $(n-1)^2$ entries of E are 0.

(b) Conclude from (a) that one row of E, say, row i, must have $n - 1$ zero entries, and hence the remaining entry, say, the ij-entry, equals 1, and this is the only nonzero entry in row i and the column j of E.

(c) Show that the matrix E' obtained from E by deleting row i and column j is extreme in the convex polyhedron of $(n-1) \times (n-1)$ doubly stochastic matrices.

(d) Conclude by induction that E is a permutation matrix. This gives a geometric proof of Lemma 2.6.2, and hence Birkhoff's theorem.

III. Eigenvalues of normal matrices.[39]

Recall that a matrix U (with entries over the complex numbers) is *unitary* if $UU^T = I$; a matrix A is *normal* if there exists a unitary matrix U such that UAU^T is a diagonal matrix. Let (a_{ij}) and (b_{ij}) be normal $n \times n$ matrices, $\alpha_1, \alpha_2, \ldots, \alpha_n$ be the eigenvalues of A, and $\beta_1, \beta_2, \ldots, \beta_n$ be the eigenvalues

[38] Hoffman and Wielandt (1953). [39] Hoffman and Wielandt (1953).

of B. Show that there exists a permutation π of $\{1, 2, \ldots, n\}$ such that

$$\sum_{i=1}^{n} |\alpha_i - \beta_{\pi(i)}| \leq \|A - B\|,$$

where the matrix norm $\|C\|$ is defined by

$$\|C\| = \sqrt{\sum_{i,j=0}^{n} |c_{ij}|^2}.$$

In other words, for a pair of normal matrices, the "minimum" ℓ_1-distance between their eigenvalues is at most the ℓ_2-distance between the matrices themselves.

IV. An assignment problem.[40]

Let (c_{ij}) be a matrix with nonnegative entries. Show that the linear program

$$\text{minimize} \sum_{i,j=0}^{n} c_{ij} x_{ij}$$

subject to

$$x_{ij} \geq 0, \quad \sum_{i=0}^{n} x_{ij} = 1, \quad \sum_{j=0}^{n} x_{ij} = 1$$

has an integer solution.

V. Quantifying Birkhoff's theorem.[41]

Show that every doubly stochastic matrix is the convex combination of $n^2 - 2n + 2$ or fewer permutation matrices and that this bound is sharp.

2.6.5. (a) Show that there is a 3×3 doubly stochastic matrix which is not the product of transfers.[42]

(b) (Research problem) Study the set of doubly stochastic matrices which are products of transfers.

2.6.6. Let J be the $n \times n$ matrix with all entries equal to 1. Show that $\frac{1}{n}J$ is an idempotent doubly stochastic matrix. (A matrix J is *idempotent* if $J^2 = J$.) Describe all idempotent doubly stochastic matrices.

2.6.7. Doubly substochastic matrices[43]: A square matrix is *doubly substochastic* if it has nonnegative real entries and all its row and column sums are less than or equal to 1.

[40] Dantzig (1963, section 15-1). [41] Farahat and Mirsky (1960).
[42] Marshall and Olkins (1979, p. 40). [43] Mirsky (1959) and von Neumann (1953).

(a) Using the argument in Lemma 2.6.2, show that if (c_{ij}) is doubly substochastic, then there exists a doubly stochastic matrix (d_{ij}) such that $c_{ij} \leq d_{ij}$ for all i, j.

(b) Show that every $n \times n$ doubly substochastic matrix is a convex combination of $n \times n$ matrices that have at most one 1 in each row and each column and all other entries equal to 0.

2.6.8. A doubly stochastic matrix is *even* if it is a convex combination of matrices of even permutations. Show that $d_1, d_2, \ldots, d_n$ are the diagonal entries of an $n \times n$ even doubly stochastic matrix if and only if[44]

$$\sum_{i=1}^{n} d_i \leq n - 3 + 3\min\{d_1, d_2, \ldots, d_n\}.$$

2.6.9. (a) Let $\underline{r} \preceq \underline{s}$. Show that the set of doubly stochastic matrices D such that $\underline{r} = \underline{s}D$ is convex.

(b) (Research problem) For a given pair of vectors $\underline{r}$, $\underline{s}$, study the convex set of doubly stochastic matrices D such that $\underline{r} = \underline{s}D$.

2.6.10. (a) Prove the following result[45]: let D be the $n \times n$ doubly stochastic matrix (d_{ij}). Then either D is the identity matrix or D satisfies the condition (*) there exists a permutation π, not equal to the identity, such that all the off-diagonal entries, $d_{i,\pi(i)}$, $i \neq \pi(i)$, are positive.

(b) Show that for any matrix (a_{ij}),

$$\sup_{(d_{ij})} \sum_{i,j=1}^{n} a_{ij} d_{ij} = \max_{\pi} \sum_{i=1}^{n} a_{i,\pi(i)},$$

where the supremum is taken over all $n \times n$ doubly stochastic matrices (d_{ij}) and the maximum is taken over all permutations of $\{1, 2, \ldots, n\}$.

2.6.11. *Symmetric doubly stochastic matrices.*[46]

(a) Determine the extreme points of the convex set $\mathcal{S}_n$ of doubly stochastic symmetric $n \times n$ matrices.

(b) Determine the extreme points of the convex set of doubly substochastic symmetric $n \times n$ matrices.

(c) Using the exponential formula (see Section 4.1), find exponential generating functions for the number of extreme points of $\mathcal{S}_n$.

2.6.12. Show that equality occurs in Muirhead's inequality if and only if $x_1 = x_2 = \cdots = x_n$ or $\underline{a} = \underline{b}$.

[44] Mirsky (1961). [45] Mirsky (1958).
[46] Converse and Katz (1975), Katz (1970, 1972), and Stanley (1999, exercise 5.24, p. 122).

2.6.13. A function $f : [0, \infty) \to [0, \infty)$ is *completely monotonic* if $(-1)^n f^{(n)}(x) \geq 0$ for all x in $[0, \infty)$.[47] Prove the following theorem: let $m \leq n$, $0 \leq p_1 \leq p_2 \leq \cdots \leq p_n$, and $0 \leq z_1 \leq z_2 \leq \cdots \leq z_m$. Then the rational function

$$\prod_{i=1}^{m}(x + z_i) \Big/ \prod_{i=1}^{n}(x + p_i)$$

is completely monotonic if for $1 \leq k \leq m$,

$$\sum_{i=1}^{k} z_i \geq \sum_{i=1}^{k} p_i.$$

2.6.14. *Inequalities among symmetric functions.*[48]

Muirhead's inequality is about evaluating symmetric functions at *nonnegative* real numbers. We may also ask for inequalities that hold for *all* real numbers. Since some fractional powers are not defined over all real numbers, this question is usually restricted to symmetric polynomials. A polynomial $p(x_1, x_2, \ldots, x_n)$ is *symmetric* if for every permutation π of $\{1, 2, \ldots, n\}$,

$$p(x_1, x_2, \ldots, x_n) = p(x_{\pi(1)}, x_{\pi(2)}, \ldots, x_{\pi(n)}).$$

(a) *Newton's inequalities:* Let

$$a_k = \frac{k!(n-k)!}{n!} \sum_{\{i,j,\ldots,l\}} x_i x_j \cdots x_l,$$

where the sum ranges over all k-element subsets of $\{1, 2, \ldots, n\}$. Show that for all real numbers,

$$a_{k-1} a_{k+1} \leq a_k^2.$$

(b) *MacLaurin's inequality*: Show that for all positive real numbers $x_1, x_2, \ldots, x_n$,

$$a_1 \geq a_2^{1/2} \geq a_3^{1/3} \geq \cdots \geq a_n^{1/n}.$$

A polynomial is *positive* if it can be written as a sum of squares of rational functions. The context is Hilbert's seventeenth problem. If a polynomial is a sum of squares, then it is nonnegative when evaluated at any n-tuple of real numbers.

[47] Ball (1994). [48] Procesi (1978).

(c) *Procesi's theorem*: Let

$$h_s(x_1, x_2, \ldots, x_n) = \sum_{i=1}^{n} x_i^s.$$

The polynomials $h_s(x_1, x_2, \ldots, x_n)$ are the *power-sum symmetric functions*. Let Δ_m be the determinant of the submatrix formed by the first m rows and columns of the matrix

$$\begin{pmatrix} h_0 & h_1 & h_2 & \ldots & h_{n-1} \\ h_1 & h_2 & h_3 & \ldots & h_n \\ \vdots & & & & \\ h_{n-1} & h_n & h_{n+1} & \ldots & h_{2n-1} \end{pmatrix}.$$

Then a symmetric polynomial is positive if and only if it is a sum of the form

$$\sum_{i,j,\ldots,l} \sigma_M \Delta_i \Delta_j \cdots \Delta_l,$$

where the indices $i, j, \ldots, l$ need not be distinct and the coefficient σ_M of the monomial $\Delta_i \Delta_j \cdots \Delta_l$ is a linear combination with positive coefficients of squares of symmetric polynomials.

2.6.15. Give a proof of König's theorem (2.1.1) by considering the $m \times m$ matrix with ij-entry equal to $|A_i \cap B_j|$.[49]

2.6.16. *Permanents of doubly stochastic matrices.*[50]

Recall that the *permanent* per (a_{ij}) of an $n \times n$ matrix (a_{ij}) is defined by

$$\text{per}\,(a_{ij}) = \sum_{\pi} a_{1,\pi(1)} a_{2,\pi(2)} \cdots a_{n,\pi(n)}.$$

(a) Suppose that $P(X_{ij})$ is a general matrix function (see Exercies 2.3.3). Show that $P(X_{ij})$ is the permanent if and only if $P(X_{ij})$ is invariant under permutations of rows or columns and $P(\delta_{ij})$, P evaluated at the identity matrix, equals 1.

(b) Show that if (a_{ij}) is an $n \times n$ matrix with nonnegative real entries,

$$\text{per}\,(a_{ij}) \leq \prod_{i=1}^{n} (a_{i1} + a_{i2} + \cdots + a_{in});$$

[49] van der Waerden (1927). This is another rediscovery of methods in König's 1916 paper.
[50] Marcus and Newman (1959).

that is, the permanent is at most the product of the row sums. Similarly, the permanent is at most the product of the column sums.

(c) Show that if A and B are matrices with nonnegative entries, then

$$\text{per}\,(A+B) \geq \text{per}\,(A) + \text{per}\,(B).$$

Using this, prove that for any $n \times n$ doubly stochastic matrix D,

$$\text{per}\,(D) \geq (n^2 - 2n + 2)^{-(n-1)}.$$

2.6.17. *Van der Waerden's permanent conjecture.* Show that if (d_{ij}) is an $n \times n$ doubly stochastic matrix, then

$$\text{per}\,(d_{ij}) \leq \frac{n!}{n^n}.$$

Equality holds if and only if (d_{ij}) is the matrix with all entries equal to $1/n$.

2.6.18. *Infinite-dimensional doubly stochastic matrices.*[51]

Consider the vector space of real matrices $(a_{ij})_{1 \leq i,j < \infty}$ with countably many rows and columns with the norm $\|\cdot\|$ defined by

$$\|(a_{ij})\| = \max\left\{\sup_i \sum_{\alpha=1}^{\infty} |a_{i\alpha}|,\ \sup_j \sum_{\alpha=1}^{\infty} |a_{\alpha j}|\right\}.$$

A matrix (a_{ij}) is *boundedly line-summable* if its norm $\|(a_{ij})\|$ is finite. Bounded line-summable matrices form a real Banach space $\mathbb{B}$ under the aforementioned norm. A matrix in $\mathbb{B}$ with nonnegative entries is *doubly stochastic* if for all i and j, $\sum_{\alpha=1}^{\infty} a_{j\alpha} = 1$ and $\sum_{\alpha=1}^{\infty} a_{\alpha i} = 1$.

(a) Let H be the closure (in the $\|\cdot\|$-topology) of all finite convex combinations of permutation matrices in $\mathbb{B}$. Show that this is the set of doubly stochastic matrices satisfying the additional condition: for every positive ϵ, there exists a positive integer N (depending only on ϵ) such that in each row or column, the sum of the N largest entries is at least $1 - \epsilon$.

(b) Let $t_{ij} : \mathbb{B} \to \mathbb{R}$, $A \mapsto a_{ij}$, and $\mathcal{T}$ be the weakest topology under which all the linear functionals t_{ij}, $\sum_{\alpha=0}^{\infty} t_{i\alpha}$, $\sum_{\alpha=0}^{\infty} t_{\alpha j}$, $1 \leq i, j \leq \infty$, are continuous. Show that the $\mathcal{T}$-closure of all finite convex combinations of permutation matrices is the set of all doubly stochastic matrices.

[51] Isbell (1955) and Kendall (1960).

(c) Birkhoff's Problem 111. Find infinite or continuous versions of Theorem 2.6.1. See *Matching theory* for some attempts.

2.7 The Gale–Ryser Theorem

When does there exist a relation with given marginals? That is, given two sequences $\underline{r}$ and $\underline{s}$, where $\underline{r} = (r_1, r_2, \ldots, r_m)$ and $\underline{s} = (s_1, s_2, \ldots, s_n)$, of nonnegative integers, does there exist an $m \times n$ matrix (a_{ij}) with entries equal to 0 or 1 satisfying the row sum conditions, for a given row index i,

$$\sum_{j=1}^{n} a_{ij} = r_i,$$

and the column sum conditions, for a given column index j,

$$\sum_{i=1}^{m} a_{ij} = s_j?$$

Suppose there is such a matrix. Then

$$r_1 + r_2 + \cdots + r_m = s_1 + s_2 + \cdots + s_n.$$

Permuting the columns if necessary, we may assume that $s_1 \geq s_2 \geq \cdots \geq s_n$. The number s_1 of 1s in the first column is at most the number of rows that are nonzero; that is,

$$s_1 \leq |\{i\colon r_i \geq 1\}|,$$

the number of nonzero coordinates r_i in $\underline{r}$. In general, we must have

$$s_1 + s_2 + \cdots + s_k \leq |\{i\colon r_i \geq 1\}| + |\{i\colon r_i \geq 2\}| + \cdots + |\{i\colon r_i \geq k\}|.$$

These necessary conditions can be stated in terms of conjugates and majorizations.

Let $\underline{c}$ be a nonincreasing sequence $(c_1, c_2, \ldots, c_n)$. Then for $l \geq c_1$, the *length-l conjugate* of $\underline{c}$, denoted by $\underline{c}^*$ regardless of the length, is the sequence defined by

$$c_k^* = |\{i\colon c_i \geq k\}|.$$

Then a necessary condition for a (0-1)-matrix to exist with row sums $\underline{r}$ and column sums $\underline{s}$ is

$$\underline{s} \preceq \underline{r}^*,$$

where $\preceq$ is the majorization order defined in the previous section and $\underline{r}^*$ is chosen to have the same length as $\underline{s}$. This necessary condition is also sufficient.

2.7.1. The Gale–Ryser theorem. Let $(r_1, r_2, \ldots, r_m)$ and $(s_1, s_2, \ldots, s_n)$ be two sequences of nonnegative integers. Then an $m \times n$ (0-1)-matrix exists with row sums $\underline{r}$ and column sums $\underline{s}$ if and only if $\underline{s} \preceq \underline{r}^*$.

As with the marriage theorem, there are several proofs of the sufficiency portion of this theorem. We present in detail a proof based on a matching theorem.

Let $R : S \to X$ be a relation, m a positive integer, and $(d_1, d_2, \ldots, d_m)$ a sequence of nonnegative integers. A *Higgins matching with defects* $(d_1, d_2, \ldots, d_m)$ is a sequence of partial matchings $M_1, M_2, \ldots, M_m$ in R such that

$$|M_i(S)| \geq |S| - d_i$$

and

$$M_i(S) \cap M_j(S) = \emptyset \text{ whenever } i \neq j.$$

We use the following notation: if x is a real number, then $x^+ = \max(x, 0)$.

2.7.2. Higgins' theorem on disjoint partial matchings.[52] Let $R : S \to X$ be a relation. Then R has a Higgins matching with defects $(d_1, d_2, \ldots, d_m)$ if and only if the *Higgins–Hall condition,*

$$\text{for all } A \subseteq S, \quad |R(A)| \geq \sum_{i=1}^{m} (|A| - d_i)^+,$$

holds.

Proof. We construct a new relation from R. Let X^* be the union

$$X \cup D_1 \cup D_2 \cup \cdots \cup D_m,$$

where D_i are new sets (mutually disjoint and disjoint from X) such that $|D_i| = d_i$. Let $R' : S \times \{1, 2, \ldots, m\} \to X^*$ be the relation defined as follows: if $a \in$

[52] Higgins (1959). However, instead of Higgins' proof, we will use the proof given by Mirsky (1971, chapter 5).

S, $i \in \{1, 2, \ldots, m\}$, and $x \in X$,

$$((a, i), x) \in R' \text{ if } (a, x) \in R,$$

and if $a \in S$, $i \in \{1, 2, \ldots, m\}$, and $y \in D_i$,

$$((a, i), y) \in R' \text{ for all } a \in S,$$

and none of the edges $((a, i), y)$ is in R' if $y \in D_j$ and $i \neq j$. It is readily verified that R has a Higgins matching with defects $(d_1, d_2, \ldots, d_m)$ if and only if R' has a matching.

We prove next that if R contains a Higgins matching with defects $(d_1, d_2, \ldots, d_m)$, then the Higgins–Hall condition holds. Let $M_1, M_2, \ldots, M_m$ be partial matchings forming the Higgins matching and let S_i be the subsets of those elements in S matched by the partial matching M_i (so that the restriction $M_i : S_i \to M_i(S)$ is a bijection). Then, the ranges $M_i(S)$ are pairwise disjoint and $|S \backslash S_i| \leq d_i$. Thus, for $A \subseteq S$,

$$\begin{aligned} |R(A)| &\geq \sum_{i=1}^{m} |M_i(A)| \\ &= \sum_{i=1}^{m} |A \cap S_i| \\ &= \sum_{i=1}^{m} |A| - |A \cap (S \backslash S_i)| \\ &\geq \sum_{i=1}^{m} (|A| - d_i)^+. \end{aligned}$$

We shall finish the proof using the marriage theorem. To do this, we need to show that the Higgins–Hall condition for R implies the Hall condition for R'. Let B be a subset of $S \times \{1, 2, \ldots, m\}$. Let A be the *projection* of B onto S; that is to say, A is the subset of elements in S occurring as a first coordinate of some element in B. Let I be the projection of B onto $\{1, 2, \ldots, m\}$ and J be the complement $\{1, 2, \ldots, m\} \backslash I$. Then

$$|B| \leq |I||A|$$

and

$$R'(B) = R(A) \cup \bigcup_{i \in I} D_i.$$

Using the Higgins–Hall condition and the fact that

$$(a - d)^+ + d \geq a,$$

we have

$$\begin{aligned}
|R'(B)| &= |R(A)| + \sum_{i \in I} d_i \\
&\geq \sum_{i=1}^{m} (|A| - d_i)^+ + \sum_{i \in I} d_i \\
&= \sum_{i \in I} [(|A| - d_i)^+ + d_i] \; + \; \sum_{j \in J} (|A| - d_j)^+ \\
&\geq \sum_{i \in I} |A| \\
&= |I||A| \\
&\geq |B|.
\end{aligned}$$

This completes the proof of Theorem 2.7.2. □

To arrive at the special case of Higgins' theorem that yields the Gale–Ryser theorem, we need an easy technical lemma. Rearrange the sequence $(d_1, d_2, \ldots, d_m)$ so that it is nondecreasing. Let $n \geq d_m$ and

$$r_i = n - d_i.$$

Then $(r_1, r_2, \ldots, r_m)$ is nonincreasing. Let $(r_1^*, r_2^*, \ldots)$ be the length-r_1 conjugate sequence.

2.7.3. Lemma. Suppose $1 \leq a \leq n$. Then

$$\sum_{i=1}^{n} (a - d_i)^+ = \sum_{j=n-a+1}^{r_1} r_j^*.$$

Proof. Consider the $m \times n$ array in which the ith row consists of r_i 1s followed by d_i 0s. For example, if $n = 9$ and $\underline{r} = (6, 4, 3, 1, 0)$, then we have the 5×9 matrix

$$\begin{array}{lccccccccc}
d_1 = 3,\ r_1 = 6 & 1 & 1 & 1 & 1 & 1 & 1 & 0 & 0 & 0 \\
d_2 = 5,\ r_2 = 4 & 1 & 1 & 1 & 1 & 0 & 0 & 0 & 0 & 0 \\
d_3 = 6,\ r_3 = 3 & 1 & 1 & 1 & 0 & 0 & 0 & 0 & 0 & 0. \\
d_4 = 8,\ r_4 = 1 & 1 & 0 & 0 & 0 & 0 & 0 & 0 & 0 & 0 \\
d_5 = 9,\ r_5 = 0 & 0 & 0 & 0 & 0 & 0 & 0 & 0 & 0 & 0
\end{array}$$

Then both sides of the equation in the lemma equal the number of 1s in columns $n - a + 1, n + a + 2, \ldots, n$ in the array. □

2.7.4. Corollary. Let $X_1, X_2, \ldots, X_n$ be mutually disjoint sets such that $|X_i| = s_i$ and $G : \{1, 2, \ldots, n\} \to X_1 \cup X_2 \cup \cdots \cup X_n$ be the relation defined by $G(i) = X_i$. Then G has a Higgins matching with defects $(d_1, d_2, \ldots, d_m)$ if $\underline{s} \preceq \underline{r}^*$, where $\underline{r} = (n - d_1, n - d_2, \ldots, n - d_m)$.

Proof. Observe first that since

$$\sum_{i=1}^{n} s_i = \sum_{j=1}^{r_1} r_j^*,$$

$\underline{s} \preceq \underline{r}^*$ implies that for all k, $1 \le k \le n$,

$$\sum_{i=k}^{n} s_i' \ge \sum_{j=k}^{r_1} r_j^*,$$

where $(s_1', s_2', \ldots, s_n')$ is a rearrangement of $(s_1, s_2, \ldots, s_n)$ into a nonincreasing sequence. Using this observation, we will check that the Higgins–Hall condition holds for the relation G. Let $A \subseteq \{1, 2, \ldots, n\}$. Then

$$\begin{aligned} |G(A)| &= \sum_{a: a \in A} s_a \\ &\ge \sum_{i=n-|A|+1}^{n} s_i'. \\ &\ge \sum_{j=n-|A|+1}^{r_1} r_j^* \\ &= \sum_{i=1}^{n} (|A| - d_i)^+. \end{aligned}$$

Hence, a Higgins matching with defects $(d_1, d_2, \ldots, d_m)$ exists in G. □

Finally, we construct an $m \times n$ matrix (a_{ij}) from a Higgins matching $M_1, M_2, \ldots, M_m$ with defects $(d_1, d_2, \ldots, d_m)$ by the rule

$$a_{ij} = \begin{cases} 1 & \text{if } M_i(\{1, 2, \ldots, n\}) \cap X_j \ne \emptyset \\ 0 & \text{otherwise.} \end{cases}$$

The matrix (a_{ij}) has row sums (r_i) and column sums (s_j). This completes the first proof of the Gale–Ryser theorem.

Ryser's proof is sketched in Exercise 2.7.3. Gale derives the theorem from a network flow theorem. He also proves the following more general form of the Gale–Ryser theorem.

2.7.6. Gale's theorem.[53] Let $(r_1, r_2, \ldots, r_m)$ and $(s_1, s_2, \ldots, s_n)$ be two vectors with nonnegative integer entries. Then there exists an $m \times n$ (0, 1)-matrix (a_{ij}) satisfying the row sum condition

$$\sum_{j=1}^{n} a_{ij} \le r_i,$$

and the column sum condition,

$$\sum_{i=1}^{m} a_{ij} \le s_j,$$

if and only if the rearrangements $(r'_1, r'_2, \ldots, r'_m)$ and $(s'_1, s'_2, \ldots, s'_m)$ into non-increasing vectors satisfy the following inequality: for all positive integers k,

$$\sum_{i=1}^{k} s'_i \le \sum_{i=1}^{k} r^*_i, \tag{G}$$

where $(r^*_1, r^*_2, \ldots, r^*_m)$ is a conjugate of $(r'_1, r'_2, \ldots, r'_m)$.

Sketch of proof. Show by induction that the following algorithm produces a matrix (a_{ij}) with the required properties: let (r'_i) and (s'_j) be nondecreasing rearrangements of (r_i) and (s_j). Let π be a permutation such that $r_{\pi(i)} = r'_i$. Construct the first column of the matrix by setting a_{1i} equal to 1 if $i = \pi(1)$, $i = \pi(2), \ldots,$ or $i = \pi(s'_1)$, and 0 otherwise. Now construct the second column with the updated vector $(r_1 - a_{11}, r_2 - a_{12}, \ldots, r_m - a_{1m})$. □

Gale gives the following concrete illustration of his algorithm: suppose n families are going on a picnic in m buses, where the jth family has s_j members and the ith bus has r_i seats. Provided the condition (G) is satisfied, then it is possible to seat all passengers so that no two members of the same family are in the same bus. Just use the simple rule: at each stage, distribute the largest unseated family among those buses having the greatest number of vacant seats.

Exercises

2.7.1. It follows from the Gale–Ryser theorem that $\underline{r} \preceq \underline{s}^*$ if and only if $\underline{s} \preceq \underline{r}^*$; that is, conjugation is an order-reversing involution of the majorization order. Give an independent proof.

2.7.2. Prove the following weaker version of the Gale–Ryser theorem: given nonnegative integers $r_1, r_2, \ldots, r_m$ and $s_1, s_2, \ldots, s_n$, there exists an $m \times n$

[53] Gale (1957).

matrix with nonnegative integer entries satisfying the row and column sum conditions if and only if

$$r_1 + r_2 + \cdots + r_m = s_1 + s_2 + \cdots + s_n.$$

2.7.3. *Ryser's proof.*[54]

Let m and n be given positive integers, $(r_1, r_2, \ldots, r_m)$ be a given row sum vector, $\underline{r}^*$ be a length-n conjugate of $\underline{r}$, and (a_{ij}) be the $m \times n$ Ferrers matrix defined by

$$a_{ij} = \begin{cases} 1 & \text{if } j \leq r_i, \\ 0 & \text{otherwise.} \end{cases}$$

Show that if $(s_1, s_2, \ldots, s_m) \preceq \underline{r}^*$, then one can rearrange the 1s in (a_{ij}) into a matrix with row sums $\underline{r}$ and column sums $\underline{s}$ by *interchanges* of the form

$$\begin{matrix} 1 & \ldots & 0 \\ \vdots & & \vdots \\ 0 & \ldots & 1 \end{matrix} \quad \leftrightarrow \quad \begin{matrix} 0 & \ldots & 1 \\ \vdots & & \vdots \\ 1 & \ldots & 0 \end{matrix}$$

Show the more general result: let A and B be two (0-1)-matrices with row sums $\underline{r}$ and column sums $\underline{s}$. Then A can be transformed into B by interchanges.

2.7.4. *A submodular function proof.*

Define the *excess function* $\eta(A)$ by the formula

$$\eta(A) = |R(A)| - \left[\sum_{i=1}^{k} (|A| - d_i)^+ \right].$$

(a) Show that η is submodular.

(b) Prove Higgins' theorem by induction, using the proof of the matroid marriage theorem as a guide.

2.7.5. (Research problem) Find an analog of the Gale–Ryser theorem for symmetric (0-1)-matrices.

2.7.6. (Research problem) *An extension of Birkhoff's theorem.*

Let $(r_1, r_2, \ldots, r_m)$ and $(s_1, s_2, \ldots, s_n)$ be vectors with nonnegative integer entries. The set Matrix$(\underline{r}, \underline{s})$ of all $m \times n$ matrices with real nonnegative entries with row sums $\underline{r}$ and column sums $\underline{s}$ is convex. Determine the extreme points of Matrix$(\underline{r}, \underline{s})$.

[54] Ryser (1957).

2.8 Matching Theory in Higher Dimensions

"The possibility of extending the marriage theorem to several dimensions does not seem to have been explored. Thinking rather crudely, one might replace a matrix by a tensor." In this section, we explain this remark, made by Harper and Rota in *Matching theory* (p. 211).

We begin with some *informal* tensor algebra. Let $V_1, V_2, \ldots, V_k$ be vector spaces of dimension $d_1, d_2, \ldots, d_k$ over a field $\mathbb{F}$. A *decomposable k-tensor* in the *tensor product* $V_1 \otimes V_2 \otimes \cdots \otimes V_k$ is a formal product $v_1 \otimes v_2 \otimes \cdots \otimes v_k$, where the ith vector v_i is in V_i. A *k-tensor* is a linear combination of decomposable k-tensors. Tensors are *multilinear*; that is, they satisfy the property

$$\begin{aligned} &v_1 \otimes \cdots \otimes (\lambda v_i + \mu u_i) \otimes \cdots \otimes v_m \\ &= \lambda(v_1 \otimes \cdots \otimes v_i \otimes \cdots \otimes v_m) \\ &\quad + \mu(v_1 \otimes \cdots \otimes u_i \otimes \cdots \otimes v_m) \end{aligned}$$

and all relations implied by this property.

A *covector* or *dual vector* of V is a linear functional or a linear transformation $V \to \mathbb{F}$. The covectors form the vector space V^* dual to V. Let $e_1, e_2, \ldots, e_d$ be a chosen basis for V. Then $e_1^*, e_2^*, \ldots, e_d^*$ is the basis for the dual V_i^* defined by $e_i^*(e_j) = 0$ if $i \neq j$ and 1 if $i = j$; in other words, $e_i^*(e_j) = \delta_{ij}$.

Consider the tensor product $\mathbb{F}^m \otimes (\mathbb{F}^n)^*$. The tensor $e_i \otimes e_j^*$ defines a linear transformation $\mathbb{F}^n \to \mathbb{F}^m$ by $u \mapsto e_j^*(u)e_i$. The matrix of this linear transformation has ij-entry equal to 1 and all other entries equal to 0. Hence, the matrix (a_{ij}) can be regarded as the tensor

$$\sum_{i=1}^{n}\sum_{j=1}^{m} a_{ij} e_i \otimes e_j^*.$$

Generalizing this, and choosing bases for V_i, we can think of a k-tensor as a k-dimensional array of numbers from $\mathbb{F}$.

2.8.1. Lemma. The rank of a matrix A equals

$$\min\left\{s\colon A = \sum_{i=1}^{s} v_i \otimes u_i^*\right\},$$

the minimum number of decomposable 2-tensors in an expression of A as a sum of decomposable 2-tensors.

Proof. Suppose $\sum_{i=1}^{s} v_i \otimes u_i^*$. Then the image of A is spanned by the vectors $v_1, v_2, \ldots, v_s$. Hence, the rank of A, which equals the dimension of the image of A, is at most s.

Let A be an $m \times n$ matrix of rank ρ. Let $r_1, r_2, \ldots, r_\rho$ be ρ linearly independent rows in A, and

$$r_i = \sum_{j=1}^{\rho} b_{ij} r_j, \text{ for } \rho + 1 \leq i \leq m.$$

Then A can be written as a sum

$$\left(e_1 + \sum_{h=\rho+1}^{m} b_{h1} e_h\right) \otimes r_1^* + \left(e_2 + \sum_{h=\rho+1}^{m} b_{h2} e_h\right) \otimes r_2^*$$
$$+ \cdots + \left(e_\rho + \sum_{h=\rho+1}^{m} b_{h1} e_h\right) \otimes r_\rho^*$$

of s decomposable tensors. Hence, $\rho \geq s$. □

An example might make the second part of the proof clearer. Consider the matrix

$$\begin{pmatrix} 0 & 1 & 2 & 3 \\ 2 & 3 & 1 & 5 \\ 6 & 7 & 1 & 9 \end{pmatrix}.$$

Then

$$r_1^* : (x_1, x_2, x_3, x_4) \mapsto x_2 + 2x_3 + 3x_4 \quad \text{and}$$
$$r_2^* : (x_1, x_2, x_3, x_4) \mapsto 2x_1 + 3x_2 + x_3 + 5x_4,$$

and the matrix can be written as the sum of two decomposable tensors by the following computation:

$$e_1 \otimes r_1^* + e_2 \otimes r_2^* + e_3 \otimes (-2r_1^* + 3r_2^*) = (e_1 - 2e_3) \otimes r_1^* + (e_2 + 3e_3) \otimes r_2^*.$$

Motivated by Lemma 2.8.1, we define the *rank* of a tensor as the minimum number of decomposable tensors in an expression of A as a sum of decomposable tensors.[55] In the case of matrices, the rank can be calculated (efficiently) by triangulation or Gaussian elimination. There is no known analogous algorithm for tensors. For matrices, there is also the determinant, which determines whether a square matrix has full rank. There is no invariant for tensors which is as explicit or useful as the determinant for matrices. To oversimplify, one

[55] This is the commonly accepted definition of the rank of a tensor for algebraists.

of primary aims of "classical invariant theory" is to find concepts analogous to triangulations and determinants for general tensors.[56] The research problem here is to develop an elegant (and hence useful) theory. Whatever this theory is, it should contain a higher-dimensional matching theory as a special case.

Let $A \subseteq \{1, 2, \ldots, d_1\} \times \{1, 2, \ldots, d_2\} \times \cdots \times \{1, 2, \ldots, d_k\}$. The *free tensor* supported by A is the k-tensor

$$\sum_{\{i_1, i_2, \ldots, i_k\} \subseteq A} x_{i_1, i_2, \ldots, i_k} e_{i_1} \otimes e_{i_2} \otimes \cdots \otimes e_{i_k}$$

in $\mathbb{F}^{d_1} \otimes \mathbb{F}^{d_2} \otimes \cdots \otimes \mathbb{F}^{d_k}$, where $\mathbb{F}$ is a sufficiently large field so that the nonzero entries $x_{i_1, i_2, \ldots, i_k}$ are algebraically independent over some given subfield of $\mathbb{F}$. The problem of higher-dimensional matching theory is to define, for a given free tensor, a combinatorial object whose "size" equals its rank. Such a definition should indicate the right explicit definition for the determinant of a tensor. Optimistically, one would also hope for an analog of the König–Egerváry theorem, where the rank equals the minimum over some "cover" of A.

Related to this problem is the question of higher-dimensional submodular inequalities (see Section 2.4). We will need to assume knowledge of matroid theory. A main axiom of matroid theory is the submodular inequality for the rank function rk:

$$\mathrm{rk}(A) + \mathrm{rk}(B) \geq \mathrm{rk}(A \cup B) + \mathrm{rk}(A \cap B).$$

This inequality is satisfied by the rank function of vectors, that is, 1-tensors. With an easy argument (see Exercise 2.8.1), this inequality can be adapted to matrices or 2-tensors. If $M[T|S]$ is a matrix, let $M[B|A]$ be the submatrix obtained by restricting M to the rows B and columns A and $\mathrm{rank}(B, A)$ be the rank of the submatrix $M[B|A]$. Then

$$\mathrm{rank}(B, A) + \mathrm{rank}(D, C) \geq \mathrm{rank}(B \cup D, A \cap C) + \mathrm{rank}(B \cap D, A \cup C).$$

This inequality is the *bimatroid submodular inequality*.[57] It compares ranks of *rectangular* sets of the form $B \times A$. There is no known analog of the submodular inequality for higher-dimensional tensors. Indeed, no "new" rank inequality for higher tensors is known. This would be a step toward a higher-dimensional matroid theory.

[56] See Grosshans (2003) and Rota (2002); for a mainstream account, see Gelfand et al. (1994).
[57] Kung (1978) and Schrijver (1979).

Exercises

2.8.1. (a) Prove the bimatroid submodular inequality.

(b) (Research problem) The bimatroid inequality involves "rectangular" sets of the form $B \times A$. Are there rank inequalities involving arbitrary subsets of $T \times S$ for free matrices?

2.8.2. *Rank inequalities for vectors.*[58]

(a) Prove that the rank function rk of vectors satisfies *Ingleton's inequality:* for four subsets X_1, X_2, X_3, and X_4 of vectors,

$$\begin{aligned}\operatorname{rk}(X_1) + \operatorname{rk}(X_2) + \operatorname{rk}(X_1 \cup X_2 \cup X_3) + \operatorname{rk}(X_1 \cup X_2 \cup X_4) + \operatorname{rk}(X_3 \cup X_4)\\ \leq \operatorname{rk}(X_1 \cup X_2) + \operatorname{rk}(X_1 \cup X_3) + \operatorname{rk}(X_1 \cup X_4) + \operatorname{rk}(X_2 \cup X_3)\\ + \operatorname{rk}(X_2 \cup X_4).\end{aligned}$$

Show that this inequality does not follow from the submodular inequality.

(b) (Research problem posed by A. W. Ingleton) Find other inequalities for rank functions of vectors.

(c) (Research problem) "Describe" all the inequalities satisfied by rank functions of vectors. (It is known that the set of forbidden minors for representability over the real or complex numbers is infinite. However, this does not preclude a reasonable description of all rank inequalities. For example, are all rank equalities for vectors consequences of Grassmann's equality?)

2.8.3. *Rank inequality for matrices.*

There are many matrix rank inequalities involving products of matrices. The simplest is

$$\operatorname{rank}(AB) \leq \min\{\operatorname{rank}(A), \operatorname{rank}(\mathrm{B})\}.$$

Another is the Frobenius rank inequality

$$\operatorname{rank}(AB) + \operatorname{rank}(BC) \leq \operatorname{rank}(B) + \operatorname{rank}(ABC).$$

Develop a theory of matrix rank inequalities. For example, are they all consequences of a finite set of inequalities?

2.8.4. *Higher-dimensional permanents.*[59]

Let A be an $n_1 \times n_2 \times \cdots \times n_d$ array of numbers with entries $a_{i_1,i_2,\ldots,i_d}$. Then a reasonable definition of the *permanent* per A is

$$\operatorname{per} A = \sum \prod_{i=1}^{n_1} a_{i,\sigma_1(i),\sigma_2(i),\ldots,\sigma_d(i)},$$

[58] Ingleton (1971). [59] Dow and Gibson (1987) and Muir and Metzler (1933, chapter 1).

where the sum ranges over all d-tuples $(\sigma_1, \sigma_2, \ldots, \sigma_d)$ of one-to-one functions $\sigma_i : \{1, 2, \ldots, n_1\} \to \{1, 2, \ldots, n_i\}$. Extend as many of the properties of two-dimensional permanents as possible. In particular, prove an analog of Exercise 2.6.16(a).

2.9 Further Reading

There are probably as many approaches to matching theory as there are areas of mathematics. The survey paper *Matching theory* shows some of these connections. Approaches we have completely ignored are those of graph theory, combinatorial optimization, polyhedral combinatorics, probabilistic and asymptotic combinatorics, and analysis of algorithms. For further reading, we recommend the following books or survey papers:

R. Brualdi and H.J. Ryser, *Combinatorial Matrix Theory*, Cambridge University Press, Cambridge, 1991.

R. Brualdi and B.L. Shader, *Matrices of Sign-Solvable Linear Systems*, Cambridge University Press, Cambridge, 1995.

G.H. Hardy, J.E. Littlewood, and G. Pólya, *Inequalities*, 2nd edition, Cambridge University Press, Cambridge, 1952.

L. Lovász and M.D. Plummer, *Matching Theory*, North-Holland, Amsterdam and New York, 1986.

A.W. Marshall and I. Olkins, *Inequalities: Theory of Majorization and Its Applications*, Academic Press, New York and London, 1979.

H. Minc, *Permanents*, Addison-Wesley, Reading, MA, 1978.

H. Minc, *Nonnegative Matrices*, Wiley, New York, 1988.

L. Mirsky, Results and problems in the theory of doubly-stochastic matrices, *Z. Wahrscheinlichkeittheor. Verwandte Geb.* 1 (1962–1963) 319–334.

L. Mirsky, *Transversal Theory*, Academic Press, New York, 1971.

3
Partially Ordered Sets and Lattices

3.1 Möbius Functions

Rota wrote, in the introduction to *Foundations I*,

> It often happens that a set of objects to be counted possesses a natural ordering, in general only a partial order. It may be unnatural to fit the enumeration of such a set into a linear order such as the integers: instead, it turns out in a great many cases that a more effective technique is to work with the natural order of the set. One is led in this way to set up a "difference calculus" relative to an arbitrary partially ordered set.

From this tentative beginning grew the research program of Möbius functions on partially ordered set. There are many expositions of this program.[1] We give a selective account of this theory.

A partially ordered set P is *locally finite* if every interval in P is finite. Let P be a locally finite partially ordered set and $\mathbb{A}$ be a commutative ring with identity. Consider the collection of functions $P \times P \to \mathbb{A}$ such that

$$f(x, y) = 0 \text{ if } x \not\leq y. \qquad (*)$$

This collection forms an $\mathbb{A}$-module and it is made into an $\mathbb{A}$-algebra, the *incidence algebra $\mathcal{I}(P)$ of the partially ordered set P*, by *convolution*

$$(f * g)(x, y) = \sum_{z: x \leq z \leq y} f(x, z) g(z, y).$$

If P is finite, then we can choose a linear extension of P (see Section 1.2) and represent a function f in $\mathcal{I}(P)$ by the $|P| \times |P|$ matrix with rows and columns indexed by P and xy-entry $f(x, y)$. By $(*)$, such matrices are upper-triangular, and taking into account entries that must be zero, convolution corresponds to

[1] See, for example, Aigner (1979) and Stanley (1986).

matrix multiplication. In particular, incidence algebras are algebras of upper-triangular matrices.

The identity of $\mathcal{I}(P)$ is the *delta function*

$$\delta(x, y) = 1 \text{ if } x = y \text{ and } 0 \text{ otherwise.}$$

The *zeta function* is the function

$$\zeta(x, y) = 1 \text{ if } x \leq y \text{ and } 0 \text{ otherwise.}$$

The *Möbius function* μ is the convolutional inverse of ζ; that is,

$$\zeta * \mu = \mu * \zeta = \delta.$$

Explicitly,

$$\sum_{z: x \leq z \leq y} \mu(z, y) = \sum_{z: x \leq z \leq y} \mu(x, z) = \delta(x, y).$$

The explicit definition shows that μ exists and gives two recursions with initial conditions: for all x in P, $\mu(x, x) = 1$. If $x < y$, the bottom-up recursion is

$$\mu(x, y) = - \sum_{z: x \leq z < y} \mu(x, z)$$

and the top-down recursion is

$$\mu(x, y) = - \sum_{z: x < z \leq y} \mu(z, y).$$

From either recursion, it is immediate that if y covers x, then $\mu(x, y) = -1$. Further, if C is the chain $x_0 < x_1 < \cdots < x_n$, then

$$\mu_C(x_0, x_0) = 1, \ \mu_C(x_0, x_1) = -1, \ \text{ and } \mu_C(x_0, x) = 0 \text{ if } x \neq x_0, x_1.$$

Finally, we define the *elementary matrix functions*. If $u \leq v$, let ϵ_{uv} be the function defined by $\epsilon_{uv}(x, y) = 1$ if $u = x$ and $v = y$, and 0 otherwise. As a matrix, the function ϵ_{uv} has all entries equal to zero, except for the uv-entry, which equals 1. If f is a function in $\mathcal{I}(P)$, then

$$\epsilon_{ux} * f * \epsilon_{yv} = f(x, y)\epsilon_{uv}$$

and, in particular,

$$\epsilon_{xx} * f * \epsilon_{yy} = f(x, y)\epsilon_{xy}.$$

When u and v range over all pairs such that $u \leq v$, the functions ϵ_{uv} form a basis for $\mathcal{I}(P)$ as an $\mathbb{A}$-module.

An easy computation (using the recursions) yields the Möbius function of a Cartesian product of partial orders in terms of the Möbius functions of its components.

3.1.1. The product formula. Let P and Q be partially ordered sets. Then

$$\mu_{P\times Q}((x,u),(y,v)) = \mu_P(x,y)\mu_Q(u,v).$$

Finite Boolean algebras are products of chains of length 1. Thus, for subsets A and B of S such that $A \subseteq B$,

$$\mu(A,B) = (-1)^{|B|-|A|}$$

in the Boolean algebra of all subsets of S. If n is a positive integer having prime factorization $n = p_1^{a_1} p_2^{a_2} \cdots p_k^{a_k}$, then its lattice of divisors under the divisibility order is isomorphic to a product of k chains, the first having length a_1, the second having length a_2, and so on. Further, if m divides n, then the interval $[m,n]$ is isomorphic to the lattice of divisors of n/m. Hence, $\mu(m,n) = \mu(1,n/m) = \mu(n/m)$, where $\mu(k)$ is the single-variable number-theoretic Möbius function defined by $\mu(k) = (-1)^r$ if k is the product of r distinct primes and 0 otherwise.

There are two classic inversion formulas. One is the principle of inclusion and exclusion:

$$\begin{aligned} f(A) &= \sum_{B:\,B\subseteq A} g(B) \text{ for all } A \subseteq S \;\Leftrightarrow \\ g(A) &= \sum_{B:\,B\subseteq A} (-1)^{|A|-|B|} f(B) \text{ for all } A \subseteq S. \end{aligned}$$

The other is the number-theoretic Möbius inversion formula:

$$f(n) = \sum_{d:\,d\,|\,n} g(d) \text{ for all } n \;\Leftrightarrow\; g(n) = \sum_{d:\,d\,|\,n} f(d)\mu(n/d) \text{ for all } n.$$

These formulas generalize immediately to locally finite partially ordered sets.

3.1.2. The Möbius inversion formula. Let P be a locally finite partially ordered set. Let f and g be functions from P to $\mathbb{A}$. Then

$$f(x) = \sum_{y:\,y\le x} g(y) \text{ for all } x \in P \;\Leftrightarrow\; g(x) = \sum_{y:\,y\le x} f(y)\mu(y,x) \text{ for all } x \in P.$$

Dually,

$$f(x) = \sum_{y:\, y \geq x} g(y) \text{ for all } x \in P \ \Leftrightarrow\ g(x) = \sum_{y:\, y \geq x} \mu(x, y) f(y) \text{ for all } x \in P.$$

There are many easy proofs. One way is to think of f and g as column or row vectors, and ζ and μ as matrices. Then the theorem just says $f = g * \zeta$ if and only if $f * \mu = g$, or in the dual version, $f = \zeta * g$ if and only if $\mu * f = g$.

To use the Möbius inversion formula, one needs to calculate Möbius functions. There are several theorems about the values of Möbius functions. They can be grouped into four kinds: product theorems, closure or Galois connection theorems, chain-counting theorems, and complementation theorems.

The basic product theorem is Formula 3.1.1. Let Interval(P) be the set of intervals (including the empty interval) in a partially ordered set P. This set may be ordered by set-containment; that is, $[u, v] \leq [x, y]$ if $[u, v] \subseteq [x, y]$. When $[u, v]$ and $[x, y]$ are nonempty, then $[u, v] \leq [x, y]$ if and only if $u \geq x$ and $v \leq y$.

3.1.3. Theorem.[2] Let μ be the Möbius function of Interval(P). If $[x, y]$ is nonempty, then

$$\mu(\emptyset, [x, y]) = -\mu_P(x, y),$$

and if both $[u, v]$ and $[x, y]$ are nonempty, then

$$\mu([u, v], [x, y]) = \mu_P(x, u)\mu_P(v, y).$$

Proof. When $[u, v]$ is nonempty, the interval $[[u, v], [x, y]]$ is isomorphic to $[x, u] \times [v, y]$. Hence, if $[u, v]$ is nonempty, the theorem follows from the product theorem. Further, by the top-down recursion and the earlier case where $[u, v]$ is nonempty,

$$\begin{aligned}
\mu(\emptyset, [x, y]) &= -\sum_{[u,v]:\, \emptyset < [u,v] \leq [x,y]} \mu([u, v], [x, y]) \\
&= -\sum_{u,v:\, x \leq u \leq v \leq y} \mu_P(x, u)\mu_P(v, y) \\
&= -[\mu * \zeta * \mu](x, y) = -\mu(x, y).
\end{aligned}$$ □

Next, we discuss theorems relating the Möbius function of a partially ordered set with the Möbius function of a smaller "quotient." The most useful theorem is the following closure theorem.[3]

[2] Crapo (1968a). [3] This theorem is distilled from *Foundations I*.

3.1.4. The closure theorem. Let $x \mapsto \overline{x}$ be a closure operator on a finite partially ordered set P and $\mathrm{Cl}(P)$ the partially ordered set of closed elements. If x is an element of P and y is a closed element, then

$$\sum_{z:\,\overline{z}=y} \mu_P(x,z) = \begin{cases} \mu_{\mathrm{Cl}(P)}(x,y) & \text{if } x \text{ is closed,} \\ 0 & \text{otherwise.} \end{cases}$$

An incidence algebra proof. One way to prove this is by a formal manipulation:

$$\begin{aligned}
\sum_{z:\,z\in P,\,\overline{z}=y} \mu_P(x,z) &= \sum_{z:\,z\in P} \mu_P(x,z)\delta_{\mathrm{Cl}(P)}(\overline{z},y) \\
&= \sum_{z,t:\,z\in P,\,t\in\mathrm{Cl}(P),\,x\le\overline{z}\le t\le y} \mu_P(x,z)\zeta_{\mathrm{Cl}(P)}(\overline{z},t)\mu_{\mathrm{Cl}(P)}(t,y) \\
&= \sum_{z,t:\,z\in P,\,t\in\mathrm{Cl}(P),\,x\le z\le t\le y} \mu_P(x,z)\zeta_P(\overline{z},t)\mu_{\mathrm{Cl}(P)}(t,y) \\
&= \sum_{z\in P,\,t\in\mathrm{Cl}(P):\,x\le z\le t\le y} \mu_P(x,z)\zeta_P(z,t)\mu_{\mathrm{Cl}(P)}(t,y) \\
&= \sum_{t\in\mathrm{Cl}(P):\,x\le t\le y} \delta_P(x,t)\mu_{\mathrm{Cl}(P)}(t,y),
\end{aligned}$$

where, in the fourth line, we made use of the fact that for a closure operator, $\overline{z} \le t$ if and only if $z \le t$. The final manifestation of the sum equals zero if x is not closed and $\mu_{\mathrm{Cl}(P)}(x,y)$ if x is closed. □

A more intuitive proof uses a partition–recursion argument. This method was used by Philip Hall and Louis Weisner,[4] two pioneers in the study of Möbius functions. Observe that the interval $[x,y]$ in P is partitioned into the subsets $\{z\colon \overline{z} = u\}$, where u ranges over all closed elements in $[x,y]$. Hence,

$$\begin{aligned}
0 &= \sum_{z:\,z\in[x,\overline{y}] \text{ in } P} \mu_P(x,z) \\
&= \sum_{u:\,u\in[x,\overline{y}] \text{ in } \mathrm{Cl}(P)} \left[\sum_{z\in P:\,\overline{z}=u} \mu_P(x,z) \right].
\end{aligned}$$

There are two cases of the theorem. Consider first the case when x is closed. We induct on the length of the longest chain from x to y in $\mathrm{Cl}(P)$. If the length is zero, then $x = y$, $\overline{z} = x$ if and only if $z = x$, and $\mu_{\mathrm{Cl}(P)}(x,x) = 1 = \mu_P(x,z)$. By induction, we may assume that the theorem holds for all elements u in

[4] Hall (1936) and Weisner (1935).

$\mathrm{Cl}(P)$ such that $x \le u < y$; that is, for all such elements u,

$$\sum_{z:\, z \in [x,y] \text{ in } P,\, \overline{z}=u} \mu_P(x,z) = \mu_{\mathrm{Cl}(P)}(x,u).$$

Thus,

$$0 = \sum_{z:\, z \in P,\, \overline{z}=y} \mu_P(x,z) \;+ \sum_{u:\, u \in \mathrm{Cl}(P),\; x \le u < y} \mu_{\mathrm{Cl}(P)}(x,u).$$

Hence, the sum

$$\sum_{z: z \in P,\, \overline{z}=y} \mu_P(x,z)$$

satisfies the bottom-up recursion for $\mu_{\mathrm{Cl}(P)}(x,y)$ and the two quantities are equal.

A similar argument works when $x < \overline{x}$. In this case, since every element z in $[x,\overline{x}]$ has closure $\overline{x}$, the base case is

$$\sum_{z \in P:\, x \le z \le \overline{x}} \mu_P(x,z) = \sum_{z \in P:\, z \in [x,\overline{x}]} \mu_P(x,z) = 0.$$

Induction now yields $\sum_{z \in P:\, \overline{z}=y} \mu_P(x,z) = 0$. This completes the second proof of Theorem 3.1.4.

If a is a fixed element in a lattice L, then $x \mapsto x \vee a$ is a closure operator on L. Thus, we obtain the following very useful special case.[5]

3.1.5. Weisner's theorem. Let L be a finite lattice and a be a fixed element in L. Then

$$\mu(\hat{0},\hat{1}) = -\sum_{x:\, x \vee a = \hat{1},\, x \ne \hat{1}} \mu(\hat{0},x).$$

Dually,

$$\mu(\hat{0},\hat{1}) = -\sum_{x:\, x \wedge a = \hat{0},\, x \ne \hat{0}} \mu(x,\hat{1}).$$

Another special case of Theorem 3.1.4 is the case of a closure operator on a set.

3.1.6. The set-closure theorem. Let $A \mapsto \overline{A}$ be a closure operator on the finite set S such that the empty set is closed and L be the lattice of closed sets.

[5] Weisner (1935).

If U is a closed set, then

$$\mu_L(\emptyset, U) = \sum_{A:\,\overline{A}=U} (-1)^{|A|}.$$

The next theorem was first proved by Philip Hall.[6] This theorem is essentially the second part of the closure theorem.

3.1.7. Theorem. Let L be a finite lattice. Then $\mu(\hat{0}, \hat{1}) = 0$ unless the meet of all the coatoms is $\hat{0}$ and the join of all the atoms is $\hat{1}$.

Proof. If L is a lattice, then

$$x \mapsto \bigwedge \{c\colon c \text{ is a coatom and } c \geq x\}$$

is a coclosure operator and the dual

$$x \mapsto \bigvee \{a\colon a \text{ is an atom and } x \geq a\}$$

is a closure operator. □

3.1.8. The Galois coconnection theorem. Let $\varphi : P \to Q$ and $\psi : Q \to P$ be a Galois coconnection, $x \mapsto \overline{x}$ the closure operators induced on P and Q, and $C(\varphi, \psi)$ the partially ordered set of closed elements. Let $x \in P$ and $y \in Q$. If both x and y are closed, then

$$\sum_{a:\,\varphi(a)=y} \mu_P(a, x) = \mu_{C(\varphi,\psi)}(y, x) = \sum_{s:\,\psi(s)=x} \mu_Q(y, s).$$

If at least one of x and y is not closed, then

$$\sum_{a:\,\varphi(a)=y} \mu_P(a, x) = 0 = \sum_{s:\,\psi(s)=x} \mu_Q(y, s).$$

Let L be a lattice. A subset $A \subset L$ is a *lower crosscut* satisfying two conditions:

LCC1. $\hat{0} \notin A$.
LCC2. If $x \in L$ and $x \neq \hat{0}$, then there exists an element a in A such that $x \geq a$.

An *upper crosscut* is a lower crosscut in the dual order.

[6] Hall (1936).

For a nonempty subset A in a finite lattice L, define the functions

$$\varphi : L \to 2^A,\ x \mapsto \{a \in A\colon a \le x\},$$
$$\psi : 2^A \to L,\ E \mapsto \bigvee E,$$

where $\bigvee E$ is defined to be the join of all the elements in E and, for use later on, $\bigwedge E$ is the meet of all the elements in E. The pair φ, ψ of functions defines a Galois coconnection between L and the Boolean algebra 2^A of all subsets of A. Since A is a lower crosscut, $\varphi(x) = \emptyset$ if and only if $x = \hat{0}$. In addition, $\psi(\varphi(\hat{0})) = \psi(\emptyset) = \hat{0}$, and so $\hat{0}$ is closed. Applying Theorem 3.1.4 to this Galois coconnection with $y = \emptyset$, $x = \hat{1}$, we obtain the following result.

3.1.9. A crosscut theorem. Let A be a lower crosscut in a finite lattice L. Then

$$\mu(\hat{0}, \hat{1}) = \sum_{E\colon E \subseteq A,\ \bigvee E = \hat{1}} (-1)^{|E|},$$

where the sum ranges over all subsets $E \subseteq A$ such that $\bigvee A = \hat{1}$. Dually, if B is an upper crosscut, then

$$\mu(\hat{0}, \hat{1}) = \sum_{E\colon E \subseteq B,\ \bigwedge E = \hat{0}} (-1)^{|E|}.$$

In a lattice L, the set of all atoms (that is, elements covering $\hat{0}$) is a lower crosscut. Thus, the formula for $\mu(\hat{0}, \hat{1})$ given in Theorem 3.1.9 depends only on the joins of atoms. Explicitly, $\mu(\hat{0}, \hat{1}) = 0$ if $\hat{1}$ is not a join of atoms. If $\hat{1}$ is a join of atoms, then $\mu(\hat{0}, \hat{1}) = \mu_{L^*}(\hat{0}, \hat{1})$, where L^* is the join-sublattice of L consisting of all joins of atoms. Joins in L^* agree with joins in L. The meet in L^* is given by

$$a \wedge b = \bigvee_{c \in L^*\colon c \le a \text{ and } c \le b} c.$$

For a distributive lattice L, the join-sublattice generated by the atoms is a Boolean algebra. Hence, if $\hat{1}$ is not a join of atoms, $\mu(\hat{0}, \hat{1}) = 0$, and if $\hat{1}$ is a join of atoms, then $\mu(\hat{0}, \hat{1}) = (-1)^m$, where m is the number of atoms in L.

A set C is a *crosscut* if it satisfies the following three conditions:

CC1. $\hat{0} \notin C$ and $\hat{1} \notin C$.

CC2. C is an antichain.

CC3. Every maximal chain from $\hat{0}$ to $\hat{1}$ intersects C at (exactly) one element.

Note that a crosscut need not be an upper or lower crosscut.

If C is a nonempty subset, then the pair

$$\varphi : \text{Interval}(L) \to 2^C,\ \ [x, y] \mapsto [x, y] \cap C,$$
$$\psi : 2^C \to \text{Interval}(L),\ \ E \mapsto \begin{cases} [\bigwedge E, \bigvee E] & \text{if } E \neq \emptyset \\ \emptyset & \text{if } E = \emptyset \end{cases}$$

of functions defines a Galois coconnection between the Boolean algebra 2^C and the lattice of intervals of L. If A is a crosscut, then $\psi(\varphi(\emptyset)) = \psi(\emptyset) = \emptyset$, and hence the empty interval is closed.

3.1.10. Another crosscut theorem. Let C be a crosscut in the finite lattice L. Then

$$\mu_L(\hat{0}, \hat{1}) = \sum_{E:\, E \subseteq A,\, \bigvee E = \hat{1},\, \bigwedge E = \hat{0}} (-1)^{|E|}.$$

Proof. We apply Theorem 3.1.8 to the Galois coconnection φ, ψ : Interval $(L) \leftrightarrow 2^C$ with $y = \emptyset$ and $x = [\hat{0}, \hat{1}]$, obtaining

$$\sum_{I:\, \varphi(I) = \emptyset} \mu_{\text{Interval}(L)}(I, [\hat{0}, \hat{1}]) = \sum_{E:\, \psi(E) = [\hat{0}, \hat{1}]} (-1)^{|E|}. \tag{CC}$$

The left-hand sum can be simplified using Theorem 3.1.3. There are two cases. When I is the empty interval, we use

$$\mu_{\text{Interval}(L)}(\emptyset, [\hat{0}, \hat{1}]) = -\mu_L(\hat{0}, \hat{1});$$

when I equals the nonempty interval $[u, v]$, we use

$$\mu_{\text{Interval}(L)}([u, v], [\hat{0}, \hat{1}]) = \mu_L(\hat{0}, u)\mu_L(v, \hat{1}).$$

Substituting this into the left-hand sum, we obtain

$$-\mu_L(\hat{0}, \hat{1}) + \sum_{u, v:\, u \leq v, [u, v] \cap C = \emptyset} \mu_L(\hat{0}, u)\mu_L(v, \hat{1}).$$

This can be rewritten as

$$-\mu_L(\hat{0}, \hat{1}) + \sum_{u, v:\, [u, v] \cap C = \emptyset} \mu_L(\hat{0}, u)\zeta_L(u, v)\mu_L(v, \hat{1}).$$

When $[u, v] \cap C = \emptyset$, then either $u \leq v < C$ (that is, $u \leq v < a$ for all $a \in C$) or $C < u \leq v$ (that is, $a < u \leq v$ for all $a \in C$). Hence, the sum can be broken up into two nonempty sums: one containing the interval $[\hat{0}, \hat{0}]$ and the other containing $[\hat{1}, \hat{1}]$. Each sum can be can be simplified as follows:

$$\sum_{u,v:u\leq v<C} \mu_L(\hat{0}, u)\zeta_L(u, v)\mu_L(v, \hat{1}) = \sum_{v:v<C} \left(\sum_{u:u\leq v} \mu_L(\hat{0}, u)\zeta_L(u, v) \right) \mu_L(v, \hat{1})$$
$$= \sum_{v:v<C} \delta_L(\hat{0}, v)\mu_L(v, \hat{1})$$
$$= \mu_L(\hat{0}, \hat{1})$$

and dually,

$$\sum_{u,v:C<u\leq v} \mu_L(\hat{0}, u)\zeta_L(u, v)\mu_L(v, \hat{1}) = \mu_L(\hat{0}, \hat{1}).$$

Hence, the left-hand sum in Equation (CC) equals $\mu_L(\hat{0}, \hat{1})$. □

3.1.11. Philip Hall's theorem on chains.[7] Let $x < y$ in a partially ordered set P. Then

$$\mu(x, y) = -c_1 + c_2 - c_3 + c_4 - \cdots,$$

where c_i is the number of length-i chains $x < x_1 < x_2 < \cdots < x_{i-1} < y$ stretched from x to y.

Proof. We give two proofs; the first is a simple formal manipulation. Let $\eta = \zeta - \delta$. Then

$$\mu = (\delta + \eta)^{-1}$$
$$= \delta - \eta + \eta * \eta - \eta * \eta * \eta + \cdots.$$

Since $x < y$, Hall's formula follows.

Hall's original proof uses a partition–recursion argument. It begins by observing that

$$\sum_C (-1)^{\text{length}(C)} = 0,$$

where C ranges over all chains stretched from x to an element z in the interval $[x, y]$. This follows since we can pair the chain $x < x_1 < \cdots < x_{n-1} < y$ with the chain $x < x_1 < \cdots < x_{n-1}$. To finish the proof, we use induction, starting from the fact that $\mu(x, x)$ equals 1, the number of length-0 chains starting from x. □

[7] Hall (1936).

Expressions of Möbius functions as alternating sums of the number of chains suggest there is a homological interpretation. The *chain complex* Chain(P) of the partially ordered set P is the simplicial complex with the chains of P as simplices. Then Theorem 3.1.11 says that the Euler characteristic χ(Chain(P)) equals $\mu_{P^+}(\bar{0}, \bar{1}) + 1$, where P^+ is obtained from P by adding a new minimum $\bar{0}$ and a new maximum $\bar{1}$. There is also a homological interpretation of crosscuts and Theorem 3.1.10.[8]

The introduction of methods from algebraic topology into the theory of Möbius functions and, more generally, into the theory of partially ordered sets is an important legacy of *Foundations I.* We refer the reader to the survey Björner (1995).

The next result is a variation on the closure theorem. It also leads to the complementation theorem. Let $f : P \to Q$ be an order-preserving function. Let $[a, b]$ be an interval in P such that $f(a) < f(b)$. Define $[a, b]_f$ to be the partially ordered set obtained by identifying all the elements z in the interval $[a, b]$ in P such that $f(z) = f(b)$; that is, $[a, b]_f$ is the set

$$\{x \in [a, b]\colon f(x) < f(b)\} \cup \{b\}$$

in P, with the order induced from $[a, b]$.

3.1.12. Theorem.[9] Let $a < b$, $f(a) < f(b)$, and $\mu_f(a, b)$ be the value of Möbius function from a to b in $[a, b]_f$. Then the following three equations, equivalent to each other, hold:

$$\mu_P(a, b) = \sum_{z\colon z \in [a,b],\, f(z)=f(b)} \mu_f(a, z)\mu_P(z, b), \tag{Mf1}$$

$$\mu_f(a, b) = \sum_{y\colon y \le z,\, f(y)=f(z)} \mu_P(a, b), \tag{Mf2}$$

$$\mu_P(a, b) = \sum_{y,z\colon y,z \in [a,b],\, f(y)=f(z)=f(b)} \mu_P(a, y)\zeta_P(y, z)\mu_P(z, b). \tag{Mf3}$$

Proof. We begin by showing that Equations (Mf1) and (Mf2) are equivalent. To do this, regard $\mu_f(a, y)$ and $\mu_P(a, y)$ as functions of y defined on P and use the Möbius inversion formula 3.1.2,

Next, we prove (Mf1) using a chain-counting argument. If R is a partially ordered set with a minimum $\hat{0}$ and a maximum $\hat{1}$, we define the *chain polynomial* $C(R; \lambda)$ *(in the variable* λ*)* by

$$C(R; \lambda) = \sum_{i\colon i \ge 0} c_i(R)\lambda^i,$$

[8] Folkman (1966). [9] Crapo (1966).

where $c_i(R)$ is the number of chains of length i stretched from $\hat{0}$ to $\hat{1}$ in R. By Theorem 3.1.11, $C(R; -1) = \mu_R(\hat{0}, \hat{1})$.

Consider an interval $[a, b]$ in P such that $f(a) < f(b)$. Each chain stretched from a to b in $[a, b]$ can be divided into two nonempty segments, $a < z_1 < \cdots < z_{i-1} < z_i$ and $z_i < z_{i+1} < \cdots < b$, where $f(z_{i-1}) \neq f(b)$ and $f(z_i) = f(z_{i+1}) = \cdots = f(b)$. Thus,

$$C([a, b]; \lambda) = \sum_{z: z \in [a,b],\ f(z)=f(b)} C([a, z]_f; \lambda) C([z, b]; \lambda).$$

Setting $\lambda = -1$, we obtain Equation (Mf1) and also (Mf2). Equation (Mf2) can also be proved directly using a partition–recursion argument similar to that in the proof of Theorem 3.1.4.

To finish the proof, we use (Mf2), with $b = z$, to substitute $\mu_f(a, z)$ into (Mf1), obtaining

$$\mu_P(a, b) = \sum_{z:\, z \in [a,b],\ f(z)=f(b)} \left(\sum_{y:\, y \leq z,\ f(y)=f(z)} \mu_P(a, y) \right) \mu_P(z, b).$$

Thus, (Mf3) is equivalent to (Mf1) and (Mf2), and it holds as well. □

Recall that in a lattice, c is a complement of a if $c \wedge a = \hat{0}$ and $c \vee a = \hat{1}$.

3.1.13. Crapo's complementation theorem. Let L be a finite lattice, a be any element in L, and $a^\perp$ be the set of complements of a in L. Then

$$\mu(\hat{0}, \hat{1}) = \sum_{c,d:\, c,d \in a^\perp} \mu(\hat{0}, c) \zeta(c, d) \mu(d, \hat{1}),$$

where the sum is over all ordered pairs (c, d) in $a^\perp \times a^\perp$ (so that $c = d$ is allowed). Put another way,

$$\mu(\hat{0}, \hat{1}) = \sum_{c,d:\, c,d \in a^\perp \text{ and } c \leq d} \mu(\hat{0}, c) \mu(d, \hat{1}),$$

Proof. The theorem holds trivially if $a = \hat{0}$. Thus we can assume that $a > \hat{0}$. We apply Theorem 3.1.12 to the order-preserving function $f : L \to [a, \hat{1}]$, $x \mapsto x \vee a$, obtaining

$$\mu(\hat{0}, \hat{1}) = \sum_{z:\, z \vee a = \hat{1}} \mu_f(\hat{0}, z) \mu(z, \hat{1}). \tag{M}$$

To finish the proof, we will show that $\mu_f(\hat{0}, z) = 0$ unless $z \wedge a = \hat{0}$. Suppose $a \vee z = \hat{1}$. The infimum of two elements in the partial order $[\hat{0}, z]_f$ is the same as their infimum in the lattice $[\hat{0}, z]$. Thus, $[\hat{0}, z]_f$ is a lattice. Let m be a coatom in $[\hat{0}, z]_f$. Then, as $m \neq z$, $m \vee a < \hat{1}$. Further,

$$[m \vee (z \wedge a)] \vee a = m \vee [(z \wedge a) \vee a] = m \vee a < \hat{1}.$$

Hence, $m \vee (z \wedge a)$ is in $[\hat{0}, z]_f$ and does not equal z. Since m is a coatom, it follows that $m \vee (z \wedge a) = m$; that is to say, $m \geq z \wedge a$. We conclude that $z \wedge a$ is a lower bound for all coatoms in $[\hat{0}, z]_f$. By Theorem 3.1.7, $\mu_f(\hat{0}, z) = 0$ unless the meet of all the coatoms is $\hat{0}$, or $z \wedge a = \hat{0}$. We can now restrict the sum in Equation (M) to those z such that $z \wedge a = \hat{0}$ (as well as $z \vee a = \hat{1}$). □

Another argument showing that $\mu_f(\hat{0}, z) = 0$ unless $z \wedge a = \hat{0}$ can be found in the solution to Exercise 3.1.13.

3.1.14. Corollary. Let L be a finite lattice. Suppose there exists an element a in L which does not have a complement. Then $\mu_L(\hat{0}, \hat{1}) = 0$.

Exercises

Incidence algebras as algebras.

Much work has been done on the algebraic aspects of incidence algebras. A comprehensive account can be found in Spiegel and O'Donnell (1997). We give six examples of such results. To avoid technicalities, we consider incidence algebras over a field $\mathbb{A}$.

3.1.1. *The functor $\mathcal{I}$.*[10]

A function $\sigma : P \to Q$ between partially ordered sets P and Q is *proper* if

PM1. σ is injective.
PM2. $\sigma(x_1) \leq \sigma(x_2)$ implies that $x_1 \leq x_2$.
PM3. If $y_1 \leq y_2$ in Q and both y_1 and y_2 are in the image of σ, then every element y in the interval $[y_1, y_2]$ is also in image of f.

[10] *Foundations VI*, pp. 267–318.

If $\sigma : P \to Q$ is proper, we define a function from $\mathcal{I}(Q) \to \mathcal{I}(P)$ as follows: if g is a function in $\mathcal{I}(Q)$, then

$$\mathcal{I}(\sigma)(g)(x_1, x_2) = g(\sigma(x_1), \sigma(x_2)). \tag{C}$$

Show that $\mathcal{I}(\sigma)$ is an $\mathbb{A}$-algebra homomorphism if and only if σ is a proper map. (We assume that $\mathbb{A}$-algebra homomorphisms preserve identities.) Conclude that the construction $\mathcal{I}$ is a contravariant functor from the category of partially ordered sets and proper maps to the category of $\mathbb{A}$-algebras and $\mathbb{A}$-algebra homomorphisms.

3.1.2. *Ideals in incidence algebras.*

(a) If J is a (two-sided) ideal in $\mathcal{I}(P)$, let

$$\text{support}(J) = \{\epsilon_{xy}\colon\ \epsilon_{xy} \in J\}.$$

Show that J consists of all functions f in $\mathcal{I}(P)$ such that $f(x, y) = 0$ whenever $\epsilon_{xy} \notin \text{support}(J)$.

(b) Let zero(J) be the set of intervals $[x, y]$ such that $f(x, y) = 0$ for all functions f in J. Show that if $[x, y] \in \text{zero}(J)$ and $[u, v] \leq [x, y]$, then $[u, v] \in \text{zero}(J)$.

(c) Show that there is an order-reversing bijection between the lattice of ideals in $\mathcal{I}(P)$ and the lattice of (order) ideals in Interval(P). Conclude that the lattice of ideals in $\mathcal{I}(J)$ is distributive. In addition, conclude that the maximal ideals in $\mathcal{I}(P)$ are ideals of the form

$$\{f\colon\ f(a, a) = 0\},$$

where a is a fixed element of P.

3.1.3. *Characterization of incidence algebras.*[11]

(a) Let $\mathbb{A}$ be a field and $\mathcal{I}$ be a subalgebra of the algebra of $n \times n$ matrices over $\mathbb{A}$. Then there is a partially ordered set P such that $\mathcal{I}$ is $\mathbb{A}$-algebra isomorphic to the incidence algebra $\mathcal{I}(P)$ if and only if the following two conditions hold:

IA1. $\mathcal{I}$ contains n pairwise orthogonal idempotents.

IA2. The quotient algebra $\mathcal{I}/J(\mathcal{I})$ is commutative, where $J(\mathcal{I})$ is the Jacobson radical of $\mathcal{I}$.

(b) Let $\mathbb{A}$ be a field and P and Q be partially ordered sets. Then $\mathcal{I}(P)$ and $\mathcal{I}(Q)$ are isomorphic $\mathbb{A}$-algebras if and only if P and Q are isomorphic partial orders.

[11] Stanley (1970) and Feinberg (1977).

3.1.4. *The Amitsur–Levitski theorem for incidence algebras.*[12]

The *standard identity* I_N in the N variables $x_1, x_2, \ldots, x_N$ is

$$\sum_{\sigma} \operatorname{sign}(\sigma) x_{\sigma(1)} x_{\sigma(2)} \cdots x_{\sigma(N)} = 0,$$

where the sum ranges over all permutations σ of $\{1, 2, \ldots, N\}$ and $\operatorname{sign}(\sigma)$ is the parity of σ. For example, the standard identity I_2 is $x_1x_2 - x_2x_1 = 0$. The Amitsur–Levitski theorem says that the algebra of $n \times n$ matrices over a commutative ring satisfies the standard identity I_{2n} in $2n$ variables and does not satisfy the standard identity I_k in any smaller number k of variables.[13]

Let P be a locally finite partially ordered set. Show that if P contains a chain of length $n-1$, then the incidence algebra $\mathcal{I}(P)$ does not satisfy the standard identity I_n. Show that if m is the maximum length of a chain in P, then $\mathcal{I}(P)$ satisfies I_{2m+2}.

3.1.5. *Cartesian and tensor products.*

Show that $\mathcal{I}(P \times Q) = \mathcal{I}(P) \otimes \mathcal{I}(Q)$. In particular, the matrix of $\zeta_{P \times Q}$ is the tensor or Kronecker product of the matrices ζ_P and ζ_Q. Using this, give another proof of Theorem 3.1.2.

3.1.6. *The standard topology.*

The *standard topology* on an incidence algebra of a locally finite partially ordered set is the topology defined by a sequence $f_1, f_2, f_3, \ldots$ converges to a function f if for every ordered pair (x, y), $x \leq y$ in P, there is an index N (which may depend on (x, y)) such that $f_n(x, y) = f(x, y)$ for all $n \geq N$. Extend the results in Exercises 3.1.1–3.1.3 to locally finite partially ordered sets.

3.1.7. *The interval semigroup of a partially ordered set.*[14]

Define a binary operation on the set Interval(P) of intervals of a partially ordered set P by

$$[a, b][c, d] = \begin{cases} [a, d] & \text{if } b = c, \\ 0 & \text{otherwise.} \end{cases}$$

This binary operation makes Interval(P) into a semigroup. Show that the semigroup algebra (over a commutative ring $\mathbb{A}$) is isomorphic to the incidence algebra $\mathcal{I}(P)$ over $\mathbb{A}$.

[12] Feinberg (1976).
[13] See Rowen (1980); a graph-theoretic proof can be found in Swan (1963, 1969).
[14] Ward (1939).

3.1.8. *Zeta polynomials*[15]

Let P be a finite partially ordered set. A *multichain* is a multiset $\{x_1, x_2, \ldots, x_n\}$ of elements in P that can be arranged so that $x_1 \leq x_2 \leq \cdots \leq x_n$. The *zeta polynomial* $Z(P; n)$ is the number of multichains of size n in P.

(a) Show that

$$Z(P; n) = \sum_{i=2}^{d} b_i \binom{n-2}{i-2}, \qquad \text{(Z)}$$

where b_i is the number of multichains of length $i - 2$ in P and d is the length of the longest chain in P. Using this result, we can think of Z as a polynomial in the variable n.

(b) Show that if P has a minimum $\hat{0}$ and a maximum $\hat{1}$, then

$$Z(P; n) = \zeta^n(\hat{0}, \hat{1}),$$

where ζ^n is the n-fold convolution of the zeta function in the incidence algebra. If n is negative and $n = -m$, then ζ^n is interpreted to be μ^m.

(c) Show that the order polynomial (discussed in Exercise 1.2.4) $\Omega(P, n)$ equals the zeta polynomial $\zeta(I(P), n)$ of the lattice of order ideals of P.

(d) Show that $Z(\text{Interval}(P); n) = Z(P; 2n - 1)$.

(e) Let $\varphi : P \to Q$ and $\psi : Q \to P$ be a Galois coconnection and suppose that Q has a minimum. Let y be an element in P. Then

$$\sum_{a: a \leq y} Z([\hat{0}, \varphi(a)]; n + 1)\mu_P(a, y) = \sum_{x: \psi(x) = y} Z([\hat{0}, x]; n).$$

In particular, set $n = -1$ to obtain the Galois coconnection theorem (3.1.8).

Let E be a subset of a partially ordered set P. An element x in P is *ordered relative to* E if $x \geq s$ for all $s \in E$ or $x \leq s$ for all $s \in E$; the element x is *strictly ordered relative to* E if x is ordered relative to E but not ordered relative to any larger set D containing E. Let P_E be the subset of elements in P ordered relative to E, with the partial order induced by P. If both the infimum $\bigwedge E$ and the supremum $\bigvee E$ of E exist, then P_E is the union of the principal filter $F(\bigvee E)$ and the principal ideal $I(\bigwedge E)$.

(f) A subset A of a partially ordered set P is a *cutset* if it intersects every maximal chain in P. Show that if A is a cutset of P, then

$$Z(P; n) = - \sum_{E: E \subseteq A,\, E \neq \emptyset} (-1)^{|E|} Z(P_E; n).$$

[15] Edelman (1980) and Stanley (1986, p. 129).

(g) From (f), derive the following crosscut theorem: let L be a finite lattice and A be a cutset such that $A \neq \{\hat{0}\}$ and $A \neq \{\hat{1}\}$. Then

$$\mu_L(\hat{0}, \hat{1}) = \sum_{S:\, S \subseteq A,\, \bigvee S = \hat{1},\, \bigwedge S = \hat{0}} (-1)^{|S|}.$$

3.1.9. *A formula for permanents.*[16]

(a) Prove *Ryser's formula:* Let $X = [x_{ij}]_{1 \leq i,j \leq n}$. Then

$$\operatorname{per} X = \sum_{E:\, E \subseteq \{1,2,\ldots,n\}} (-1)^{n-|E|} \prod_{i=1}^{n} \sum_{j:\, j \in E} x_{ij}.$$

For example, when $n = 2$, Ryser's formula is

$$x_{11}x_{22} + x_{12}x_{21} = (x_{11} + x_{12})(x_{21} + x_{22}) - x_{11}x_{21} - x_{12}x_{22}.$$

(b) Use Ryser's formula to give a reasonable way to calculate the permanent.

(c) Let $X = [x_{ij}]_{1 \leq i \leq m, 1 \leq j \leq n}$, where $n \geq m$. Extend the definition of the permanent by

$$\operatorname{per} X = \sum_{A:\, A \subseteq \{1,2,\ldots,m\}, |A| = n} \operatorname{per} [x_{ij}]_{1 \leq i \leq m, j \in A}.$$

Show that

$$\operatorname{per} X = \frac{1}{(n-m)!} \sum_{E:\, E \subseteq \{1,2,\ldots,m\}} (-1)^{m-|E|} |E|^{n-m} \prod_{i=1}^{n} \sum_{j:\, j \in E} x_{ij}.$$

(d) (Research problem) Find analogs of Ryser's formula for higher-dimensional permanents, as defined in Exercise 2.8.5.

3.1.10. *The Eulerian function of a closure operator.*[17]

Let $A \to \overline{A}$ be a closure operator on a finite set S. A subset A of S *spans* if $\overline{A} = S$; an s-tuple $(a_1, a_2, \ldots, a_s)$ of elements a_i in S *spans* if the underlying set $\{a_1, a_2, \ldots, a_s\}$ spans S. The *Eulerian function* $\phi(S; s)$ of the closure is the Dirichlet polynomial defined by

$$\phi(S; s) = \sum_{X:\, X \in L} \mu(X, S) |X|^s,$$

where L is the lattice of closed subsets of S.

(a) Show that when s is a nonnegative integer, $\phi(S; s)$ equals the number of spanning s-tuples.

Eulerian functions were first studied by P. Hall in his 1936 paper. Let G be a group. The function sending a subset $A \subseteq G$ to the subgroup generated by

[16] Crapo (1968b) and Ryser (1963).

[17] Gaschütz (1959), Hall (1936), and Kung (1996b, Sections 3.2 and 3.3).

A is a closure operator on G. The lattice of closed sets is $L(G)$, the lattice of subgroups of G.

(b) An *automorphism* of the closure operator $A \mapsto \overline{A}$ on S is a permutation α on S such that for all $A \subseteq S$, $\alpha(\overline{A}) = \overline{\alpha(A)}$. Show that if $\mathcal{A}$ is the automorphism group of the closure operator, then the order $|\mathcal{A}|$ divides $\phi(G; n)$ for every nonnegative integer n. In particular, if G is a group with automorphism group $\mathrm{Aut}(G)$, then $|\mathrm{Aut}(G)|$ divides $\phi(G; n)$.

(c) The *Frattini subgroup* $\Phi(G)$ of the group G is the intersection of all the maximal subgroups of G. Show that

$$\begin{aligned}\phi(G; s) &= \sum_{H:\, \Phi(G) \leq H \leq G} \mu(H, G)|H|^s \\ &= |\Phi(G)|^s \phi(G/\Phi(G); s).\end{aligned}$$

(d) Show that if $|H|$ and $|K|$ are relatively prime and $H \times K$ is the direct product, then

$$\phi(H \times K; s) = \phi(H; s)\phi(K; s).$$

(e) Prove *Gaschütz's factorization theorem*. If N is a normal subgroup of G, then

$$\phi(G; s) = \phi(G/N; s)\phi(G \downarrow N; s),$$

where

$$\phi(G \downarrow N; s) = \sum_{H:\, H \leq G \text{ and } NH = G} \mu(H, G)|N \cap H|^s.$$

(f) Let p be a prime. A finite group G is a p-group if $|G| = p^m$ for some nonnegative integer m. Show that

$$\phi(G; s) = p^{m-r} \prod_{i=0}^{r-1} (p^s - p^i)$$

for some positive integer r, $r \leq m$.

(g) Show that if Z_n is the cyclic group of order n, then

$$\phi(Z_n; s) = \sum_{i:\, i \mid n} \mu(n/i) i^s,$$

where μ is the number-theoretic Möbius function.

(h) Calculate the Eulerian functions of dihedral groups.

(i) Calculate the Eulerian function of the projective special linear groups $\mathrm{PSL}(2, p)$ when p is a prime.

(j) Prove that if G is a finite simple group, then $\phi(G; 2) \neq 0$.

3.1.11. *The Lindström–Wilf determinantal formula.*[18]

A *meet-semilattice* is a partially ordered set in which the infimum $x \wedge y$ of any two elements exists. (The existence of a maximum is not assumed.) Let P be a finite meet-semilattice, $\mathbb{A}$ be a commutative ring with identity, and $F : P \times P \to \mathbb{A}$ be a function, not necessarily in the incidence algebra $\mathcal{I}(P)$.

(a) Show that

$$\det[F(x \wedge y, y)]_{x,y \in P} = \prod_{x : x \in P} \left(\sum_{z : z \leq x} F(z, x)\mu(z, x) \right).$$

We note the dual form. Let P be a join-semilattice, that is, a finite partially ordered set in which the supremum $x \vee y$ of any two elements exists. Then

$$\det[F(y, x \vee y)]_{x,y \in P} = \prod_{x : x \in P} \left(\sum_{z : z \geq x} \mu(x, z)F(x, z) \right).$$

(b) Show that

$$\det[\operatorname{sign} \mu(x \wedge y, y)]_{x,y \in P} = \prod_{y : y \in P} \left(\sum_{x : x \in P} |\mu(x, y)| \right).$$

(c) Show that

$$\det[\gcd(i, j)]_{1 \leq i,j \leq n} = \prod_{i=1}^{n} \phi(i),$$

where ϕ is the Euler totient function.

(d) Let P be a finite meet-semilattice such that $\mu(\hat{0}, x) \neq 0$ for all x in P. Show that there exists a permutation $\phi : P \to P$ such that $x \wedge \phi(x) = \hat{0}$ for all $x \in P$.

(e) Let b be a positive integer greater than 1. If i is a nonnegative integer, let $e_0(i) + e_1(i)b + e_2(i)b^2 + \cdots$ be the b-ary expansion of i, $\alpha(i) = \sum_{s : s \geq 0} e_s(i)$, and $\eta(i)$ be the number of b-ary digits $e_s(i)$ that are nonzero. If i and j are positive integers, define $i \wedge j$ to be the (nonnegative) integer

$$\min\{e_0(i), e_0(j)\} + \min\{e_1(i), e_1(j)\}b + \min\{e_2(i), e_2(j)\}b^2 + \cdots.$$

Note that when $b = 2$, $\min\{e_s(i), e_s(j)\} = e_s(i)e_s(j)$. Show that

$$\det[(-1)^{\alpha(i \wedge j)}]_{0 \leq i,j \leq n} = 2^{\eta(0)+\eta(1)+\cdots+\eta(n)}.$$

[18] Lindström (1969) and Wilf (1968).

3.1.12. *The Redheffer matrix.*[19]

Let P be a finite partially ordered set with a minimum $\hat{0}$. The (*generalized*) *Redheffer matrix* $R(P)$ the matrix with rows and columns indexed by P (linearly ordered by a chosen linear extension) with $R(P)$ the matrix obtained from the (upper-triangular) matrix of the zeta function by replacing the first column (labeled by $\hat{0}$) by a column with all entries equal to 1. Show that the permanent of $R(P)$ equals the number of chains in P (of any length) with bottom $\hat{0}$ and the determinant of $R(P)$ equals the sum

$$\sum_{x:x\in P} \mu(\hat{0}, x).$$

In particular, $R(P)$ is singular if P has a maximum. A special case is the number-theoretic Redheffer matrix R_n, the matrix with rows and columns indexed by $\{1, 2, \ldots, n\}$ with ij-entry equal to 1 if $j = 1$ or i divides j, and 0 otherwise. Then

$$\det R_n = \sum_{k=1}^{n} \mu(k),$$

where μ is the number-theoretic Möbius function.

3.1.13. *A useful inversion formula.*[20]

Prove the following inversion formula: let L be a lattice in which $\mu(x, \hat{1}) \neq 0$ for all $x \in L$. Let f and g be functions from L to a ring $\mathbb{A}$. Then

$$g(x) = \sum_{y:\, y\vee x=\hat{1}} f(y) \text{ for all } x \iff$$

$$f(x) = \sum_{y:\, y\in L} \left(\sum_{z:\, 0\leq z\leq x\wedge y} \frac{\mu(z, x)\mu(z, y)}{\mu(z, \hat{1})} \right) g(y) \text{ for all } x.$$

For future reference, we record the dual version. Let L be a lattice in which $\mu(\hat{0}, x) \neq 0$ for all $x \in L$ and $f, g : L \to \mathbb{A}$ be functions. Then

$$g(x) = \sum_{y:\, y\wedge x=\hat{0}} f(y) \text{ for all } x \iff$$

$$f(x) = \sum_{y:\, y\in L} \left(\sum_{z:\, x\vee y\leq z\leq \hat{1}} \frac{\mu(x, z)\mu(y, z)}{\mu(\hat{0}, z)} \right) g(y) \text{ for all } x.$$

[19] Redheffer (1977) and Wilf (2004–2005). Other results about the Redheffer's matrix, including its relation to the Riemann hypothesis, can be found in papers cited in Wilf's paper.

[20] Doubilet, *Foundations VII* and Dowling and Wilson (1975).

3.1.14. *A complementing permutation.*[21]

Let L be a finite lattice in which $\mu(\hat{0}, x) \neq 0$ and $\mu(x, \hat{1}) \neq 0$ for all $x \in L$.

(a) Show that there exists a bijection $\phi : L \to L$ such that for all $x \in L$, $\phi(x) \wedge x = \hat{0}$ and $\phi(x) \vee x = \hat{1}$; that is, $\phi(x)$ is a complement of x.

(b) (Research problem) Suppose $\mu(\hat{0}, x) \neq 0$ and $\mu(x, \hat{1}) \neq 0$ for all $x \in L$. Is the matrix $[\delta(\hat{0}, x \wedge y)\delta(\hat{1}, x \vee y)]_{x,y \in L}$ (with xy-entry equal to 1 if x and y are complements and 0 otherwise) nonsingular?

(c) (Research problem) Find natural explicit complementing permutations for special lattices or classes of lattices?

3.1.15. *Lattices of intervals.*[22]

The set $L(\mathcal{C})$ of unions of subsets from a finite collection $\mathcal{C}$ of subsets forms a lattice with join equal to union and meet defined by

$$A \wedge B = \bigcup_{C : C \subseteq A \text{ and } C \subseteq B} C.$$

An (*integer*) *interval* is a subset of integers of the form $\{a, a+1, a+2, \ldots, b-1, b\}$, where $a \geq 1$. An *interval lattice* is a lattice of the form $L(\mathcal{C})$, where $\mathcal{C}$ is a collection of intervals.

(a) Show that an interval in an interval lattice is an interval lattice.

(b) Show that if L is an interval lattice, then $\mu(A, B) = -1$, 0, or 1.

3.2 Chains and Antichains

Let P be a partially ordered set. A subset A in P is an *antichain* if no two elements in it are comparable, or equivalently, $|A \cap C| \leq 1$ for any chain C in P. The *width* $w(P)$ of a partially ordered set P is the maximum size of an antichain in P. If P is partitioned into m chains, then $m \geq w(P)$. Dilworth's chain decomposition theorem says that equality can be achieved.

3.2.1. Dilworth's chain decomposition theorem.[23] A finite partially ordered set of width k can be partitioned into k chains.

As well as a fundamental theorem in the theory of partially ordered sets, Dilworth's theorem is "perhaps the most elegant matching theorem" (*Matching theory*). It also implies the following Ramsey-theoretic result.

[21] Dowling (1977). [22] Greene (1988). [23] Dilworth (1950).

3.2.2. Corollary. A finite partially ordered set of size at least $mn+1$ contains a chain of length $m+1$ or an antichain of size $n+1$.

Another interpretation, using perfect graphs, is given in Exercise 3.2.3.

We will give several proofs, beginning with Dilworth's proof in his 1950 paper. In Dilworth's own words,[24] the idea of the proof is that "if ... a portion of the ordered set was split into n chains and x was an element not in any of the n chains, then it should be possible to include x and reorganize the chains so that the enlarged set was again a set join of n chains."

Dilworth's proof. Let P be a partially ordered set of width k, $C_1, C_2, \ldots, C_k$ be k disjoint chains in P, and $C = C_1 \cup C_2 \cup \cdots \cup C_k$. Suppose there is an element a in P, not in C. For each chain C_i, let

$$\begin{aligned} U_i &= \{x \in C_i \colon x > a\}, \\ N_i &= \{x \in C_i \colon x \text{ is incomparable with } a\}, \\ L_i &= \{x \in C_i \colon x < a\}, \end{aligned}$$

$U = U_1 \cup U_2 \cup \cdots \cup U_k$, $N = N_1 \cup N_2 \cup \cdots \cup N_k$, and $L = L_1 \cup L_2 \cup \cdots \cup L_k$. Then, there exists an index m such that $(U \backslash U_m) \cup N$ has width strictly less than k. To prove this, suppose the contrary; that is, for each index m, there is an antichain A_m of size k in $(U \backslash U_m) \cup N$. Let $A = A_1 \cup A_2 \cup \cdots \cup A_k$ and s_i the minimum element in $A \cap C_i$. Then

(a) $\{s_1, s_2, \ldots, s_k\}$ is an antichain, and
(b) $\{s_1, s_2, \ldots, s_k\} \subseteq N$.

To prove (a), suppose that $s_i \geq s_j$. Let $s_j \in A_r$ and let t_i be the element in $A_r \cap C_i$. Then $t_i \geq s_i \geq s_j$. In addition, $t_i \neq s_j$ since t_i and s_j are in two different chains, contradicting the assumption that A_r is an antichain. To prove (b), observe that $A_i \cap U_i = \emptyset$ by construction, and hence there is at least one element of A in N_i.

It follows from (a) and (b) that $\{a, s_1, s_2, \ldots, s_k\}$ is an antichain in P of size $k+1$, a contradiction. Using the same argument in the order-dual, we can also prove that there exists an index l such that $(L \backslash L_l) \cup N$ has width strictly less than k.

To finish the proof, observe that if $s \in U_i$ and $t \in L_j$, then $s \geq a \geq t$. Hence, an antichain A in $C \backslash (U_m \cup L_l)$ is contained in either $(U \backslash U_m) \cup N$ or $(U \backslash L_l) \cup N$. It follows that $C \backslash (U_m \cup L_l)$ has width at most $k-1$. By induction,

[24] Background to chapter 1, in Dilworth (1990).

it can be partitioned into $k - 1$ chains. These chains, together with the chain $L_l \cup \{a\} \cup U_m$, give a partition of $C \cup \{a\}$ into k chains. □

The first proof of a fundamental theorem is often not efficient, because, in part, the discoverer has to conjecture the theorem and decide whether the conjecture is true. In addition, the conjecture comes from some concrete situation and the essence has to be extracted. Dilworth's proof is a good example of this phenomenon. There are many simpler (inductive) proofs of Dilworth's theorem. We shall give four of them.

Perles' proof.[25] Suppose that there exists a maximum-size antichain A not equal to the set of maximal elements or the set of minimal elements. Consider the ideal $I(A)$ and filter $F(A)$ generated by A, defined by

$$I(A) = \{x\colon x \leq a \text{ for some } a \text{ in } A\},$$
$$F(A) = \{x\colon x \geq a \text{ for some } a \text{ in } A\}.$$

Then by our assumption, $|I(A)| < |P|$ and $|F(A)| < |P|$. As A has maximum size, $I(A) \cup F(A) = P$, $I(A) \cap F(A) = A$, and both $I(A)$ and $F(A)$ have width $w(P)$. By induction, there exist chain partitions with $w(P)$ chains for $I(A)$ and $F(A)$. Each chain in the partition for $I(P)$ ends at an element a in A and each chain in the partition for $F(A)$ starts at an element a in A. Hence, the chains in the two partitions can be spliced together to obtain a partition for P with $w(P)$ chains.

It remains to consider the case when A equals the set of maximal elements or the set of minimal elements. Let a be a maximal element and b a minimal element such that $a \geq b$. (It may happen that $a = b$.) Then $P \backslash \{a, b\}$ has width $w(P) - 1$. By induction, $P \backslash \{a, b\}$ has a chain partition with $w(P) - 1$ chains. Adding the chain $\{a, b\}$ to it, we obtain a chain partition for P with $w(P)$ chains. □

Tverberg's proof.[26] Let C be a maximal chain in P. If $P \backslash C$ has width $w(P) - 1$, then C added to a partition of $P \backslash C$ into $w(P) - 1$ chains yields a chain partition of P into $w(P)$ chains. We can now suppose that $P \backslash C$ has width $w(P)$. Let A be an antichain of size $w(P)$ in P disjoint from C. The maximum element in C is not in $I(A)$, and hence, $|I(A)| < |P|$. Dually, the minimum element in C is not in $F(A)$ and $|F(A)| < |P|$. By induction, there are chain partitions in $I(A)$ and $F(A)$ of size $w(P)$. We can now apply the splicing argument in Perles' proof. □

Galvin's proof.[27] Let a be a maximal element of P and suppose that $P \backslash \{a\}$ has width k, so that P has width k or $k + 1$. By induction, $P \backslash \{a\}$ has a

[25] Perles (1963). [26] Tverberg (1967). [27] Galvin (1994).

partition into k chains $C_1, C_2, \ldots, C_k$. If A is an antichain with k elements, then $|A \cap C_i| = 1$. Let a_i be the maximum element in the chain C_i that is in some size-k antichain. Then $\{a_1, a_2, \ldots, a_k\}$ is an antichain. (To see this, suppose that $a_2 > a_1$, say. Let $\{b_1, a_2, b_3, \ldots, b_k\}$ be a size-k antichain containing a_2 so that $b_1 \in C_1$; then $a_2 > a_1 \geq b_1$, a contradiction.)

If $\{a, a_1, a_2, \ldots, a_k\}$ is an antichain, then P has width $k+1$ and $\{a\}, C_1, C_2, \ldots, C_k$ is a partition into $k+1$ chains. If not, $a > a_i$ for some i. Then $\{y: y \in C_i, y \leq a_i\} \cup \{a\}$ is a chain K. Since every size-k antichain in $P \backslash \{a\}$ contains one of the elements in $\{y: y \in C_i, y \leq a_i\}$, $P \backslash K$ does not contain any size-k antichain. By induction, there is a partition $D_1, D_2, \ldots, D_{k-1}$ of $P \backslash K$ into $k-1$ antichains. Adding K to this partition, we obtain a partition of P into k chains. □

Dilworth's motivation[28] for studying chain partitions is to determine the order dimension of a (finite) distributive lattice. Recall from Section 1.2 that the order dimension $\dim(P)$ of a partially ordered set P is the minimum number d such that there exists an order-preserving function from P into a Cartesian product of d chains. In a product of d chains, an element covers at most d elements: hence, $\dim(P)$ is at least the maximum number of elements covered by an element in P.

3.2.3. Lemma. An element x in a finite distributive lattice L covers exactly k elements if and only if x is the join of an antichain of k join-irreducibles.

Proof. Let $j_1 \vee j_2 \vee \cdots \vee j_k$ be the unique decomposition of x into an antichain of join-irreducibles. By Birkhoff's representation theorem (1.3.3), we may think of x as the ideal I_x in the partially ordered set $J(L)$ with maximal elements $j_1, j_2, \ldots, j_k$.

Observe that if j is a maximal element in an ideal I, then $I \backslash \{j\}$ is an ideal and I covers $I \backslash \{j\}$. Conversely, if I covers J, then there exists at least one maximal element h in I, not in J, $J \subseteq I \backslash \{h\}$, and hence, $J = I \backslash \{h\}$. Hence, the ideal I_x covers exactly k ideals, $I_x \backslash \{j_1\}, I_x \backslash \{j_2\}, \ldots, I_x \backslash \{j_k\}$. □

By the lemma, we conclude that $\dim(L) \geq w(J(L))$. In fact, equality holds.

3.2.4. Dilworth's dimension theorem. The order dimension of a finite distributive lattice L equals the width of its partially ordered set $J(L)$ of join-irreducibles.

[28] This is recounted in the background to chapter 1 of Bogart et al. (1991).

Proof. We only need to prove that the width w of $J(L)$ is an upper bound for $\dim(L)$. By the chain partition theorem, there exists a partition $C_1, C_2, \ldots, C_w$ of $J(L)$ into w (pairwise disjoint) chains. Think of a chain not as a totally ordered set $x_1 < x_2 < \cdots < x_s$, but as a chain of subsets

$$\emptyset \subset \{x_1\} \subset \{x_1, x_2\} \subset \cdots \subset \{x_1, x_2, \ldots, x_s\},$$

and let $\tilde{C}_i$ be the chain of subsets corresponding to C_i. In addition, think of an element x in L as an ideal in $J(L)$, so that the intersection $x \cap C_i$ is a subset occurring in $\tilde{C}_i$. Consider the function $L \to \tilde{C}_1 \times \tilde{C}_2 \times \cdots \times \tilde{C}_w$ defined by

$$x \mapsto (x \cap C_1, x \cap C_2, \ldots, x \cap C_w).$$

If $x \subseteq x'$, then $x \cap C_i \subseteq x' \cap C_i$. In addition,

$$x = \bigcup_{i=1}^{w} (x \cap C_i).$$

Hence, this function is an injective order-preserving function. We conclude that $w(J(L)) \geq \dim(L)$. □

The idea of using a maximum-size antichain formed from minimal or maximal elements of subsets of the partially ordered set occurs in Dilworth's proof. This idea led to a structural result about maximum-size antichains. Recall from Section 1.2 that the antichains in a partially ordered set P form a distributive lattice $\mathcal{D}(P)$ under the order $A \leq B$ if for all elements b in B, there exists an element a in A such that $a \leq b$, or, equivalently, $I(A) \subseteq I(B)$. In terms of ideals, meets and joins are set intersection and unions. In terms of the antichains themselves, the join is the set $\max A \cup B$, the subset of maximal elements in the set $A \cup B$. The meet $A \wedge B$ is somewhat more complicated. The obvious description, from the definition of meet in $\mathcal{D}(P)$, is

$$A \wedge B = \max I(A) \cap I(B).$$

There are two other plausible candidates. Let

$$A \,\Delta\, B = ((A \cup B) \backslash (\max A \cup B)) \cup (A \cap B)$$

and

$$A \,\nabla\, B = \min A \cup B,$$

where $\min A \cup B$ is the set of minimal elements in $A \cup B$. All three sets are antichains. It is possible that they are all different. For example, take $A = \{a\}$ and $B = \{b\}$ in the partially ordered set $\{a, b, c\}$, with $c < a$ and

$c < b$. However, it is immediate from the definitions that $A \,\Delta\, B \subseteq A \wedge B$, $A \,\Delta\, B \subseteq A \,\nabla\, B$, and

$$|A| + |B| = |A \vee B| + |A \,\Delta\, B|.$$

From these, we conclude that

$$|A| + |B| \leq |A \vee B| + |A \wedge B|,$$
$$|A| + |B| \leq |A \vee B| + |A \,\nabla\, B|.$$

In particular, if A and B are maximum-size antichains, then $A \vee B$ and $A \wedge B$ are also maximum-size antichains. This proves the following theorem of Dilworth.[29]

3.2.5. Theorem. The maximum-size antichains form a sublattice $\mathcal{S}(P)$ of the lattice $\mathcal{D}(P)$ of all antichains (or ideals).

Since $\mathcal{D}(P)$ is distributive, $\mathcal{S}(P)$ is distributive. Note that when A and B are maximum-size antichains, $A \wedge B = \min A \cup B$.

Theorem 3.2.5 yields yet another proof of the decomposition theorem.

Pretzel's proof.[30] A chain C is *saturated* in the partially ordered set P if it has nonempty intersection with every maximum-size antichain in P. We can use Theorem 3.2.4 to construct a saturated chain inductively. Let a_1 be an element in the meet A_1 of all the maximum-size antichains in the sublattice $\mathcal{S}(P)$. Consider the meet A_2 of all the maximum-size antichains not containing a_1. Since $A_1 \leq A_2$, the antichain A_2 contains an element a_2 such that $a_2 \geq a_1$. Choose a_2 to obtain the length-2 chain $\{a_1, a_2\}$. Continue.

If C is a saturated chain, then $w(P \backslash C) = w(P) - 1$. We can now prove Dilworth's theorem by induction. □

Theorem 3.2.5 has an important corollary.[31] An *order automorphism* of P is an order-preserving bijection $P \to P$.

3.2.6. Corollary. Let P be a finite partially ordered set. There exists a maximum-size antichain in P invariant under all order automorphisms of P.

Proof. An order automorphism of P induces an order automorphism of the lattice $\mathcal{S}(P)$. Hence, the maximum antichain in $\mathcal{S}(P)$ is invariant under all order automorphisms. The minimum antichain is also invariant. □

[29] Dilworth (1960) and Freese (1974). [30] Pretzel (1979).
[31] Freese (1974) and Kleitman et al. (1971).

Dilworth's theorem has been extended to unions of antichains by Greene and Kleitman.[32] A subset A in a partially ordered set P is a *k-family* if for every chain C in P, $|A \cap C| \leq k$, or equivalently, A is a union of k (or fewer) antichains. Let $d_k(P)$ be the maximum size of a k-family in P. If $C_1, C_2, \ldots, C_e$ is a partition of P into e chains, let

$$E(C_1, C_2, \ldots, C_e) = \sum_{i=1}^{e} \min\{k, |C_i|\}.$$

Each chain C_i intersects a k-family at most $\min\{k, |C_i|\}$ times. Hence,

$$E(C_1, C_2, \ldots, C_e) \geq d_k(P).$$

Equality can always be achieved.

3.2.7. Theorem (Greene and Kleitman). Let P be a finite partially ordered set. Then

$$d_k(P) = \min\{E(C_1, C_2, \ldots, C_e)\},$$

the minimum being taken over all partitions $C_1, C_2, \ldots, C_e$ of P into chains.

A proof will be sketched in Exercise 3.2.6.

Exercises

3.2.1. *The Erdős–Szekeres theorem.*[33]

(a) A *monotone* sequence is a sequence that is nondecreasing or nonincreasing. Using Corollary 3.2.2., show that any finite integer sequence of length $n^2 + 1$ contains a monotone subsequence of length at least $n + 1$.

(b) Give an independent proof of Corollary 3.2.2 using the pigeon-hole principle.

3.2.2. *A dual of Dilworth's theorem.*[34]

Let P be a partially ordered set. Then the maximum length of a chain in P equals the minimum number of antichains in a partition of P into antichains.

3.2.3. *Chain decompositions and matchings.*[35]

A relation $R : S \to T$ can be made into a rank-2 partially ordered set by putting S "below" T, so that $a < b$ if $(a, b) \in R$.

(a) Apply Dilworth's theorem to obtain the König–Egerváry theorem (in Section 2.3).

[32] Greene and Kleitman (1976). [33] Erdös and Szekeres (1935).
[34] This is an easy result. See *Matching theory* (p. 212) and Mirsky (1971a) and elsewhere.
[35] Fulkerson (1956).

If P is a partially ordered set, let $R : P_1 \to P_2$ be the relation on two copies of P defined by $(a, b) \in R$ if $a < b$.

(b) Show that partial matching M of R gives a chain partition of P into c chains, where $c + |M| = |P|$.

(c) Show that if $U_1 \cup U_2$, where $U_1 \subseteq P_1$ and $U_2 \subseteq P_2$, is a vertex cover of R, then ignoring which copy an element is from, $P \backslash (U_1 \cup U_2)$ is an antichain, and hence the maximum size of an antichain equals $|P| - d$, where d is the minimum size of a vertex cover of R.

(d) Deduce Dilworth's theorem from the König–Egerváry theorem.

3.2.4. *Incomparability graphs of partially ordered sets.*

Let G be a graph on the vertex set V. A *proper coloring* with *c colors* of G is a function $f : V \to \{1, 2, \ldots, c\}$ with the following property: $f(a) \neq f(b)$ if $\{a, b\}$ is an edge in G. The *chromatic number* $\chi(G)$ of G is the minimum nonnegative integer c such that there is a proper coloring of G with c colors. A *clique* of G is a subset $A \subseteq V$ such that every pair $\{a, b\}$ with $a, b \in A$ is an edge; that is, a clique is an *induced* complete subgraph. The *clique number* $\omega(G)$ is the maximum size of a clique in G. It is immediate that $\chi(G) \geq \omega(G)$. An *independent* or *stable set* is the opposite of a clique: it is a vertex subset C such that no pair $\{a, b\}$ with $a, b \in C$ is an edge. A graph G is *perfect* if for all induced subgraphs of G (including G itself), the chromatic number equals the clique number.

The *comparability graph* of P is the graph on the vertex set P with $\{a, b\}$ an edge whenever a and b are comparable; that is, $a \leq b$ or $a \geq b$. The *incomparability graph* is the complement of the comparability graph; that is, $\{a, b\}$ is an edge in the incomparability graph if and only if $\{a, b\}$ is not an edge in the comparability graph. Observe that a chain is a clique and an antichain is an independent set in the comparability graph and the roles are switched in the incomparability graph.

(a) Show that Dilworth's theorem is equivalent to the statement that the incomparability graph is perfect.

(b) Using Exercise 3.2.2, show that the comparability graph of a partially ordered set is perfect. Together with Lovász's theorem that the complement of a perfect graph is perfect;[36] this yields an indirect proof of Dilworth's theorem.

[36] Lovász (1983).

3.2.5. *Width and order dimension.*[37]

(a) Prove Dilworth's lemma. Let P be a partially ordered set. Then

$$w(P) \geq \dim(P).$$

Dilworth observed this result without proof in a footnote in his 1950 paper.

(b) Let A be the antichain of maximal elements in a partially ordered set P. Show that if $A \neq P$, then

$$\dim(P) \leq w(P \backslash A) + 1.$$

(c) Let A be an antichain in P such that $A \neq P$. Then

$$\dim(P) \leq 2w(P \backslash A) + 1.$$

(d) All three inequalities are tight. Find examples of partially ordered sets for which equality holds.

3.2.6. *Lattices of k-families and k-saturated partitions.*[38]

Let $\mathcal{A}_k(P)$ be the collection of k-families of P. In particular, $\mathcal{A}_1(P)$ is $\mathcal{A}(P)$, the collection of antichains of P. The collection $\mathcal{A}(P)$ forms a distributive lattice isomorphic to the lattice of order ideals of P.

The *depth* $\delta_A(a)$ of an element a in a k-family A is the length of the longest chain in A with bottom a. Let

$$A_i = \{a\colon a \in A \text{ and } \delta_A(a) = i\}.$$

The subsets A_i are antichains. The k-tuple $[A_1, A_2, \ldots, A_k]$ is the *canonical partition* of the k-family A (even though some of the sets may be empty).

(a) Show that the canonical partition of a k-family A is the unique k-tuple such that

- A_i are antichains,
- $A_i \cap A_j = \emptyset$, if $i \neq j$, and
- $A_k < A_{k-1} < \cdots < A_1$ in $\mathcal{A}(P)$.

If A and B are k-families, define

$$A \vee B = \bigcup_{i=1}^{k} A_i \vee B_i,$$

where $A_i \vee B_i$ is the join in $\mathcal{A}(P)$.

[37] Dilworth (1960) and Trotter (1975).
[38] Frank (1980), Greene and Kleitman (1976), Hoffman and Schwartz (1977), and Saks (1979).

(b) Show that $[A_1 \vee B_1, A_2 \vee B_2, \ldots, A_k \vee B_k]$ is the canonical partition of $A \vee B$ and that $A \vee B$ is indeed the least upper bound of A and B. In particular, the canonical partition gives an injection γ from $\mathcal{A}_k(P)$ to $\mathcal{A}(P)^k$, the Cartesian product of k copies of the antichain lattice $\mathcal{A}(P)$ such that $\gamma(A \vee B) = \gamma(A) \vee \gamma(B)$

If $S \subseteq P$, let S^* be the set of maximal elements of the set $\{a: a < s \text{ for some } s \in S\}$. Note that the inequality in the condition is strict. If $[A_1, A_2, \ldots, A_k]$ is a k-tuple of antichains, define $[\overline{A_1}, \overline{A_2}, \ldots, \overline{A_k}]$ inductively by

$$\overline{A_1} = A_1,\ \overline{A_2} = A_2 \wedge (\overline{A_1})^*, \ldots,$$
$$\overline{A_i} = A_i \wedge (\overline{A_{i-1}})^*, \ldots, \overline{A_k} = A_k \wedge (\overline{A_{k-1}})^*.$$

(c) Show that $[\overline{A_1}, \overline{A_2}, \ldots, \overline{A_k}]$ is the canonical partition of a k-family. Define

$$A \wedge B = \bigcup_{i=1}^{k} \overline{A_i \wedge B_i}.$$

(d) Show that $A \wedge B$ is the greatest lower bound of A and B in $\mathcal{A}_k(P)$.

Parts (b) and (d) show that $\mathcal{A}_k(P)$ is a lattice. A lattice L is *locally distributive* if for every element $x \in L$, the interval $[x, x^*]$, where x^* is the join of all the elements covering x, is a Boolean algebra.

(e) Show that $\mathcal{A}_k(P)$ is locally distributive. However, $\mathcal{A}_k(P)$ need not be distributive or modular.

(f) (Research problem) Is every finite locally distributive lattice isomorphic to $\mathcal{A}_k(P)$ for some partial order P and some integer k? This will extend Birkhoff's representation theorem.

(g) Let $l(P)$ be the length of the longest chain in P. Show that an $l(P)$-family A is a join-irreducible in $\mathcal{A}_{l(P)}(P)$ if and only if A is a chain $x_1, x_2, \ldots, x_k$, where for $1 \leq i \leq k-1$, x_{i+1} covers x_i. (The description of meet-irreducibles is more complicated; see the paper of Greene and Kleitman.)

Let $d_k(P)$ be the maximum size of a k-family in P. A *Sperner k-family* is a k-family having the maximum size $d_k(P)$. Let $\mathcal{S}_k(P)$ be the collection of Sperner k-families in P.

(h) Show that $\mathcal{S}_k(P)$ is a sublattice of $\mathcal{A}_k(P)$. Using this, show that there exists a maximum-size k-family in P invariant under all order automorphisms of P.

A chain partition $C_1, C_2, \ldots, C_e$ of P is *k-saturated* if

$$E(C_1, C_2, \ldots, C_e) = \sum_{i=1}^{e} \min\{k, |C_i|\} = d_k(P).$$

(i) Show that for a given k, there exists a chain partition that is simultaneously k-saturated and $(k+1)$-saturated. This gives a proof of Theorem 3.2.6.

(j) Show that $\mathcal{S}_k(P)$ is a distributive lattice.

(k) (Research problem posed by Greene and Kleitman) Describe the join- and meet-irreducibles in the lattice $\mathcal{S}_k(P)$.

3.2.7. *Greene's extension of Dilworth's theorem.*[39]
Let $\hat{d}_k(P)$ be the size of the largest subset of P obtained by taking the union of k chains. There exists an integer partition $\Delta = \Delta_1 + \Delta_2 + \cdots + \Delta_l$ of $|P|$ such that

$$d_k(P) = \Delta_1 + \Delta_2 + \cdots + \Delta_k$$

and

$$\hat{d}_k(P) = \Delta_1^* + \Delta_2^* + \cdots + \Delta_k^*,$$

where Δ^* is the partition conjugate to Δ.

3.3 Sperner Theory

Sperner theory is about maximum-size antichains in partially ordered sets. In this section, we discuss Sperner theory in Boolean algebras of subsets of a finite set. Sperner theory was founded by the following result.[40]

3.3.1. Sperner's theorem. Let $\mathcal{A}$ be an antichain in the Boolean algebra $2^{\{1,2,\ldots,n\}}$. Then

$$|\mathcal{A}| \leq \binom{n}{\lfloor n/2 \rfloor}.$$

[39] Fomin (1978) and Greene (1974, 1976).
[40] Sperner (1928). There are two Sperner theorems: the other is about triangulations.

Equality occurs only if all the subsets in $\mathcal{A}$ have the same size, $\lfloor n/2 \rfloor$ or $\lceil n/2 \rceil$.

Note that by Stirling's approximation for $n!$,

$$\binom{n}{\lfloor n/2 \rfloor} \approx \frac{2^n}{\sqrt{\pi n/2}}.$$

We will give several proofs of Sperner's theorem.

Sperner's proof. Let $\mathcal{A}$ be a maximum-size antichain of subsets in $2^{\{1,2,\ldots,n\}}$. Let m and M be the minimum and maximum size of a subset in $\mathcal{A}$, $\mathcal{A}_m$ be the subfamily of all subsets in $\mathcal{A}$ of size m, and $\mathcal{C}$ be the family of subsets A such that $|A| = m + 1$ and A contains a subset in $\mathcal{A}_m$. Observe that $\mathcal{C} \cap \mathcal{A} = \emptyset$ and $(\mathcal{A} \backslash \mathcal{A}_m) \cup \mathcal{C}$ is an antichain.

It follows from counting incidences that

$$|\mathcal{A}_m||n - m| \leq |\mathcal{C}||m + 1|. \tag{NM}$$

If $m \leq \frac{n-1}{2}$,

$$\frac{\mathcal{C}}{\mathcal{A}_m} \geq \frac{n - m}{m + 1} \geq 1,$$

with equality at the second inequality if and only if $m = \frac{n-1}{2}$. Hence, if $m < \frac{n-1}{2}$, then $(\mathcal{A} \backslash \mathcal{A}_m) \cup \mathcal{C}$ is an antichain of larger size. We conclude that if $\mathcal{A}$ is a maximum-size antichain, then $m \geq \frac{n-1}{2}$. Arguing in the order-dual, we can also conclude that $M \leq \frac{n+1}{2}$.

If n is even, then $m = M$ and the theorem follows. If n is odd, then $m + 1 = M$. It may happen $m < M$ and one needs to rule out the possibility of constructing a larger antichain taking a mixture of subsets of size $\lfloor \frac{n-1}{2} \rfloor$ and $\lceil \frac{n-1}{2} \rceil$. This can be done easily using (NM). □

The idea of counting incidences in two different ways is used more efficiently in the next proof, due independently to Yamamoto, Mešalkin, and Lubell (in chronological order of publication).[41]

Proof using the LYM inequality. A chain in $2^{\{1,2,\ldots,n\}}$ is *maximal* if it has length n or equivalently, it is constructed by starting with the empty set and adding elements, one at a time. There are $n!$ maximal chains. Given a subset $A \subseteq \{1, 2, \ldots, n\}$ of size k, a maximal chain containing A can be constructed

[41] Lubell (1966), Mešalkin (1963), and Yamamoto (1954).

by starting at the empty set, adding elements from A one at a time until one obtains A, and then adding elements in the complement of A one at a time. Thus, there are $k!(n-k)!$ maximal chains containing A.

Let $\mathcal{A}$ be an antichain, $\mathcal{A}_k$ be the subfamily of subsets in $\mathcal{A}$ of size k, and $\alpha_k = |\mathcal{A}_k|$. Since an antichain and a chain can have at most one subset in common, we obtain the *LYM inequality*:

$$\sum_{k=0}^{n} \alpha_k k!(n-k)! \leq n!.$$

Using the fact that $\max\{\binom{n}{k}: 0 \leq k \leq n\} = \binom{n}{\lfloor n/2 \rfloor}$, we have

$$\frac{\sum_{k=0}^{n} \alpha_k}{\binom{n}{\lfloor n/2 \rfloor}} \leq \sum_{k=0}^{n} \frac{\alpha_k}{\binom{n}{k}} \leq 1.$$

This proves the inequality in Sperner's theorem. If equality occurs, then $\alpha_k = 0$ for all k such that $\binom{n}{k} < \binom{n}{\lfloor n/2 \rfloor}$. In other words, $\alpha_k = 0$ unless $k = \lfloor n/2 \rfloor$ or $\lceil n/2 \rceil$. If n is even, this implies that $\mathcal{A}$ is the antichain of all elements of size $\lfloor n/2 \rfloor$. If n is odd, then it could happen that $\mathcal{A}$ consists of a mixture of subsets of size $\lfloor n/2 \rfloor$ or $\lceil n/2 \rceil$. This possibility also occurs in Sperner's proof and can be excluded by the same easy argument. □

Third proof. The third proof uses Corollary 3.2.6. This result says that there exists a maximum-size antichain invariant under all order automorphisms of P. Put another way, there is a maximum-size antichain that is a union of orbits of the automorphism group of P. An order automorphism of $2^{\{1,2,\ldots,n\}}$ restricts to a permutation of the one-element subsets, and hence the automorphism group of $2^{\{1,2,\ldots,n\}}$ is the symmetric group on $\{1, 2, \ldots, n\}$. Hence, the orbits are $\{A: |A| = k\}$. It follows that $\{A: |A| = \lfloor n/2 \rfloor\}$ is a maximum-size antichain. □

Corollary 3.2.6 extends easily to k-families (see Exercise 3.2.7(h)). Thus, the third proof yields the following extension first observed by Erdős.

3.3.2. An extension of Sperner's theorem. Let $\mathcal{A}$ be a k-family in the Boolean algebra $2^{\{1,2,\ldots,n\}}$. Then

$$|\mathcal{A}| \leq \sum_{j=-\lfloor k/2 \rfloor}^{\lfloor k/2 \rfloor} \binom{n}{\lfloor n/2 \rfloor + j}.$$

Equality occurs if and only if $\mathcal{A}$ is the collection of all the subsets in $\mathcal{A}$ of size $\lfloor n/2 \rfloor + j$, where $-\lfloor k/2 \rfloor \le j \le \lfloor k/2 \rfloor$ if n is even, and $-\lfloor k/2 \rfloor \le j \le \lfloor k/2 \rfloor$ or $-\lfloor k/2 \rfloor + 1 \le j \le \lfloor k/2 \rfloor + 1$ if n is odd.

A *symmetric chain* in the Boolean algebra $2^{\{1,2,\ldots,n\}}$ is a chain $E_k, E_{k+1}, \ldots, E_{n-k}$ such that $|E_i| = i$ (and, of course, $E_i \subset E_{i+1}$). Symmetric chains are centered at a set of size $n/2$ if n is even and two sets of equal size $\lfloor n/2 \rfloor$ and $\lceil n/2 \rceil$ if n is odd. A *symmetric chain partition* is a partition of $2^{\{1,2,\ldots,n\}}$ into symmetric chains. Thus, if n is even, there are $\binom{n}{n/2} - \binom{n}{n/2+1}$ chains of length 1, $\binom{n}{n/2+1} - \binom{n}{n/2+2}$ chains of length 3, and so on. Similarly, if n is odd, there are $\binom{n}{\lceil n/2 \rceil} - \binom{n}{\lceil n/2 \rceil+1}$ chains of length 2, $\binom{n}{\lceil n/2 \rceil+1} - \binom{n}{\lceil n/2 \rceil+2}$ chains of length 4, and so on. This can be compactly stated as follows: there are

$$\binom{n}{\lceil (n+k)/2 \rceil} - \binom{n}{\lceil (n+k+1)/2 \rceil}$$

chains of length k. For example, when $n = 4$, a symmetric chain decomposition has 1 chain of length 4, 3 chains of length 2, and 2 chains of length 0.

3.3.3. Lemma.[42] The Boolean algebra $2^{\{1,2,\ldots,n\}}$ has a symmetric chain partition.

Proof. We proceed by induction. When $n = 1$, $2^{\{1\}}$ can be partitioned into one chain of size 2 and the lemma holds. Suppose that $2^{\{1,2,\ldots,n-1\}}$ has been partitioned into symmetric chains. Let

$$E_k, E_{k+1}, \ldots, E_{n-k}$$

be a length-l symmetric chain in this partition, where $l = n - 2k + 1$. We define two new chains. The first is constructed by removing $E_{(n-1)-k}$ and adding the element n to each of the remaining sets, giving the length-$(l-1)$ symmetric chain

$$E_k \cup \{n\}, E_{k+1} \cup \{n\}, \ldots, E_{n-k+1} \cup \{n\}$$

in $2^{\{1,2,\ldots,n\}}$. The new chain is empty when the original chain has length 0, that is, when n is even and $k = n/2$. The second chain is constructed by putting the subset $E_{n-k} \cup \{n\}$ on the top of the chain, giving the length-$(l+1)$ symmetric chain

$$E_k, E_{k+1}, \ldots, E_{n-k}, E_{n-k} \cup \{n\}.$$

[42] de Bruijn et al.(1949).

The new chains that are nonempty yield a symmetric chain decomposition of $2^{\{1,2,\ldots,n\}}$. □

The chain partition constructed inductively in Lemma 3.3.3 can be described explicitly by a *bracket* notation. Represent the subsets in $\{1, 2, \ldots, n\}$ by a length-n sequence of 0s and 1s and replace each 1 by "(" and each 0 by ")". For example, for the subset $\{1, 3, 4, 6, 7, 8, 9\}$ in $\{1, 2, \ldots, 10, 11\}$, we have

1 0 1 1 0 1 1 1 1 0 0
( ) (( ) (((( )).

Close brackets where possible, so that for our example, we obtain

() (() (((()).

The subsets with the same closed brackets (at the same coordinates) form a chain. For our example, there are four such subsets:

() (() (((())	1 0 1 1 0 1 1 1 1 0 0
()) () (((())	1 0 0 1 0 1 1 1 1 0 0
()) ()) ((())	1 0 0 1 0 0 1 1 1 0 0
()) ())) (())	1 0 0 1 0 0 0 1 1 0 0.

These subsets form the chain

$$\{1, 4, 8, 9\} \subset \{1, 4, 7, 8, 9\} \subset \{1, 4, 6, 7, 8, 9\} \subset \{1, 3, 4, 6, 7, 8, 9\}.$$

We end this section with a discussion of the following problem, posed by Littlewood and Offord: *Let $\alpha_1, \alpha_2, \ldots, \alpha_n$ be complex numbers of absolute value strictly greater than* 1. *How many among the sums $\sum_{i=1}^{n} \epsilon_i \alpha_i$, where $(\epsilon_1, \epsilon_2, \ldots, \epsilon_n)$ are the 2^n vectors, with ϵ_i equal to -1 or $+1$, have absolute value less than or equal to* 1*?* Although there is an application to zeros of random polynomials, the problem is appealing on its own.[43]

The answer is in the following theorem, proved independently by Katona and Kleitman.[44]

[43] Littlewood and Offord showed that if $\alpha_0, \alpha_1, \ldots, \alpha_n$ are complex numbers, then "most" of the polynomials

$$\alpha_0 + \epsilon_1\alpha_1 z + \epsilon_2\alpha_2 z^2 + \cdots + \epsilon_n\alpha_n z^n$$

have "few" real zeros. Further applications can be found in Tao and Vu (2006).

[44] Katona (1966) and Kleitman (1965, 1970).

3.3.4. Theorem. Let $\alpha_1, \alpha_2, \ldots, \alpha_n$ be complex numbers of absolute value greater than 1 and U be a region in the complex plane satisfying the following property: if $x, y \in U$, then $|x - y| \leq 2$. Then at most $\binom{n}{\lfloor n/2 \rfloor}$ of the sums

$$\sum \epsilon_i \alpha_i, \quad \epsilon_i = \pm 1,$$

can lie in U. In particular, at most $\binom{n}{\lfloor n/2 \rfloor}$ of the sums lie inside any circle of radius 1 in the complex plane.

The ingenious idea is to turn the analytic data into combinatorial data. This is best understood by looking at the one-dimensional or real case, proved by Erdős in 1945.[45]

3.3.5. Theorem. Let $\alpha_1, \alpha_2, \ldots, \alpha_n$ be real numbers of absolute value at least 1 and I be an half-open interval of length 2. Then at most $\binom{n}{\lfloor n/2 \rfloor}$ of the sums $\sum_i \epsilon_i \alpha_i, \epsilon = \pm 1$, are in I.

Proof. Since the multiset of sums $\sum \epsilon_i \alpha_i$ is unaffected if we substitute $-\alpha_i$ for α_i, we may assume that $\alpha_i > 0$. If $A \subseteq \{1, 2, \ldots, n\}$, let

$$S(A) = \sum_{j:\, j \in A} \alpha_j - \sum_{j:\, j \notin A} \alpha_j. \tag{S}$$

This associates each sum $\sum \epsilon_i \alpha_i$ with a subset A in $\{1, 2, \ldots, n\}$. If $A \subset B$, then $|S(B) - S(A)| \geq 2$, and hence the two sums $S(A)$ and $S(B)$ cannot both be in a half-open length-2 interval. Hence if the sums $S(A)$ lie in a half-open length-2 interval, the sets A form an antichain. We can now apply Sperner's theorem. □

Just as in the real case, we may replace α_i by $-\alpha_i$ in the complex case. Thus, we can assume that all the complex numbers α_i lie in the upper half-plane.

Given a region U satisfying the property in Kleitman's theorem, let $\mathcal{F}$ be the family of subsets A such that the sum $S(A)$ is in U, and T be the subset of indices j such that α_j lies in the first quadrant; that is, both the real and imaginary parts of α_j are nonnegative. Let A and B be subsets in $\mathcal{F}$ such that $B \subset A$, $S(A) = \sum \epsilon_i \alpha_i$, and $S(B) = \sum \bar{\epsilon}_i \alpha_i)$ be the vectors determined by A

[45] Erdős (1945).

and B. Then,

$$\begin{aligned}\epsilon_i - \bar{\epsilon}_i &= 2 \text{ if } i \in A\backslash B,\\ \epsilon_i - \bar{\epsilon}_i &= 0 \text{ if } i \in A \cap B,\\ \epsilon_i - \bar{\epsilon}_i &= 0 \text{ if } i \in \{1,2,\ldots,n\}\backslash A.\end{aligned}$$

Thus,

$$\begin{aligned}S(A) - S(B) &= \sum_{i=1}^{n} (\epsilon_i - \bar{\epsilon}_i)\alpha_i\\ &= 2 \sum_{i:i \in A\backslash B} \alpha_i.\end{aligned}$$

If, in addition, $A\backslash B \subseteq T$, then the last sum ranges over complex numbers of absolute value greater than 1 in the first quadrant, and hence the sum also has absolute value greater than 1 (and is in the first quadrant). Hence, $|S(A) - S(B)| \geq 2$, and in particular, $S(A)$ and $S(B)$ cannot both lie in the region U. We conclude that the family $\mathcal{F}$ satisfies the property:

(K_1) If A and B are subsets in $\mathcal{F}$, $A \subseteq B$, and the difference set $B\backslash A$ is contained in T, then $A = B$.

We can repeat the argument on the second quadrant to conclude that $\mathcal{F}$ also satisfies the property:

(K_2) If A and B are subsets in $\mathcal{F}$, $A \subseteq B$, and $B\backslash A$ is contained in the complement $\{1, 2, \ldots, n\}\backslash T$, then $A = B$.

We can now finish the proof of Theorem 3.3.4 with the following "two-part Sperner theorem."

Theorem 3.3.6. Let $T \subseteq \{1, 2, \ldots, n\}$ and $\mathcal{F}$ be a family of subsets of $\{1, 2, \ldots, n\}$ satisfying properties (K_1) and (K_2). Then $\mathcal{F}$ has at most $\binom{n}{\lfloor n/2 \rfloor}$ subsets.

Proof. Let $t = |T|$. We shall consider the case when both n and t are even. The other three cases are almost the same. Once this case is understood, a single unified proof for all four cases can easily be written using floor and ceiling functions.

Fix a symmetric chain partition of the Boolean algebra 2^T of subsets of T. Consider a chain C of odd length $2k + 1$ consisting of the subsets

$$D_{t/2-k}, D_{t/2-k+1}, \ldots, D_{t/2}, \ldots, D_{t/2+k}.$$

For an index j, $t/2 - k \le j \le t/2 + k$, let $\mathcal{F}_j(\mathcal{C})$ be the family of subsets E such that E is contained in the complement $\{1, 2, \ldots, n\}\backslash T$ and $E \cup D_j$ is in the family $\mathcal{F}$. By (K_2), the families $\mathcal{F}_j(\mathcal{C})$ are antichains, and hence the union $\mathcal{F}_{t/2-k}(\mathcal{C}) \cup \mathcal{F}_{t/2-k+1}(\mathcal{C}) \cup \cdots \cup \mathcal{F}_{t/2+k}(\mathcal{C})$ is a k-family in $2^{\{1,2,\ldots,n\}\backslash T}$. By Theorem 3.3.2,

$$\left|\bigcup_{j=-k}^{k} \mathcal{F}_j(\mathcal{C})\right| \le \sum_{j=-k}^{k} \binom{n-t}{(n-t)/2 + j}.$$

By (K_1), the sets $\mathcal{F}_j(\mathcal{C})$, where $\mathcal{C}$ ranges over all symmetric chains in the chosen partition and j ranges over an appropriate integer interval, are pairwise disjoint. In particular, the unions $\bigcup_j \mathcal{F}_j(\mathcal{C})$ are pairwise disjoint. Hence,

$$|\mathcal{F}| \le \sum_{\mathcal{C}} \left|\bigcup_{j} \mathcal{F}_j(\mathcal{C})\right|.$$

Using the fact that in a symmetric chain partition of 2^T, there are

$$\binom{t}{t/2 + k} - \binom{t}{t/2 + k + 1}$$

chains of (odd) length $2k + 1$ (and no chains of even length), we obtain

$$|\mathcal{F}| \le \sum_{k=0}^{t/2} \left[\binom{t}{t/2 + k} - \binom{t}{t/2 + k + 1}\right] \sum_{j=-k}^{k} \binom{n-t}{(n-t)/2 + j}.$$

To simplify this inequality, we change the order of summation, obtaining

$$\begin{aligned} |\mathcal{F}| \le & \sum_{j=-t/2}^{0} \binom{n-t}{(n-t)/2 + j} \sum_{k=-t/2}^{j} \left[\binom{t}{t/2 + k} - \binom{t}{t/2 + k - 1}\right] \\ & + \sum_{j=1}^{t/2} \binom{n-t}{(n-t)/2 + j} \sum_{k=j}^{t/2} \left[\binom{t}{t/2 + k} - \binom{t}{t/2 + k + 1}\right]. \end{aligned}$$

The inner sums (over k) are telescoping. Thus,

$$\sum_{j=-t/2}^{0} \binom{n-t}{(n-t)/2+j} \sum_{k=-t/2}^{j} \left[\binom{t}{t/2+k} - \binom{t}{t/2+k-1} \right]$$

$$= \sum_{j=-t/2}^{0} \binom{n-t}{(n-t)/2+j} \binom{t}{t/2+j}$$

$$= \sum_{j=-t/2}^{0} \binom{n-t}{(n-t)/2+j} \binom{t}{t-(t/2+j)}$$

$$= \sum_{j=-t/2}^{0} \binom{n-t}{n/2-(t/2-j)} \binom{t}{t/2-j}$$

$$= \sum_{i=t/2}^{t} \binom{n-t}{n/2-i} \binom{t}{i},$$

where, in the last step, we change the index of summation from j to $i = t/2 - j$. Similarly, the second sum simplifies to

$$\sum_{i=0}^{t/2-1} \binom{n-t}{n/2-i} \binom{t}{i}.$$

To finish the proof, we use the van der Monde–Chu identity

$$\binom{n}{n/2} = \sum_{i=0}^{t} \binom{n-t}{n/2-i} \binom{t}{i}$$

to conclude that $|\mathcal{F}| \leq \binom{n}{n/2}$. □

The proof we have given is Kleitman's 1965 proof. There are other ways to prove Theorem 3.3.4. In particular, it turns out that the separate steps to convert analysis to combinatorics and then to use the existence of a symmetric chain decomposition are unnecessary. One can construct an analog of a symmetric chain decomposition on the sums directly, yielding a general d-dimensional analog (see Exercise 3.3.5).

Exercises

3.3.1. Let $\mathcal{C}$ be a collection of subsets of $\{1, 2, \ldots, n\}$ such that for every pair A and B of subsets in $\mathcal{C}$, $A \cap B \neq \emptyset$.

(a) Show that $|\mathcal{C}| \leq 2^{n-1}$.

(b) Describe those collection $\mathcal{C}$ such that $|\mathcal{C}| = 2^{n-1}$.

3.3.2. *Mešalkin's extension of Sperner's theorem.*[46]

Two partitions $\{A_1, A_2, \ldots, A_s\}$ and $\{B_1, B_2, \ldots, B_s\}$ with the same number s of blocks of $\{1, 2, \ldots, n\}$ are *comparable* if for some indices i and j, $A_i \subseteq B_j$ or $A_i \supseteq B_j$. Show that the maximum size of a family of incomparable partitions equals the following maximum of multinomial coefficients:

$$\max\left\{\binom{n}{n_1, n_2, \ldots, n_s} : n_1 + n_2 + \cdots + n_s = n\right\}.$$

3.3.3. *The probability that one set contains another.*[47]

Put a probability measure on $2^{\{1,2,\ldots,n\}}$. Show that if two subsets A and B are chosen independently, then the probability that $A \subseteq B$ is at least $1/m$, where m is the middle binomial coefficient. (This lower bound is independent of the probability measure.)

3.3.4. Prove analogs of Theorems 3.3.1 and 3.3.3 for the lattice of divisors of a number (or a finite product of chains).

3.3.5. *The d-dimensional Littlewood–Offord problem.*[48]

Let $\alpha_1, \alpha_2, \ldots, \alpha_n$ be vectors in $\mathbb{R}^d$, with $\|\alpha_i\| \geq 1$ (where $\|x\|$ is the length of the vector x). Show that at most $\binom{n}{\lfloor n/2 \rfloor}$ sums $\sum \epsilon_i \alpha_i$, $\epsilon_i = \pm 1$ lie in a unit ball.

Sperner theory for partially ordered sets.

In the 1960s, Rota suggested studying extensions, or, more accurately, nonextensions, of Sperner's theorem to arbitrary partially ordered sets. This has become an intensive area of research. We present a selection of concepts and results in general Sperner theory.

In the next two exercises, let P be a finite ranked partially ordered set having rank N, $\mathcal{W}_i(P)$ (or $\mathcal{W}_i$) be the set of elements in P of rank i, and $W_i(P) = W_i = |\mathcal{W}_i|$. The numbers W_i are the *Whitney numbers of the second kind* of P.

3.3.6. *The normalized matching property and the LYM inequality.*[49]

The partially ordered set P satisfies the *normalized matchin. property* if for every rank i and every subset $\mathcal{B} \subseteq \mathcal{W}_i$,

$$\frac{|I(\mathcal{B})|}{W_{i-1}} \geq \frac{|\mathcal{B}|}{W_i},$$

where

$$I(\mathcal{B}) = \{b\colon b \in \mathcal{W}_{i-1} \text{ and } b \leq a \text{ for some } a \text{ in } \mathcal{B}\}.$$

[46] Mešalkin (1963). [47] Baumert et al. (1980). [48] Kleitman (1976). [49] Kleitman (1974).

The partially ordered set P satisfies the *LYM inequality* if for every antichain $\mathcal{A}$ in P,

$$\sum_{a:a\in\mathcal{A}} \frac{1}{W_{\text{rank}(a)}} \leq 1,$$

or, equivalently,

$$\sum_{i=0}^{N} \frac{|\mathcal{A} \cap \mathcal{W}_i|}{W_i} \leq 1.$$

The third property is explicitly combinatorial: P has a *regular chain cover,* that is, a nonempty collection of chains $\mathcal{C}$ of P such that for each nonnegative integer i, every rank-i element occurs in the same number (depending on i) of chains in $\mathcal{C}$.

(a) Show that the three properties are equivalent.

The partially ordered set P is *Sperner* if the maximum size of an antichain in P equals $\max\{W_i(P)\colon 0 \leq i \leq n\}$.

(b) Show that if P satisfies the LYM inequality, then P is Sperner.

3.3.7. *Unimodality, matchings, and the Sperner property.*[50]

The partially ordered set P is *Sperner* if the maximum size of an antichain in P equals $\max\{W_i : 0 \leq i \leq n\}$. We can define symmetric chains and symmetric chain partitions for ranked partially ordered set exactly as for Boolean algebras.

(a) Prove the following observation of Harper and Rota. Let P be a ranked partial order satisfying the following two properties:

HR1. *Unimodality of the Whitney numbers*: There is an index m such that

$$W_0 \leq W_1 \leq W_2 \leq \cdots \leq W_m$$

and

$$W_m \geq W_{m+1} \geq \cdots \geq W_{n-1} \geq W_n.$$

HR2. *The matching condition*: For $0 \leq i \leq n-1$, the relation $\mathcal{W}_i(P) \rightarrow \mathcal{W}_{i+1}(P)$ obtained by restricting the partial order has a partial matching of maximum size $\max\{W_i, W_{i+1}\}$.

Then P is Sperner.

[50] *Matching theory* (p. 213) and Griggs (1977).

(b) Prove Griggs' theorem. A ranked partially ordered set P of rank n has a symmetric chain partition if it satisfies the following three properties: unimodality of Whitney numbers, the LYM inequality, and symmetry: For $0 \le i \le \lfloor n/2 \rfloor$, $W_i = W_{n-i}$.

3.3.8. *Dedekind's problem on free distributive lattices.*[51]

The free distributive lattice on n generators is the "largest" distributive lattice generated by n generators, in the sense that any other distributive lattice generated by n generators is an image under a lattice homomorphism. By Exercise 1.3.7, the elements of the free distributive lattice on n generators are in bijection with *monotone,* that is, surjective order-preserving functions $f : 2^{\{1,2,\ldots,n\}} \to \{0, 1\}$. Monotone functions are in bijection with antichains not equal to $\{\emptyset\}$ or $\{\{1, 2, \ldots, n\}\}$ in $2^{\{1,2,\ldots,n\}}$.

Dedekind posed the problem of finding $\psi(n)$, the number of elements in the free distributive lattice on n generators. By Exercise 1.3.8(d), the number $\psi(n) + 2$ equals the number of antichains in the Boolean algebra $2^{\{1,2,\ldots,n\}}$.

(a) Show that

$$2^{\binom{n}{\lfloor n/2 \rfloor}} \le \psi(n) \le 3^{\binom{n}{\lfloor n/2 \rfloor}}.$$

(b) Show that

$$2^{(1+\alpha_n)\binom{n}{\lfloor n/2 \rfloor}} \le \psi(n) \le 2^{(1+\beta_n)\binom{n}{\lfloor n/2 \rfloor}},$$

where $\alpha_n = ce^{-n/4}$, $\beta_n = c'(\log n)/\sqrt{n}$, and c, c' are constants.

3.4 Modular and Linear Lattices

Although the lattice axioms are abstracted from properties of set-unions and -intersections, lattice theory is not usually regarded as a generalization of the theory of Boolean algebras. Most lattice theorists regard Richard Dedekind as the founder of the subject. In his two papers,[52] Dedekind established lattice theory as the order-theoretic foundation of algebra.

For Dedekind, elements of lattices are subalgebras, ordered by set-containment. For algebras with an Abelian group operation, such lattices satisfy a modular law. The *modular law* can be stated as an identity: for all elements x, y, and z,

$$x \wedge (y \vee (x \wedge z)) = (x \wedge y) \vee (x \wedge z). \qquad \text{(M1)}$$

[51] Hansel (1966), Kahn (2002), Kleitman (1969), and Kleitman and Markowsky (1974).
[52] Dedekind (1897, 1900).

It can also be stated as a weaker form of one of the distributive axioms: for elements x, y, and z such that $x \geq z$,

$$x \wedge (y \vee z) = (x \wedge y) \vee (x \wedge z) = (x \wedge y) \vee z. \qquad \text{(M2)}$$

Since $x \geq z$ if and only if $x \wedge z = z$, it is easy to see that the two formulations of the modular law are equivalent. The inequality

$$x \wedge (y \vee (x \wedge z)) \geq (x \wedge y) \vee (x \wedge z) \qquad \text{(M3)}$$

holds in any lattice, as both x and $y \vee (x \wedge z)$ are greater than or equal to $x \wedge y$ and $x \wedge z$. Hence, the modular law is equivalent to the inequality

$$x \wedge (y \vee (x \wedge z)) \leq (x \wedge y) \vee (x \wedge z). \qquad \text{(M4)}$$

In a similar way, the modular law is also equivalent to the apparently weaker statement: for elements x, y, and z such that $x \geq z$,

$$x \wedge (y \vee z) \leq (x \wedge y) \vee z.$$

A lattice is *modular* if it satisfies the modular law. Examples of modular lattices include lattices of normal subgroups of a group, lattices of subspaces of a vector space or projective space, lattices of ideals of a ring, and lattices of submodules of a module. The partial order in all these lattices is set-containment. The meet is set-intersection, and hence meets of arbitrarily many elements exist. The join of any subset X of elements is defined by

$$\bigvee_{x:x \in X} x = \bigcap \{y\colon y \supseteq x \text{ for all } x \text{ in } X\}.$$

In algebra terminology, the join is the subgroup, subspace, ideal, or submodule *generated* by all the elements in the union $\bigcup_{x:x \in X} x$.

An easy argument shows that the modular law holds in these lattices. For example, let $L(V)$ be the lattice of subspaces of a vector space V. Then $x \wedge y = x \cap y$ and $x \vee y$ is the subspace spanned by the vectors in the union of x and y. Let x, y, and z be three subspaces and $\vec{v}$ be a vector in the subspace $x \wedge (y \vee (x \wedge z))$. Then $\vec{v} \in x$, $\vec{v} = \vec{u} + \vec{w}$, where $\vec{u} \in y$ and $\vec{w} \in x \wedge z$. The last condition implies that $\vec{w} \in x$, which, together with $\vec{u} = \vec{v} - \vec{w}$, implies that $\vec{u} \in x$, and hence, $\vec{u} \in x \wedge y$. Since $\vec{v} = \vec{u} + \vec{w}$, we conclude that $\vec{v} \in (x \wedge y) \vee (x \wedge z)$. This proves inequality (M4), and hence the modular law.[53]

[53] Kronecker called subgroups of Abelian groups "modules." This was the origin of the name "modular law."

3.4.1. Dedekind's transposition principle. If x and y are elements in a modular lattice L, the intervals $[x, x \vee y]$ and $[x \wedge y, y]$ are lattice-isomorphic under the functions

$$\varphi_y : [x, x \vee y] \to [x \wedge y, y], \quad u \mapsto u \wedge y,$$
$$\psi_x : [x \wedge y, y] \to [x, x \vee y], \quad v \mapsto x \vee v.$$

Proof. We first show that the composition $\psi_x \varphi_y$ is the identity on the interval $[x, x \vee y]$. Let $u \in [x, x \vee y]$. Then $x \leq u$ and by the modular law,

$$\psi_x \varphi_y(u) = (u \wedge y) \vee x = u \wedge (y \vee x) = u.$$

By the same argument, dualized, $\varphi_y \psi_x$ is the identity on $[x \wedge y, y]$. Hence, ψ_x and φ_y are bijections. As both functions are order-preserving, the two intervals are isomorphic as partial orders and, hence, as lattices. □

Dedekind's transposition principle implies the "third" isomorphism theorem for modules and as a special case, Abelian groups.

Let **C** be a class of lattices. A lattice F is a *free lattice* on a set X of generators for the class **C** if F is in **C** and F satisfies the universal property: if L is a lattice in **C** and $\varphi : X \to L$ is a function, then there exists a unique lattice homomorphism $\tilde{\varphi} : F \to L$ extending φ. The universal property implies that if X and Y have the same cardinality, then the free lattices on X and Y in **C** are isomorphic. Up to isomorphism, we denote by $F_{\mathbf{C}}(\aleph)$ the free lattice on a set of generators of cardinality $\aleph$, if such a lattice exists.

Let **M** be the class of modular lattices. In his 1900 paper, Dedekind constructed implicitly the free modular lattice $F_{\mathbf{M}}(3)$ on three generators as a sublattice of the lattice $L(H)$ of subspaces of an eight-dimensional vector space H. Let $\{e_i\colon 1 \leq i \leq 8\}$ be a basis of H. If X is a set of vectors, let $\langle X \rangle$ be the subspace spanned by X. Let $x = \langle e_2, e_4, e_5, e_8 \rangle$, $y = \langle e_2, e_3, e_6, e_7 \rangle$, and $z = \langle e_1, e_4, e_6, e_7 + e_8 \rangle$. Then with patience and elementary linear algebra, we find that there are 28 different subspaces formed by taking meets and joins of x, y, and z.

Dedekind's argument can be made abstractly. Let x, y, and z be three generators. In the free modular lattice they generate, the maximum $\hat{1}$ is $x \vee y \vee z$ and the minimum $\hat{0}$ is $x \wedge y \wedge z$. Let

$$\begin{aligned}
u &= (x \wedge y) \vee (y \wedge z) \vee (z \wedge x), \\
v &= (x \vee y) \wedge (y \vee z) \wedge (z \vee x), \\
x_1 &= (x \wedge v) \vee u = (x \vee u) \wedge v, \\
y_1 &= (y \wedge v) \vee u = (y \vee u) \wedge v, \\
z_1 &= (z \wedge v) \vee u = (z \vee u) \wedge v,
\end{aligned}$$

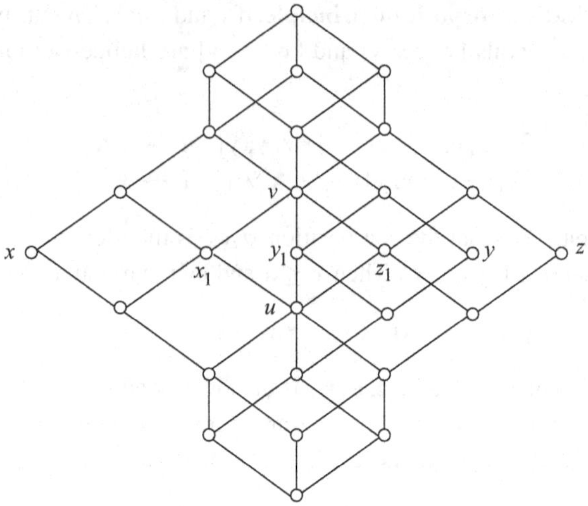

Figure 3.1 Free modular lattice on three generators x, y, and z.

as shown in Figure 3.1. These five elements are distinct. The elements $x \wedge y$, $y \wedge z$, and $z \wedge x$ generate a Boolean algebra with eight elements, giving the interval $[\hat{0}, u]$. Dually, $x \vee y$, $y \vee z$, and $z \vee x$ generate an eight-element Boolean algebra as well, giving the interval $[v, \hat{1}]$. There are six additional elements, $x \wedge x_1$, $y \wedge y_1$, $z \wedge z_1$, $x \vee x_1$, $y \vee y_1$, and $z \vee z_1$, obtainable from the lattice operations. For more details, see the solution to Exercise 3.4.3.

3.4.2. Proposition. The free modular lattice $F_{\mathrm{M}}(3)$ has 28 elements.

The free modular lattice on four generators is infinite. To show this, we exhibit an infinite number of elements in a modular lattice generated by four elements. Let PG(2, $\mathbb{R}$) be the real projective plane. We use the field of real numbers so that we can draw pictures; any infinite field will work. If a and b are points, let $\overline{a, b}$ be the line defined by a and b, and ℓ_∞ the line at infinity. Let a, b, c, and d be four points in PG(2, $\mathbb{R}$) such that the lines $\overline{a, b}$ and $\overline{c, d}$ intersect at a point v_1 on the line ℓ_∞ at infinity and $\overline{a, c} \cap \overline{b, d} = v_2$ and $v_2 \in \ell_\infty$. In particular, the two pairs $\overline{a, b}$, $\overline{c, d}$ and $\overline{a, c}$, $\overline{b, d}$ of lines are parallel in the affine subplane.

In addition, let $v = \overline{a, d} \cap \overline{b, c}$, $u = \overline{v, v_1} \cap \overline{b, d}$, $t_1 = \overline{v, v_2} \cap \overline{a, b}$, $s_1 = \overline{t_1, u} \cap \overline{c, d}$, and $v_3 = \overline{a, d} \cap \overline{t_1, s_1}$. For $i = 1, 2, 3, \ldots$, define the points recursively by

$$t_{i+1} = \overline{s_i, v_2} \cap \overline{a, b} \text{ and } s_{i+1} = \overline{t_{i+1}, v_3} \cap \overline{c, d}.$$

The points s_i and t_i are distinct, as shown in Figure 3.2.

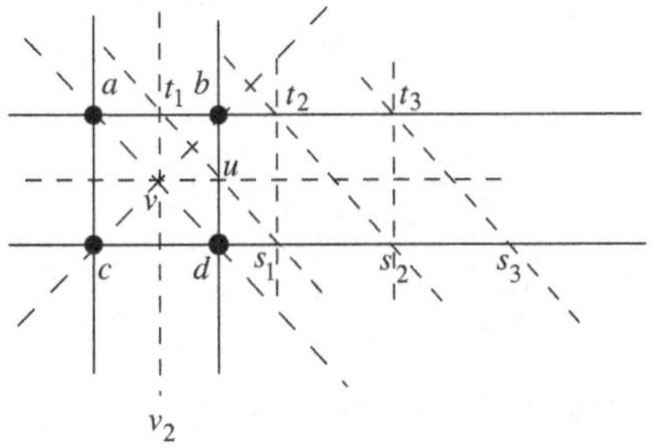

Figure 3.2 The free modular lattice on four generators is infinite.

The infinite set of points $\{s_i, t_j\colon 1 \leq i, j < \infty\}$ are lattice polynomials in a, b, c, and d in the lattice of subspaces of PG(2, $\mathbb{R}$). Hence, a, b, c, and d generate an infinite sublattice of the (modular) lattice of subspaces of PG(2, $\mathbb{R}$).

The argument just presented shows the importance of geometric ideas in studying modular lattices. The lattice operations of joins and meets of subspaces are analogs of the geometric constructions of projections and sections in synthetic projective geometry. With its focus on incidence relations and its refusal of coordinates, synthetic projective geometry can best be understood using the language of modular lattices. In particular, modular lattices offer a way to work with joins and meets of arbitrary subspaces without referring to the underlying points. We explain this point of view by studying Desargues' theorem.[54] (Desargues' theorem is a theorem for projective spaces of dimension three or higher. There are projective planes in which Desargues' theorem fails.)

3.4.3. Desargues' theorem. Let a, b, c, a', b', and c' be six points in a projective space, no three on a line, such that the lines $\overline{a, a'}$, $\overline{b, b'}$, and $\overline{c, c'}$ meet at a common point p (see Figure 3.3). Let

$$d = \overline{a, b} \wedge \overline{a', b'}, \qquad e = \overline{b, c} \wedge \overline{b', c'}, \quad \text{and} \quad f = \overline{a, c} \wedge \overline{a', c'}.$$

Then the three points d, e, and f are on the same line.

Schützenberger[55] observed that Desargues' theorem is a special case of the following lattice-theoretic condition, called the *Arguesian law*: for any elements x, y, z, x', y', and z' in L such that

$$(x \vee x') \wedge (y \vee y') = (y \vee y') \wedge (z \vee z') = (z \vee z') \wedge (x \vee x'),$$

[54] For Desargues' work, see Field and Gray (1987). [55] Schützenberger (1945).

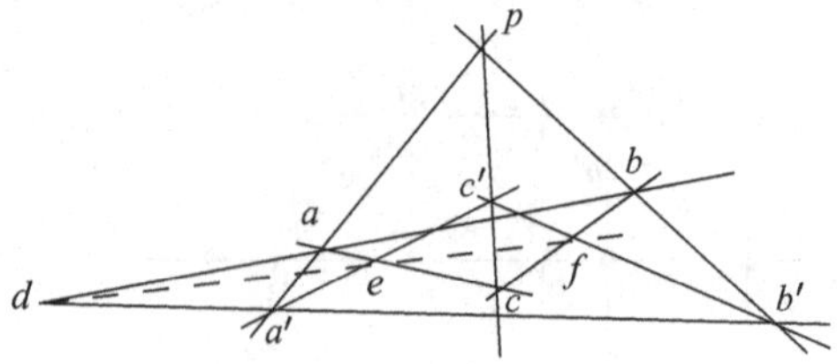

Figure 3.3 Desargues' theorem.

the following equations hold:

$$\begin{aligned}&[(x \vee y) \wedge (x' \vee y')] \vee [(x \vee z) \wedge (x' \vee z')] \\ &\quad = [(y \vee z) \wedge (y' \vee z')] \vee [(x \vee y) \wedge (x' \vee y')] \\ &\quad = [(x \vee z) \wedge (x' \vee z')] \vee [(y \vee z) \wedge (y' \vee z')].\end{aligned}$$

Since Schützenberger's paper, several equivalent versions of the Arguesian law have been found. An example is the following compact variation (CA): if[56]

$$(x \vee x') \wedge (y \vee y') \leq z \vee z',$$

then

$$(x \vee y) \wedge (x' \vee y') \leq [(x \vee z) \wedge (x' \vee z')] \vee [(y \vee z) \wedge (y' \vee z')].$$

Both versions of the Arguesian law so far are implications. Jónsson[57] has given a version which is an inequality. Let

$$u = (x \vee y) \wedge (x' \vee y') \wedge [((x \vee z) \wedge (x' \vee z')) \vee ((y \vee z) \wedge (y' \vee z'))].$$

Then the Arguesian law can be stated as

$$(x \vee x') \wedge (y \vee y') \wedge (z \vee z') \leq (x \wedge (u \vee y)) \vee (y \wedge (u \vee y')).$$

Another version is a self-dual inequality due to Day and Pickering and Haiman:[58]

$$\begin{aligned}&z \wedge [[(x \vee x') \wedge (y \vee y')] \vee z] \\ &\quad \leq x \vee [[(x \vee y) \wedge (x' \vee y')] \vee [(y \vee z) \wedge (y' \vee z')]] \wedge (x' \vee z')].\end{aligned}$$

The Arguesian law implies the modular law in the following way: if $X \geq Y$, then Jónsson's inequality with $x = y' = Z$, $y = z = x' = Y$, and $z' = X$ becomes $X \wedge (Y \vee Z) \leq Y \vee (X \wedge Z)$.

[56] Day (1983). [57] Jónsson (1953). [58] Day and Pickering (1983) and Haiman (1985b).

A lattice L is *Arguesian* if it satisfies the Arguesian law. The most natural examples of Arguesian lattices are lattices of commuting equivalence relations in combinatorics and lattices of normal subgroups in algebra. Recall from Section 1.5 that a (binary) relation R on S is a subset of the Cartesian product $S \times S$. If a and b are elements of S, we write $a\,R\,b$ instead of $(a, b) \in R$. If R and T are relations on S, then the *composition* $R \circ T$ is the relation on S defined by $a\,R \circ T\,b$ if there exists an element c in S such that $a\,R\,c$ and $c\,T\,b$. Relational composition has been studied extensively (see Section 1.5), but the general theory had no impact on combinatorics so far. However, a special case of relational composition has an elegant algebraic theory. This is the case of commuting equivalence relations.

A relation R on S is an *equivalence relation* if it is reflexive, symmetric, and transitive: in terms of relational composition, R is an equivalence relation if $I \subseteq R$, $R^{-1} = R$, and $R \circ R \subseteq R$, where I is the identity or equality relation $\{(a, a) \colon a \in S\}$ and R^{-1} is the *inverse* relation $\{(b, a) \colon (a, b) \in R\}$. Given an equivalence relation R, the equivalence classes of R form a partition of S; conversely, any partition of S determines an equivalence relation whose equivalence classes are the blocks of the partition. The notions of equivalence relations and partitions are mathematically equivalent or "cryptomorphic," but different psychologically. Let R_π be the equivalence relation associated with the partition π.

The set $\Pi(S)$ of partitions of a set S forms a lattice under the partial order of reverse refinement: $\pi \le \sigma$ when every block of π is contained in a block of σ. Under this partial order, $\Pi(S)$ is a lattice. The meet $\pi \wedge \sigma$ in $\Pi(S)$ is the partition whose blocks are $B \cap C$, where B is a block in π and C is a block in σ. In terms of their equivalence relations,

$$R_{\pi \wedge \sigma} = R_\pi \cap R_\sigma .$$

The join $\pi \vee \sigma$ is the least partition coarser than both π and σ. It does not have a simple description in terms of blocks. In terms of relations, $R_{\pi \vee \sigma}$ is the least equivalence relation containing $R_\pi \cup R_\sigma$. Since any equivalence relation containing $R_\pi \cup R_\sigma$ contains $R_\pi \circ R_\sigma$ by transitivity, both R_π and R_σ are contained in $R_\pi \circ R_\sigma$, and the transitive closure of $R_\pi \cup R_\sigma$ is an equivalence relation, we conclude that

$$R_{\pi \vee \sigma} = \bigcup_{m=1}^{\infty} \underbrace{R_\pi \circ R_\sigma \circ R_\pi \circ R_\sigma \circ \cdots \circ R_\pi \circ R_\sigma}_{m \text{ times}} .$$

Two relations R and T *commute* if $R \circ T = T \circ R$.

3.4.4. Lemma. $R_{\pi \vee \sigma} = R_\pi \circ R_\sigma$ if and only if R_π and R_σ commute.

Proof. If R and T are commuting equivalence relations, then

$$R \circ T \circ R \circ T = R \circ R \circ T \circ T = R \circ T.$$

Hence, if R_π and R_σ commute, then $R_{\pi \vee \sigma} = R_\pi \circ R_\sigma$.

Conversely, suppose that $R_{\pi \vee \sigma} = R_\pi \circ R_\sigma$. Then $R_\pi \circ R_\sigma \supseteq R_\sigma \circ R_\pi$. Taking inverses, we obtain $(R_\pi \circ R_\sigma)^{-1} \supseteq (R_\sigma \circ R_\pi)^{-1}$. But

$$(R_\pi \circ R_\sigma)^{-1} = R_\sigma^{-1} \circ R_\pi^{-1} = R_\sigma \circ R_\pi.$$

Hence, $R_\sigma \circ R_\pi \supseteq R_\pi \circ R_\pi$. We conclude that $R_\pi \circ R_\sigma = R_\sigma \circ R_\pi$. □

Since $R_{\pi \vee \sigma}$ is the smallest equivalence relation containing $R_\pi \circ R_\sigma$, we have also proved the following result.

3.4.5. Lemma. Let R and T be equivalence relations on the set S. Then R and T commute if and only if $R \circ T$ is an equivalence relation.

We next characterize commuting equivalence relations. To do this, we use a weakening of the notion of stochastic independence of partitions from information theory (see Section 1.4). Two equivalence relations R_π and R_σ are *algebraically independent* if for any block A in π and any block B in σ, $A \cap B \neq \emptyset$. If R_π and R_σ are independent, then for any two elements a and b in S, $aR_\pi b$, $aR_\sigma b$, or there exists a point c such that $aR_\pi c$ and $cR_\sigma b$. Hence,

$$R_\sigma \circ R_\pi = R_{\{S\}} = R_\pi \circ R_\sigma,$$

where $\{S\}$ is the maximum partition consisting of one block S. In particular, R_π and R_σ commute.

Let $E \subseteq S$. If R is a relation on S, then the *restriction* $R|_E$ is the relation $R \cap E \times E$. If R is an equivalence relation, then $R|_E$ is also an equivalence relation. If π is partition on S, then the *restriction* $\pi|_E$ is the partition on E whose blocks are the nonempty subset of the form $B \cap E$, where B is a block of π. Note that the partition associated with $R_\pi|_E$ is the restriction $\pi|_E$.

Let S_i be a family of disjoint subsets with a partition π_i on each subset S_i. The *disjoint union* $\sqcup_i \pi_i$ is the partition on the union $\bigcup_i S_i$ consisting of all the blocks in the partitions π_i. Suppose that for each i, σ_i is a partition of S_i such that π_i and σ_i are independent. Then $\sqcup_i \pi_i \circ \sqcup_i \sigma_i$ and $\sqcup_i \sigma_i \circ \sqcup_i \pi_i$ both equal $\sqcup_i \{S_i\}$. Hence, $\sqcup_i \pi_i$ and $\sqcup_i \sigma_i$ commute, but they are not independent. This shows one direction of the following characterization.

3.4.6. The Dubreil–Jacotin theorem.[59] The equivalence relations R_π and R_σ commute if and only if for every block C of the partition $\pi \vee \sigma$, the restrictions $\pi|_C$ and $\sigma|_C$ are algebraically independent.

Proof. We first show a special case. If the equivalence relations R_π and R_σ on S commute and $\pi \vee \sigma = \{S\}$, then $R_\pi \circ R_\sigma = R_{\{S\}}$, and hence for any two elements a and b in S, there exists an element c such that $aR_\pi c$ and $cR_\sigma b$. The element c is in the intersection of the block containing a in π and the block containing b in σ. We conclude that π and σ are algebraically independent.

To prove the general case, let C be a block of $\pi \vee \sigma$. Then $R_\pi|_C \circ R_\sigma|_C = R_{\{C\}}$, and hence, $R_\pi|_C$ and $R_\sigma|_C$ commute as equivalence relations on C. The special case holds and we conclude that $\pi|_C$ and $\sigma|_C$ are algebraically independent. □

A lattice L is *linear* if there exists a bijective lattice homomorphism $R : L \to \Pi(S)$ such that for all elements x and y in L, their images $R(x)$ and $R(y)$, thought of as equivalence relations on S, commute, so that

$$R(x \vee y) = R(x) \vee R(y) = R(x) \circ R(y).$$

3.4.7. Theorem. A linear lattice L is Arguesian.

Proof. We prove the implication (CA). Let $R : L \to \Pi(S)$ be a bijective homomorphism into a sublattice of commuting equivalence relations and x, y, z, x', y', z' be elements in L such that

$$R(x \vee x') \cap R(y \vee y') \subseteq R(z \vee z'). \tag{H}$$

Suppose $a, b \in S$ and $a\, R((x \vee y) \wedge (x' \vee y'))\, b$. Then there exist c and d such that

$$aR(x)c, \ cR(y)b, \quad aR(x')d, \ dR(y')b.$$

From this, we deduce that

$$cR(x \vee x')d, \quad cR(y \vee y')d$$

and

$$c\, R((x \vee x') \wedge (y \vee y'))d.$$

[59] Dubreil and Dubreil-Jacotin (1939).

By inequality (H), $c\,R(z \vee z')\,d$. Thus, there exists $e \in S$ such that $cR(z)e$ and $eR(z')d$.

So far, we have the following relationships involving a and e :

$$aR(x)c,\ cR(z)e,\quad aR(x')d,\ dR(z')e.$$

Hence,

$$a\,R((x \vee z) \wedge (x' \vee z'))\,e.$$

Similarly,

$$eR((y \vee z) \wedge (y' \vee z'))b.$$

Combining these two relations, we obtain

$$aR([(x \vee z) \wedge (x' \vee z')] \vee [(y \vee z) \wedge (y' \vee z')])b,$$

and thus,

$$R((x \vee y) \wedge (x' \vee y')) \subseteq R([(x \vee z) \wedge (x' \vee z')] \vee [(y \vee z) \wedge (y' \vee z')]).$$

□

We end with a discussion of free linear lattices.

3.4.8. Theorem. Free linear lattices exist.

The only known proof is indirect and uses a theorem of Birkhoff[60] that free lattices exist in a class **C** of lattices if and only if **C** is closed under forming direct products and taking sublattices.

We first observe that a sublattice of a lattice of commuting equivalence relations is also a lattice of commuting linear relation. Next, we show that a direct product of linear lattices is linear. Let I be an index set and for each $i \in I$, let L_i be a lattice represented by commuting equivalence relations on the base set S_i. Associate with each element $(x_i)_{i \in I}$ in the Cartesian product $\prod_{i:i \in I} L_i$ the equivalence relation $R((x_i))$ on the base set $\prod_{i:i \in I} S_i$ defined by

$$(a_i)\,R((x_i))\,(b_i) \quad \text{if} \quad a_i R(x_i) b_i \text{ for all } i \in I.$$

Put another way, the equivalence classes of $R((x_i))$ are all Cartesian products $\prod_{i:i \in I} B_i$, where B_i is an equivalence class in $R(x_i)$. It is routine to check that $R : \prod_{i:i \in I} L_i \to \prod_{i:i \in I} S_i$ is a representation by commuting equivalence relations.

[60] Birkhoff (1967, p. 167).

At this stage, we can finish the proof by quoting Birkhoff's theorem. It takes only a little longer to explicitly show Birkhoff's construction for linear lattices. Let X be a given set of generators. If $|X|$ is infinite, let $\mathcal{L}$ be the set of isomorphism classes of linear lattices such that $|L| \leq |X|$; if $|X|$ is finite, let $\mathcal{L}$ be the set of all isomorphism classes of finite or countable linear lattices. Let $\mathcal{H}$ be the set of all pairs (φ, L), where $L \in \mathcal{L}$ and $\varphi : X \to L$ is a function (defined up to an isomorphism of L). Let

$$\tilde{L} = \prod_{(\varphi,L):\,(\varphi,L)\in\mathcal{H}} L.$$

Then $\tilde{L}$ is linear, because it is a product of linear lattices. For each generator x in X, let $\tilde{x}$ be the element in $\tilde{L}$ with (φ, L)-component equal to $\varphi(x_k)$. Let $F_{\mathrm{L}}(X)$ be the sublattice of $\tilde{L}$ generated by the set $\{\tilde{x}\colon\ x \in X\}$. This is a linear lattice and we claim that it is free. To see this, let $\psi : X \to M$ be a function from X to a linear lattice M. Then $(\psi, M) \in \mathcal{H}$. Let $P : \tilde{L} \to M$ be the projection sending an element of $\tilde{L}$ to its (ψ, M)-component. This is a lattice homomorphism extending ψ. We conclude that $\tilde{L}$ satisfies the universal property defining a free linear lattice.

Other than existence, little is known about the structure of free linear lattices. The only case known is that $F_{\mathrm{L}}(3)$ is isomorphic to the free modular lattice $F_{\mathrm{M}}(3)$. The free modular or linear lattice on n generators contains (as a homomorphic image) the lattice generated by n subspaces in projective space. When $n = 3$, the lattice generated by n subspaces in general position equals $F_{\mathrm{L}}(3)$. Whether this equality holds for $n \geq 4$ is unknown. An explicit construction for free modular or linear lattices would yield insights into the geometric problem of finding invariants and classifying configurations of subspaces. The case of four subspaces was studied intensively by Gelfand and Ponomarev.[61]

A major research problem is whether linear lattices can be characterized by identities. By a theorem of Birkhoff (see Exercise 1.3.8), an equivalent problem is whether the class of linear lattices is closed under homomorphic images. Haiman[62] proved that linear lattices satisfy a sequence of increasingly stronger identities based on Desargues' theorem. However, he also showed that these analogs are not sufficient to characterize linearity. Haiman also developed a proof theory for universal Horn sentences (that is, sentences of the form $\bigwedge_i [P_i \leq Q_i] \Rightarrow [P \leq Q]$, where P_i, Q_i, P, Q are lattice polynomials, "$\wedge$" is logical conjunction or "and", and "$\Rightarrow$" is implication) about linear lattices. His algorithm represents a given Horn sentence as a series-parallel graph and

[61] Gelfand and Ponomarev (1972, 1974). [62] Haiman (1991).

then applies simple graph constructions that would yield either a proof or a counterexample.[63]

Exercises

3.4.1. Show that in a modular lattice, the two conditions that (I) x and y are comparable and that (II) for some element a $x \vee a = y \vee a$ and $x \wedge a = y \wedge a$, imply that $x = y$.

3.4.2. *McLaughlin's theorem.*[64]

Recall from Exercise 1.3.6 that a lattice L is modular if and only if L does not contain the pentagon N_5 as a sublattice. Prove McLaughlin's theorem: let L be a complemented lattice with minimum $\hat{0}$ and maximum $\hat{1}$. Suppose that L does not contain a pentagon with $m = \hat{0}$ and $M = \hat{1}$. Then L is modular.

3.4.3. (a) Construct the free distributive lattice $F_{\mathbf{D}}(3)$ on three generators.

(b) Do the calculations in the construction of $F_{\mathbf{M}}(3)$.

(c) Prove von Neumann's observation:[65] The sublattice generated by three elements x, y, and z in a modular lattice is distributive if and only if $x \wedge (y \vee z) = (x \wedge y) \vee (x \wedge z)$.

3.4.4. *The free modular lattice generated by two chains.*[66]

In this problem, we need the notion of a free modular lattice generated by a given partially ordered set P. Informally, this is the largest modular lattice $F_{\mathbf{M}}(P)$ containing P, with meets and joins in $F_{\mathbf{M}}(P)$ agreeing with any existing infimums and supremums in P.

(a) Define $F_{\mathbf{M}}(P)$ formally by a universal property.

Let $P(m, n)$ be the partially ordered set consisting of two disjoint chains

$$x_1 < x_2 < \cdots < x_m \quad \text{and} \quad y_1 < y_2 < \cdots < y_n,$$

the first having length $m - 1$ and the second having length $n - 1$. Let $Y(m, n)$ be the (distributive) lattice of order ideals of the Cartesian product of the two chains $\{0, 1, 2, \ldots, m\}$ and $\{0, 1, 2, \ldots, n\}$, in their natural order.

(b) Show that $|Y(m, n)| = \binom{m+n}{n}$.

(c) Prove the *Birkhoff–Schreier–Zassenhaus theorem.* The free modular lattice $F_{\mathbf{M}}(P(m, n))$ generated by the partial order $P(m, n)$ is isomorphic to a sublattice of $Y(m, n)$. In particular, it is distributive.

[63] Haiman (1985a). [64] McLaughlin (1956). [65] von Neumann (1936).
[66] Birkhoff (1967, p. 65), Schreier (1928), and Zassenhaus (1934).

3.4.5. *Modular lattices and exterior algebras.*[67]

(Research problem) Exterior algebras provide a calculus for meets and joins of subspaces. Thus, meets and joins in exterior algebras may be considered as coordinatized versions of meets and joins in modular or linear lattices. The double algebra due to Rota and his coworkers gives a workable notation for working with exterior-algebra meets and joins. Study this connection.

3.4.6. Prove that the four versions of the Arguesian law given in the text are equivalent.

3.4.7. *Three types of representations of lattices by equivalence relations.*[68]

A *representation (by equivalence relations)* of a lattice L is a bijective lattice homomorphism $R : L \to \Pi(S)$, where $\Pi(S)$ is the lattice of partitions on a set S. Partitions are thought of as equivalence relations on S. In his important 1953 paper, Jónsson defined three types of representations. A representation is *type-1* if $R(x \vee y) = R(x) \circ R(y)$ for all $x, y \in L$. Lattices with a type-1 representation are exactly the linear lattices. A representation is *type-2* if $R(x \vee y) = R(x) \circ R(y) \circ R(x)$ for all $x, y \in L$ and *type-3* if $R(x \vee y) = R(x) \circ R(y) \circ R(x) \circ R(y)$ for all $x, y \in L$.

(a) Show that a lattice L has a type-2 representation if and only if L is modular.

(b) Show that every lattice has a type-3 representation.

3.4.8. *Subgroups and equivalence relations.*

If H is a subgroup of a group G (written multiplicatively), let R_H be the equivalence relation giving the partition of G into right cosets of H. Let H and K be subgroups and $HK = \{hk\colon h \in H, k \in K\}$.

(a) Show that the equivalence relations R_H and R_K commute if and only if $HK = KH$.

(b) Show that the following lattices are linear: the lattice of normal subgroups of a group, the lattice of subgroups of an Abelian group, the lattice of subspaces of a vector space, and the lattice of submodules of a module.

3.4.9. *Generalized Arguesian Identities.*[69]

The following is Desargues' theorem in three-dimensional projective space. Let $a, a', b, b', c, c', d, d'$ be points. If the four lines $\overline{a, a'}$, $\overline{b, b'}$, $\overline{c, c'}$, and $\overline{d, d'}$ meet at one common point, then four points

$$\overline{a, b} \wedge \overline{a', b'}, \quad \overline{b, c} \wedge \overline{b', c'}, \quad \overline{c, d} \wedge \overline{c', d'}, \quad \text{and } \overline{d, a} \wedge \overline{d,' a'}$$

lie on a plane.

[67] Barnabei et al. (1985), Grassmann (1862), Hawrylycz (1994), and Mainetti and Yan (2000).
[68] Jónsson (1953). [69] Haiman (1991).

The corresponding generalized Arguesian identity is

$$\begin{aligned} &d \wedge ((d_1' \wedge d_2') \vee \{[(c_2' \wedge c_1') \vee (c_2 \wedge c_1)] \wedge [(b_2' \wedge b_1') \vee (b_2 \wedge b_1)] \\ &\quad \wedge [(a_2' \wedge a_1') \vee (a_2 \wedge a_1)]\}) \\ &\leq a_1 \vee ((a_1' \vee d_2') \wedge \{[(b_1' \vee a_2') \wedge (b_1 \vee a_2)] \vee [(c_1' \vee b_2') \wedge (c_1 \vee b_2)] \\ &\quad \vee [(d_1' \vee c_2') \wedge (d \vee c_2)]\}). \end{aligned}$$

Prove that this identity holds in all linear lattices.

(b) (Research problem) Find lattice versions of higher-dimensional theorems in projective geometry.

3.4.10. *Information theory and commuting equivalence relations.*

(a) (Research problem) Stochastic independence for partitions or equivalence relations is a key notion in information theory (see Section 1.4). Find an interesting information-theoretic interpretation of two commuting equivalence relations.

(b) (Research problem) Another problem is to generalize the notion of commuting equivalence relations to more than two relations. Motivated by probability theory, define three partitions π, σ, and τ on the same set to be *algebraically* 3-*independent* when for any three blocks $A \in \pi$, $B \in \sigma$, and $C \in \tau$, $A \cap B \cap C \neq \emptyset$ and 3-*commuting* when they are disjoint unions of algebraically 3-independent partitions. Is there a lattice-theoretic interpretation of 3-commuting equivalence relations?

3.4.11. *Stochastic independence and commutativity.*[70]

Stochastic independence is a strengthening of algebraic independence. Let $(\Omega, \mathcal{F}, \Pr)$ be a probability space, where $\mathcal{F}$ is a σ-algebra on Ω and Pr is the probability measure. Two σ-subalgebras $\mathcal{B}$ and $\mathcal{C}$ of $\mathcal{F}$ are *stochastically independent* if for subsets $B \in \mathcal{B}$ and $C \in \mathcal{C}$, $\Pr(B \cap C) = \Pr(B)\Pr(C)$. A σ-subalgebra is *strictly separable* if it can be generated by a countable collection of subsets. If $\mathcal{A}$ is a σ-subalgebra of $\mathcal{F}$, let $M_{\mathcal{A}}$ be the probability space $(\Omega, \mathcal{A}, \Pr)$.

Two theorems due to Rohlin characterize the structure of pairs of stochastically independent σ-subalgebras.

(a) Show *Rohlin's first theorem.* Let $(\Omega, \mathcal{F}, \Pr)$ be a probability space, $\mathcal{B}$ and $\mathcal{C}$ be two strictly separable σ-subalgebras of $\mathcal{F}$ and $\mathcal{A}$ be the σ-subalgebra generated by $\mathcal{B}$ and $\mathcal{C}$. Then up to a set of measure zero, the space $M_{\mathcal{A}}$ equals the direct product $M_{\mathcal{B}} \otimes M_{\mathcal{C}}$.

[70] Rohlin (1949) and Yan (1999).

(b) Show *Rohlin's second theorem.* Let $\mathcal{A}$ be a strictly separable σ-subalgebra of $\mathcal{F}$. If $\mathcal{A}$ is atomless, then $\mathcal{A}$ has a stochastically independent complement $\mathcal{B}$ and, up to a set of measure zero, the space $(\Omega, \mathcal{F}, \mathrm{Pr})$ equals the direct product $M_{\mathcal{A}} \otimes M_{\mathcal{B}}$.

The concept of commuting equivalence relations has a stochastic analog. The σ-subalgebras $\mathcal{B}$ and $\mathcal{C}$ *commute stochastically* if for subsets $B \in \mathcal{B}$ and $C \in \mathcal{C}$,

$$\mathrm{Pr}_{\mathcal{D}}(B)\mathrm{Pr}_{\mathcal{D}}(C) = \mathrm{Pr}_{\mathcal{D}}(B \cap C),$$

where $\mathcal{D} = \mathcal{B} \cap \mathcal{C}$ and $\mathrm{Pr}_{\mathcal{D}}$ is the conditional probability with respect to $\mathcal{D}$.

(c) Prove that the following statements are equivalent:

SC1. The σ-subalgebras $\mathcal{B}$ and $\mathcal{C}$ are stochastically commuting.

SC2. For all pairs of random variables X and Y measurable with respect to the σ-subalgebras $\mathcal{B}$ and $\mathcal{C}$,

$$E(X|\mathcal{B} \cap \mathcal{C})E(Y|\mathcal{B} \cap \mathcal{C}) = E(XY|\mathcal{B} \cap \mathcal{C}),$$

where $E(\cdot|\mathcal{B} \cap \mathcal{C})$ is the conditional expectation operator with respect to $\mathcal{B} \cap \mathcal{C}$.

SC3. The conditional expectation operators $E(\cdot|\mathcal{B})$ and $E(\cdot|\mathcal{C})$ commute; that is, for any random variable X,

$$E(E(X|\mathcal{B}) \,|\, \mathcal{C}) = E(E(X|\mathcal{C}) \,|\, \mathcal{B}) = E(X \,|\, \mathcal{B} \cap \mathcal{C}).$$

(d) Use Rohlin's theorems to prove the probabilistic analog of Theorem 3.4.8 for commuting equivalence relations: let $\mathcal{B}$, $\mathcal{C}$ be a pair of strictly separable σ-subalgebras which are stochastically commuting. Let $\mathcal{D} = \mathcal{B} \cap \mathcal{C}$ and $\mathcal{A}$ be the σ-subalgebra generated by $\mathcal{B}$ and $\mathcal{C}$. If $\mathcal{D}$ is atomless, then there exist strictly separable stochastically independent σ-subalgebras $\mathcal{S}$ and $\mathcal{T}$ such that the following decompositions hold, modulo sets of measure zero:

$$M_{\mathcal{B}} = M_{\mathcal{D}} \otimes M_{\mathcal{S}}, \quad M_{\mathcal{C}} = M_{\mathcal{D}} \otimes M_{\mathcal{T}}, \quad \text{and} \quad M_{\mathcal{A}} = M_{\mathcal{D}} \otimes M_{\mathcal{S}} \otimes M_{\mathcal{T}}.$$

3.5 Finite Modular and Geometric Lattices

In this section, we discuss the combinatorics of *finite* modular, semimodular, and geometric lattices. Our discussion is selective and will center on the following result.

3.5.1. Dilworth's covering theorem. Let L be a finite modular lattice. Then the number of elements covering exactly k elements equals the number of elements covered by exactly k elements.

Finite atomic modular lattices are the key to proving Dilworth's theorem. It is slightly more difficult to study finite atomic semimodular lattices and we will take this option. A lattice L is *semimodular* if for any two elements x and y, x covers $x \wedge y$ implies that $x \vee y$ covers y. By Dedekind's transposition principle (3.4.1), a modular lattice is semimodular.

A finite lattice satisfies the *Jordan–Dedekind chain condition* if for every pair x and y of elements such that $x < y$, every maximal chain from x to y has the same length.

3.5.2. Lemma. A finite semimodular lattice satisfies the Jordan–Dedekind chain condition.

Proof. We will prove the following assertion: if one maximal chain from x to y has length n, then every maximal chain from x to y has length n. We proceed by induction on n. The assertion holds if y covers x, that is, when $n = 1$. Suppose that there is a maximal chain

$$x = x_0 < x_1 < x_2 < \cdots < x_{n-1} < x_n = y$$

of length n from x to y. Let

$$x = y_0 < y_1 < y_2 < \cdots < y_{m-1} < y_m = y$$

be another maximal chain from x to y. Finally, let

$$x_1 \vee y_1 = z_0 < z_1 < \cdots < z_{k-1} < z_k = y$$

be a maximal chain from $x_1 \vee y_1$ to y. By semimodularity, $x_1 \vee y_1$ covers x_1 and hence $x_1 < z_0 < z_1 < \cdots < y$ is a maximal chain from x_1 to y. Since $x_1 < x_2 < \cdots < y$ is a maximal chain from x_1 to y having length $n - 1$, we conclude, by induction, that $k = n - 2$. On the other hand, $y_1 < z_0 < z_1 < \cdots < z_{k-1} = z_k = y$ is a maximal chain from y_1 to y. Hence, by induction, $k + 1 = m - 1$. All together, we conclude that $m = n$. □

If x is an element in a finite semimodular lattice L, let the *rank function* $\mathrm{rk}(x)$ be the length of a maximal chain from the minimum $\hat{0}$ to x. Thinking of rank as dimension, Lemma 3.5.2 asserts "invariance of dimension."

3.5.3. Lemma. The rank function of a finite semimodular lattice satisfies the submodular inequality

$$\mathrm{rk}(x) + \mathrm{rk}(y) \geq \mathrm{rk}(x \vee y) + \mathrm{rk}(x \wedge y).$$

Proof. To see this, observe that $\mathrm{rk}(x) - \mathrm{rk}(x \wedge y)$ is the length of a maximal chain from $x \wedge y$ to x. Let

$$x \wedge y = x_0 < x_1 < x_2 < \cdots < x_{m-1} < x_m = x$$

be such a maximal chain, where $m = \mathrm{rk}(x) - \mathrm{rk}(x \wedge y)$. Take the join of each element in the chain with y to obtain

$$y \leq x_1 \vee y \leq x_2 \vee y \leq \cdots \leq x_{m-1} \vee y \leq x \vee y.$$

Since $x_{i+1} \vee y$ covers or equals $x_i \vee y$, the length of a maximal chain from y to $x \vee y$ is at most m. Hence, $\mathrm{rk}(x \vee y) - \mathrm{rk}(y) \leq m$. □

A lattice is *geometric* if it is atomic and semimodular. A *matroid* M on the finite set S is defined by a closure operator $A \mapsto \overline{A}$ on the Boolean algebra 2^S of subsets of S satisfying the (*Mac Lane–Steinitz*) *exchange condition*:

If $a, b \notin \overline{A}$, then $a \in \overline{A \cup \{b\}}$ implies $b \in \overline{A \cup \{a\}}$.

A matroid is *simple* if the empty set and all one-element subsets are closed. The closed sets of a matroid are called *flats*. The flats form a lattice $L(M)$ under set-containment, with meet and join given by

$$\overline{A} \wedge \overline{B} = \overline{A} \cap \overline{B} \quad \text{and} \quad \overline{A} \vee \overline{B} = \overline{A \cup B}.$$

3.5.4. Birkhoff's theorem. The lattice $L(M)$ of flats of a finite matroid is geometric. Conversely, given a geometric lattice L, there exists a unique simple matroid M such that $L(M)$ is isomorphic to L.

Proof. Let M be a matroid on the finite set S. Then it is easy to check that the flat Y covers the flat X if and only if there exists $a \in S$ such that $X \vee \overline{\{a\}} = Y$. Thus,

$$\begin{aligned}
&X \text{ covers } X \wedge Y \\
\Leftrightarrow\ &X = (X \wedge Y) \vee \overline{\{a\}} \text{ for some } a \in S \\
\Rightarrow\ &X \vee Y = (X \wedge Y) \vee \overline{\{a\}} \vee Y = Y \vee \overline{\{a\}} \\
\Rightarrow\ &X \vee Y \text{ covers } Y.
\end{aligned}$$

We conclude that $L(M)$ is semimodular. In particular, the closure of a one-element set $\{a\}$ either is the minimum flat $\overline{\emptyset}$ or covers $\overline{\emptyset}$. Let X be a flat and let

$a_1, a_2, \ldots, a_m$ be all the elements in X not in $\overline{\emptyset}$. Then

$$X = \overline{\{a_1\}} \vee \overline{\{a_2\}} \vee \cdots \vee \overline{\{a_m\}}.$$

We conclude that $L(M)$ is atomic.

Now let L be a geometric lattice and S be the set of atoms of L. It is routine that the function $A \mapsto \overline{A}$ on subsets of S by

$$\overline{A} = \left\{c\colon c \leq \bigvee_{a:\, a \in A} a\right\}$$

is a closure operator and the lattice of closed sets of $A \to \overline{A}$ is isomorphic to L. To finish, we need to prove the exchange condition. Let $a, b \notin \overline{A}$ and $a \in \overline{A \cup \{b\}}$. Let $X = \bigvee_{e:\, e \in A} e$. Then $a \not\leq X$ but $a \leq X \vee b$. Since $b \not\leq X$, $b \wedge X = \hat{0}$, and hence, b covers $b \wedge X$. By semimodularity, $X \vee b$ covers X. As $X < X \vee a \leq X \vee b$, we conclude that $X \vee a = X \vee b$; in other words, $b \in \overline{A \cup \{a\}}$. □

We note that the rank function in the lattice $L(M)$ of flats of a matroid M on a set S induces a rank function on subsets A of S by $\mathrm{rk}(A) = \mathrm{rk}(\overline{A})$. By Lemma 3.5.3, this rank function satisfies the submodular inequality: for $A, B \subseteq S$, $\mathrm{rk}(A) + \mathrm{rk}(B) \geq \mathrm{rk}(A \cup B) + \mathrm{rk}(A \cap B)$. Thus, the lattice-induced rank function is a matroid rank function, as defined in Section 2.4.

3.5.5. Rota's positivity theorem.[71] Let L be a finite geometric lattice, x and y be elements such that $x \leq y$, and μ be the Möbius function of L. Then $\mu(x, y)$ is nonzero and has sign $(-1)^{\mathrm{rk}(y)-\mathrm{rk}(x)}$.

Proof. We proceed by induction on $\mathrm{rk}(y) - \mathrm{rk}(x)$. To begin, observe that $\mu(x, x) = 1$ and $\mu(x, y) = -1$ if y covers x. For the induction step, we use Weisner's theorem (3.1.5). Choose an element a covering x. Then

$$\mu(x, y) = - \sum_{z:\, z \in [x,y],\, z \vee a = y,\, z \neq y} \mu(x, z).$$

By the submodular inequality,

$$\mathrm{rk}(z) + \mathrm{rk}(a) \geq \mathrm{rk}(y) + \mathrm{rk}(x),$$

and hence,

$$\mathrm{rk}(z) - \mathrm{rk}(x) \geq \mathrm{rk}(y) - \mathrm{rk}(a) = [\mathrm{rk}(y) - \mathrm{rk}(x)] - 1.$$

[71] *Foundations I*, p. 357.

Since $z < y$, we have $\mathrm{rk}(z) - \mathrm{rk}(x) = [\mathrm{rk}(y) - \mathrm{rk}(x)] - 1$. By induction, $\mu(x, z)$ is nonzero and has sign $(-1)^{\mathrm{rk}(y)-\mathrm{rk}(x)-1}$. We conclude that $\mu(x, y)$ is nonzero and has sign $(-1)^{\mathrm{rk}(y)-\mathrm{rk}(x)}$. □

An easy way to prove Dilworth's covering theorem is to prove a more general result.[72] Let J and M be subsets of a finite lattice L. The subset J is *concordant* with the subset M if for every element x in L, either x is in M or there exists an element $x^\dagger$ such that

CS1. $\mu(x, x^\dagger) \neq 0$.
CS2. For every element j in J, $x \vee j \neq x^\dagger$.

If H and K are subsets of a partially ordered set, the *incidence matrix* $\mathcal{I}(H|K)$ is the matrix with rows indexed by H and columns indexed by K with the hk-entry equal to 1 if $h \leq k$ and 0 otherwise.

3.5.6. Theorem. Let J be concordant with M in a finite lattice L. Then the incidence matrix $\mathcal{I}(M|J)$ has rank $|J|$. In particular, $|J| \leq |M|$.

Proof. Let $\mathbb{Q}$ be field of rational numbers, $\mathrm{Fun}(L, \mathbb{Q})$ be the vector space of functions defined from the set L to $\mathbb{Q}$, and $\mathrm{Fun}(J, \mathbb{Q})$ be the subspace of functions *supported on* J, that is, functions such that $f(x) = 0$ unless $x \in J$. If $a \in L$, the *delta function* $\delta_a : L \to \mathbb{Q}$ is the function defined by $\delta_a(x) = 1$ if $x = a$ and 0 otherwise. The set $\{\delta_a\colon a \in L\}$ is a basis for $\mathrm{Fun}(L, \mathbb{Q})$ and the subset $\{\delta_a\colon a \in J\}$ is a basis for $\mathrm{Fun}(J, \mathbb{Q})$.

Let $T\colon \mathrm{Fun}(J, \mathbb{Q}) \to \mathrm{Fun}(L, \mathbb{Q})$ be the linear transformation defined by

$$Tf(x) = \sum_{z : z \leq x} f(z).$$

Relative to the bases of delta functions, the matrix of T is the incidence matrix $\mathcal{I}(L|J)$. The incidence matrix $\mathcal{I}(M|J)$ is a submatrix of $\mathcal{I}(L|J)$. We will show that $\mathcal{I}(M|J)$ has rank $|J|$ by showing that the linear transformation $T_M : \mathrm{Fun}(J, \mathbb{Q}) \to \mathrm{Fun}(M, \mathbb{Q})$ obtained by restricting Tf to the elements in M is injective. This will be done by showing that one can reconstruct a function f in $\mathrm{Fun}(J, \mathbb{Q})$ from the restriction $Tf|_M$ of Tf to M. We need the following lemma.

3.5.7. Lemma.

$$\sum_{y : x \leq y \leq x^\dagger} \mu(y, x^\dagger) Tf(y) = \sum_{z : z \vee x = x^\dagger} f(z).$$

[72] Kung (1987).

Proof. Let $f_x : [x, x^\dagger] \to \mathbb{Q}$ be the function defined by

$$f_x(y) = \sum_{z:\, z \vee x = y} f(z).$$

Observe first that the elements in L are partitioned into equivalence classes by the relation $a \sim b$ if and only if $a \vee x = b \vee x$, and second that $z \leq y$ for an element y in $[x, x^\dagger]$ if and only if $z \vee x \leq y$. Hence,

$$Tf(y) = \sum_{z:\, z \leq y} f(z) = \sum_{z:\, x \leq z \leq y} f_x(z).$$

Applying Möbius inversion to f_x on the interval $[x, x^\dagger]$, we obtain

$$\sum_{y:\, x \leq y \leq x^\dagger} \mu(y, x^\dagger) Tf(y) = f_x(x^\dagger) = \sum_{z:\, z \vee x = x^\dagger} f(z). \qquad \square$$

To reconstruct a function $f : J \to \mathbb{Q}$, we first reconstruct the (unrestricted) function $Tf : L \to \mathbb{Q}$ using as input the restriction $Tf : M \to \mathbb{Q}$. Once we have done this, f can be reconstructed using Möbius inversion over L.

To start the reconstruction of Tf, we note that if J is concordant with M, the maximum $\hat{1}$ must be in M. Hence, $Tf(\hat{1})$ can be read off directly from the input. We now go down the lattice inductively. If $x \in M$, then $Tf(x)$ is read off directly from the input. If $x \notin M$, then by CS2, for all $j \in J$, $x \vee j \neq x^\dagger$. Hence, $f_x(x^\dagger) = 0$ and rearranging the equation in Lemma 3.5.7, we have

$$\mu(x, x^\dagger) Tf(x) = - \sum_{y:\, x < y \leq x^\dagger} \mu(y, x^\dagger) Tf(y).$$

By induction, since $y > x$, all the values $Tf(y)$ have already been reconstructed. Hence, as $\mu(x, x^\dagger) \neq 0$, the equation yields the value of $Tf(x)$. This completes the proof of Theorem 3.5.6. $\square$

There are many examples of concordant sets.

3.5.8. The Dowling–Wilson inequalities.[73] Let L be a rank-n geometric lattice, and

$$B_k = \{x\colon\ x \in L \text{ and } \operatorname{rk}(x) \leq k\},$$
$$T^k = \{x\colon\ x \in L \text{ and } n - \operatorname{rk}(x) \leq k\}.$$

[73] Dowling and Wilson (1975).

Then B_k is concordant with T^k. In particular, if W_r is the number of rank-r elements in L and $k < n/2$, then

$$W_0 + W_1 + W_2 + \cdots + W_k \leq W_{n-k} + W_{n-k+1} + \cdots + W_{n-1} + W_n.$$

Equality holds if and only if L is modular.

Proof. For all elements x in L, let $x^\dagger = \hat{1}$. Then if $\mathrm{rk}(j) \leq k$ and $x \notin T^k$ (that is, $\mathrm{rk}(x) \leq n - k$), then the submodular inequality implies that $\mathrm{rk}(x \vee j) < n$. Thus, CS2 holds. CS1 holds by Rota's positivity theorem.

We shall not need the characterization of those geometric lattices in which equality holds and refer the reader to the paper of Dowling and Wilson. □

Dilworth's covering theorem (3.5.1) is another consequence of Theorem 3.5.6. If x is an element in a finite lattice L, let x^* be the join of all the elements covering x. If L is semimodular, then the interval $[x, x^*]$ is a geometric lattice. Dually, let x_* be the meet of all the elements covered by x.

3.5.9. Theorem. Let k be a positive integer and L be a finite modular lattice.

(a) Let J_k be the set of elements in L covered by k or fewer elements and M_k be the set of elements in L covering k or fewer elements. Then J_k is concordant with M_k, with $x^\dagger = x^*$.

(b) Let D_k be the set of elements x in L such that $\mathrm{rk}(x) - \mathrm{rk}(x_*) \leq k$ and U_k be the set of elements x in L such that $\mathrm{rk}(x^*) - \mathrm{rk}(x) \leq k$. Then D_k is concordant with U_k, with $x^\dagger = x^*$.

Proof. If $x \notin M_k$, then the interval $[x, x^*]$ has at least $k + 1$ atoms and contains the atomic geometric lattice L^x generated by the atoms in $[x, x^*]$. The lattice L^x has maximum x^*, and by the Dowling–Wilson inequalities, x^* covers at least $k + 1$ elements.

Suppose that $j \in J_k$ and $x \in L$. If $x \vee j = x^*$, then by Dedekind's transposition principle, the intervals $[x \wedge j, j]$ and $[x, x^*]$ are isomorphic. However, since x^* covers at least $k + 1$ elements, the isomorphism implies that j covers at least $k + 1$ elements, contradicting the assumption that $j \in J_k$. We conclude that for all $j \in J_k$, $x^* \neq x \vee j$. This verifies CS2. CS1 follows from Rota's positivity theorem.

We note that there are other, perhaps shorter, arguments to prove part (a). One could use the fact that $[x, x^*]$ is a finite atomic modular lattice, and in such lattices, the number of atoms equals the number of coatoms.

To prove part (b), we use a similar argument, using the fact that if $[x \wedge j, j]$ and $[x, x^*]$ are isomorphic, then $j_* \leq x \wedge j$. □

We end this section by remarking that a modular lattice is consistent in the sense given in Exercise 1.3.6. Hence, a discussion of the Kurosh–Ore theorem for decompositions of elements into join-irreducibles in modular lattices can be found there.

Exercises

3.5.1. *A lattice-theoretic version of the fundamental theorem of geometry.*

A lattice L is *reducible* if L is isomorphic to the Cartesian product $L_1 \times L_2$, where both factors L_1 and L_2 have at least two elements. A lattice is *irreducible* if it is not reducible. Show that if $n \geq 4$, then an irreducible finite rank-n atomic modular lattice is isomorphic to the lattice of subspace of a dimension-n vector space over a finite field, and if $n = 3$, such a lattice is the lattice of subspaces of a projective plane.

3.5.2. *The cycle matroid of a graph.*

Let Γ be a graph with vertex set V and edge set E. For a subset $T \subseteq E$, let $\overline{T}$ be the subset

$$T \cup \{e\colon \text{ for some subset } D \subseteq T,\ D \cup \{e\} \text{ is a cycle in } \Gamma\}.$$

(a) Show that each closed set X of edges determines a partition of the vertex set. A block of this partition consists of the vertices in a connected component of the edge subgraph $\Gamma|_X$ on V with edge set X.

(b) Show that $T \mapsto \overline{T}$ is a closure operator on E satisfying the exchange condition.

The closure operator $T \mapsto \overline{T}$ defines the *cycle matroid* of the graph Γ. Let $L(\Gamma)$ be the lattice of flats and rk its rank function.

(c) Show that $\mathrm{rk}(X) = |V| - c(X)$, where $c(X)$ is the number of connected components in $\Gamma|_X$.

If λ is a positive integer, a (*proper*) λ-*coloring* of Γ is a function $h : V \to \{1, 2, \ldots, \lambda\}$ such that $h(u) \neq h(v)$ whenever $\{u, v\}$ is an edge of Γ.

(d) Show that the number of proper λ-colorings of Γ equals a polynomial $P(\Gamma; \lambda)$ of degree $|V|$. In fact,

$$P(\Gamma; \lambda) = \sum_{X:\, X \in L(\Gamma)} \mu(\overline{\emptyset}, X)\lambda^{c(X)}.$$

The polynomial $P(\Gamma; \lambda)$ is the *chromatic polynomial* of Γ.

(e) Show that the lattice of flats of the cycle matroid of the complete graph K_n on the vertex set $\{1, 2, \ldots, n\}$ is isomorphic to the lattice of partitions on

$\{1, 2, \ldots, n\}$. Show that

$$P(K_n; \lambda) = \lambda(\lambda - 1)(\lambda - 2) \cdots (\lambda - n + 1).$$

Hence, conclude that if μ is the Möbius function of the lattice $\Pi(\{1, 2, \ldots, n\})$ of partitions, then

$$\mu(\hat{0}, \hat{1}) = (-1)^{n-1}(n-1)!$$

(f) Using Exercise 3.1.11, show that

$$\det[\lambda^{\mathrm{rk}(X \vee Y)}]_{X,Y \in L(G)} = \prod_{X:\, X \in L(G)} P(G/X; \lambda),$$

where G/X is the graph G with the edges in X contracted. Extend this theorem to matroids and their characteristic polynomials.

3.5.3. *Sperner theory.*[74]

(a) Show that the lattice $L(V)$ of subspaces of a finite vector space V is Sperner.
(b) Find an explicit symmetric chain decomposition of $L(V)$.
(c) Show that for sufficiently large n, the lattice of partitions of $\{1, 2, \ldots, n\}$ is *not* Sperner.
(d) Find other examples of geometric lattices which are not Sperner.
(e) Find continuous analogs of Sperner's theorem for lattices of subspaces over the reals.

3.5.4. *Dilworth-Hall gluing.*[75]

Let L_1 be a lattice with maximum u and L_2 be a lattice with a minimum z. Suppose that there exist elements $a_1 \in L_1$ and $a_2 \in L_2$ such that the upper interval $[a_1, u]$ in L_1 is isomorphic to the lower interval $[z, a_2]$ in L_2. The *Dilworth–Hall union* $L_1 \sqcup L_2$ *(over the intervals* $[a_1, u]$ *and* $[z, a_2]$*)* is the lattice obtained by taking the union of L_1 and L_2 and identifying isomorphic elements in $[a_1, u]$ and $[z, a_2]$, with join defined by $x \vee y$ equals $x \vee y$ in L_i if both x and y are in L_i and $x \vee y = (x \vee a_1) \vee y$ if $x \in L_1$ and $y \in L_2$. The meet is defined dually.

(a) Show that $L_1 \sqcup L_2$ is indeed a lattice.
(b) Show that $L_1 \sqcup L_2$ is modular if and only if L_1 and L_2 are modular.

[74] Canfield (1978), Dilworth and Greene (1971), Kahn (1980), Klain and Rota (1997), and Vogt and Voigt (1997).

[75] Hall and Dilworth (1944). This paper started the area of gluing constructions for modular lattices and was a motivation for studying the Arguesian law. See the survey paper Day and Freese (1990).

(c) Using this construction, show that there exists a finite modular lattice which is not a sublattice of any complemented modular lattice.

3.5.5. Show that the following are equivalent for a finite lattice L:

(a) L is semimodular.

(b) x and y cover $x \wedge y$ implies $x \vee y$ covers x and y.

(c) x covers y implies that $x \vee z$ covers or equals $y \vee z$ for every element z in L.

(d) Let $\mathrm{rk}(x)$ be the minimum length of a maximal chain from $\hat{0}$ to x. Then

$$\mathrm{rk}(x) + \mathrm{rk}(y) \geq \mathrm{rk}(x \vee y) + \mathrm{rk}(x \wedge y).$$

3.5.6. A finite lattice is *coatomic* if every element is a meet of coatoms (that is, elements covered by the maximum $\hat{1}$).

(a) Show that a finite geometric lattice is coatomic.

(b) Is it true that if a finite semimodular lattice is coatomic, then it is atomic?

3.5.7. An element x in a semimodular lattice L is *modular* if for all elements y in L,

$$\mathrm{rk}(x) + \mathrm{rk}(y) = \mathrm{rk}(x \vee y) + \mathrm{rk}(x \wedge y).$$

Let L be a finite geometric lattice and M_k be the number of rank-k modular elements in L.

(a) Show that

$$W_0 + W_1 + W_2 + \cdots + W_k \geq M_{n-k} + M_{n-k+1} + \cdots + M_{n-1} + M_n.$$

(b) (Research problem) Prove the *unimodality conjecture*. There exists an index m such that

$$W_0 \leq W_1 \leq W_2 \leq \cdots \leq W_m \text{ and } W_{m+1} \geq W_{m+2} \geq \cdots \geq W_{n-1} \geq W_n.$$

This conjecture was made by Rota in 1970, but almost no progress has been made since then. A more tractable conjecture is $W_1 < W_2 < \cdots < W_k$, where $k = \lfloor \frac{n}{2} \rfloor$.

(c) (Research problem) Show that if $k \leq n/2$, then $W_k \leq W_{n-k}$. This "top-heaviness" conjecture was made by Dowling and Wilson. It has proved to be just as intractable as the unimodality conjecture.

3.5.8. *Consistent lattices.*

Let L be a finite lattice. Recall from Exercise 1.3.4 that a join-irreducible j in a finite lattice L is consistent if for all elements x in L, $x \vee j$ equals x or is a join-irreducible in the upper interval $[x, \hat{1}]$. Let C be the set of consistent join-ireducibles and M the set of meet-irreducibles.

(a) Show that $C \cup \{\hat{0}\}$ is concordant with $M \cup \{\hat{1}\}$.

A lattice is consistent if every join-irreducible is consistent. Part (a) implies that in a finite consistent lattice, the number of join-irreducibles is at most the number of meet-irreducibles.

Let G be a finite group. A subgroup H of G is *subnormal* if there exists a chain $G = N_0 \supset N_1 \supset N_2 \supset \cdots \supset N_{r-1} \supset N_r = H$ such that N_{i+1} is a normal subgroup in the subgroup N_i.

(b) Show that the subnormal subgroups form a sublattice of the lattice of subgroups of a finite group G.[76]

(c) Let $W(G)$ be the lattice of subnormal subgroups of G. Show that $W(G)$ is consistent and dually semimodular.

3.6 Valuation Rings and Möbius Algebras

The basic construction in this section is the $\wedge$-semigroup algebra of a lattice L. Let L be a lattice, $\mathbb{A}$ be a commutative ring, and $M(L, \mathbb{A})$ be the $\mathbb{A}$-algebra consisting of (formal) linear combinations of the form $\sum a_x x$, where $a_x \in \mathbb{A}$ and $x \in L$, with all but finitely many coefficients a_x equal to zero. Multiplication in $M(L, \mathbb{A})$ is defined by $xy = x \wedge y$ if x and y are in L and extended by linearity and distributivity. Explicitly,

$$\left[\sum_{x:\, x \in L} a_x x\right]\left[\sum_{y:\, y \in L} b_y y\right] = \sum_{z:\, z \in L}\left(\sum_{x,y:\, x \wedge y = z} a_x b_y\right) z.$$

If L has a maximum $\hat{1}$, then $M(L, \mathbb{A})$ has an identity and it equals $\hat{1}$.

An $\mathbb{A}$-*valuation* v on the lattice L is a function $L \to \mathbb{A}$ satisfying

$$v(x \vee y) + v(x \wedge y) = v(x) + v(y)$$

for all elements x and y in L. Constant functions are valuations. However, unless a lattice is distributive, it does not have many valuations.

Valuations have been largely studied on Boolean algebras, where they are called measures. Although measures and valuations are defined by the same equation, the theory of valuations is richer and ranges wider. As is done in functional analysis, the study of measures can be reduced to the study of linear functionals or abstract integrals on function spaces. In analogy, we reduce the

[76] Wielandt (1939).

study of valuations on a distributive lattice to the study of linear functionals on a ring constructed from the lattice.[77]

Let S be the submodule in $M(L, \mathbb{A})$ of linear combinations of elements of the form

$$x \vee y + x \wedge y - x - y.$$

Then S is an ideal of $M(L, \mathbb{A})$. To prove this, consider the expression

$$z \wedge (x \vee y + x \wedge y - x - y), \qquad \text{(E)}$$

which equals

$$z \wedge (x \vee y) + z \wedge (x \wedge y) - z \wedge x - z \wedge y.$$

Using the distributive axioms, idempotency and commutativity, we can rewrite the first two terms to obtain

$$(z \wedge x) \vee (z \wedge y) + (z \wedge x) \wedge (z \wedge y) - z \wedge x - z \wedge y.$$

Thus, the expression (E) is in S. Since we include all linear combinations in the construction, we conclude that S is an ideal. The *valuation ring* $\mathrm{Val}(L, \mathbb{A})$ of the distributive lattice L over the ring $\mathbb{A}$ is the quotient ring $M(L, \mathbb{A})/S$.

3.6.1. Lemma. Let L be a distributive lattice and $\mathbb{A}$ a commutative ring. Then there is a natural bijection between valuations $L \to \mathbb{A}$ and $\mathbb{A}$-linear functionals $\mathrm{Val}(L, \mathbb{A}) \to \mathbb{A}$.

Proof. If $v : L \to \mathbb{A}$ is a valuation, we can extend v to a $\mathbb{A}$-linear functional on $M(L, \mathbb{A})$ by linearity; that is,

$$v\left(\sum_{x: x \in L} a_x x\right) = \sum_{x: x \in L} a_x v(x).$$

As a linear functional, $v(e) = 0$ for every element e in the ideal S. Hence, v is defined as a linear functional on $\mathrm{Val}(L, \mathbb{A})$.

Conversely, let $u : \mathrm{Val}(L, \mathbb{A}) \to \mathbb{A}$ be a linear functional. Then define a valuation $\tilde{u}$ on L by setting $\tilde{u}(x) = u(x)$, where x on the right side is the linear

[77] The basic theory of valuation rings can be found in Rota (1971). Rota intended the theory to be applied to logic and probability in a unified way. This program is sketched in *Twelve problems*. There is much to be done and clarified. The papers Rota (1973) and Ellerman and Rota (1978) give an indication of what might be done. We should say that many of proofs in Rota (1973) remain sketches. See also Geissinger (1973).

combination with one term x itself. Since u is zero for any element in the ideal S, $\tilde{u}$ is a valuation on L.

The two constructions are inverses, and hence they are both bijections. □

Let S^+ be the ideal generated by S and $\hat{0}$ and $\mathrm{Val}_0(L, \mathbb{A}) = M(L, \mathbb{A})/S^+$. An analog of Lemma 3.6.1, in which a valuation v is required to satisfy $v(\hat{0}) = 0$, holds for $\mathrm{Val}_0(L, \mathbb{A})$.

We will identify an element $x \in L$ with the linear combination x in $\mathrm{Val}(L, \mathbb{A})$. We will also assume that L has a maximum $\hat{1}$. Since any finite calculation in the valuation ring involves a finite number of elements in a distributive lattice, we can work in the (finite) sublattice generated by the elements occurring in the calculation. This sublattice always has a (relative) maximum.

In $\mathrm{Val}(L, \mathbb{A})$, we have the identity

$$x \vee y = x + y - xy = \hat{1} - (\hat{1} - x)(\hat{1} - y).$$

Iterating this, we have

$$\begin{aligned} x_1 \vee x_2 \vee \cdots \vee x_m &= \hat{1} - (\hat{1} - x_1)(\hat{1} - x_2) \cdots (\hat{1} - x_m). \\ &= \sum_{i_1, i_2, \ldots, i_k} (-1)^{k-1} x_{i_1} x_{i_2} \cdots x_{i_k}, \end{aligned}$$

the sum ranging over all nonempty subsets $\{i_1, i_2, \ldots, i_k\}$ of $\{1, 2, \ldots, m\}$. This is an algebraic version of the principle of inclusion–exclusion.

We return to the study of the $\wedge$-semigroup algebra $M(L, \mathbb{A})$, where L is a finite lattice. Wedderburn's theory of algebras suggests that one looks for idempotents. An element t in an algebra M is *idempotent* if $t^2 = t$. A set $\{t_1, t_2, \ldots, t_m\}$ is a *set of orthogonal idempotents* if

$$t_i^2 = t_i \text{ and } t_i t_j = 0 \text{ whenever } i \neq j.$$

A set of orthogonal idempotents cannot satisfy any nontrivial linear relation. The set $\{t_i\}$ is *complete* if it spans M.

If $x \in L$, let e_x be the linear combination in $M(L, \mathbb{A})$ defined by

$$e_x = \sum_{a: a \leq x} \mu(a, x) a. \qquad (ID)$$

3.5.6. Solomon's theorem.[78] The sums e_x, $x \in L$, form a complete set of orthogonal idempotents for the $\wedge$-semigroup algebra $M(L, \mathbb{A})$.

[78] Solomon (1967).

Proof. Let $\bar{e}_x$ be basis elements, one for each element x in L. Let $\bar{M}(L, \mathbb{A})$ be the $\mathbb{A}$-algebra of all formal linear combinations $\sum_{x:x\in L} a_x \bar{e}_x$, with multiplication defined by

$$\bar{e}_x^2 = \bar{e}_x \quad \text{and} \quad \bar{e}_x \bar{e}_y = 0 \text{ if } x \neq y \tag{M}$$

and extended by linearity. Let $\varphi : M(L, \mathbb{A}) \to \bar{M}(L, \mathbb{A})$ be the $\mathbb{A}$-linear map defined by

$$\varphi(x) = \sum_{a:a\leq x} \bar{e}_a.$$

Then since

$$\begin{aligned}\varphi(x)\varphi(y) &= \left(\sum_{a:a\leq x} \bar{e}_a\right)\left(\sum_{b:b\leq y} \bar{e}_b\right) \\ &= \sum_{a:a\leq x \text{ and } a\leq y} \bar{e}_a \;=\; \varphi(xy),\end{aligned}$$

φ is an $\mathbb{A}$-algebra homomorphism. By Möbius inversion, φ has an inverse, defined by

$$\varphi^{-1}(\bar{e}_x) = \sum_{a:a\leq x} \mu(a, x)a.$$

We conclude that $M(L, \mathbb{A})$ and $\bar{M}(L, \mathbb{A})$ are isomorphic. In particular, the elements e_x in $M(L, \mathbb{A})$ satisfy the same multiplication rule as their images $\bar{e}_x$ in $\bar{M}(L, \mathbb{A})$. □

It may seem strange to introduce a new algebra $\bar{M}(L, \mathbb{A})$ in the proof of Solomon's theorem. The orthogonality relation $e_x e_y = 0$ can also be proved by induction.

Many Möbius function identities follow from expanding expressions in the basis $\{x\}$ in the basis $\{e_x\}$. We give an example.

Suppose $a < \hat{1}$. Then

$$ae_{\hat{1}} = \left(\sum_{x:x\leq a} e_x\right) e_{\hat{1}} = 0$$

because $x \leq a < \hat{1}$, and hence, $e_x e_{\hat{1}} = 0$. On the other hand,

$$ae_{\hat{1}} = a\left(\sum_{x:x\leq\hat{1}} \mu(x, \hat{1})x\right) = \sum_{x:x\in L} \mu(x, \hat{1})(x \wedge a).$$

Equating coefficients of $\hat{0}$ (which equals $e_{\hat{0}}$), we obtain

$$\sum_{x:\, x \wedge a = \hat{0}} \mu(x, \hat{1}) = 0.$$

This gives another proof of Weisner's theorem (3.1.5).

Using Solomon's theorem, we can define the Möbius algebra for any finite partially ordered set P over a commutative ring $\mathbb{A}$ with identity.[79] The *Möbius algebra* $M(P, \mathbb{A})$ over the ring $\mathbb{A}$ is the dimension-$|P|$ $\mathbb{A}$-algebra with basis $x,\ x \in P$, with multiplication

$$xy = \sum_{u:\, u \in P} \left[\sum_{v:\, v \le x \text{ and } v \le y} \mu(u, v) \right] u.$$

With this multiplication, it is easy to check that the elements e_x, defined by $e_x = \sum_{u:\, u \le x} \mu(u, x) u$, form a complete set of orthogonal idempotents.

3.5.4. Theorem.[80] Let L be a finite distributive lattice and J the partially ordered set of join-irreducibles of L. Then the valuation ring $\mathrm{Val}_0(L, \mathbb{A})$ is isomorphic to the Möbius algebra $M(J, \mathbb{A})$.

Proof. By Birkhoff's theorem (1.3.3), L is isomorphic to the lattice $\mathcal{D}(J)$ of order ideals of P. Let $w : L \to M(J, \mathbb{A})$ be the function

$$w(x) = \sum_{j:\, j \in J,\, j \le x} e_j,$$

the sum ranging over all join-irreducibles j in the order ideal in J associated with x. Then w is a valuation with $w(\hat{0}) = 0$. By Lemma 3.5.1, w extends to a $\mathbb{A}$-algebra homomorphism $\tilde{w} : \mathrm{Val}_0(L, \mathbb{A}) \to M(J, \mathbb{A})$. Further, $\tilde{w}$ is surjective and both $\mathrm{Val}_0(L, \mathbb{A})$ and $M(J, \mathbb{A})$ have dimension $|J|$. Hence, $\tilde{w}$ is an isomorphism. □

Exercises

3.6.1. Show that if $a_1, a_2, \ldots, a_n$ are real numbers, then

$$\begin{aligned} &\max\{a_1, a_2, \ldots, a_n\} \\ &\quad = \sum_i a_i - \sum_{\{i,j\}} \min\{a_i, a_j\} + \sum_{\{i,j,k\}} \min\{a_i, a_j, a_k\} - \cdots \\ &\quad \pm \min\{a_1, a_2, \ldots, a_n\}. \end{aligned}$$

[79] Greene (1973). [80] Davies (1970).

3.6.2. Let L be an indecomposable finite-rank modular lattice. Show that if $v : L \to \mathbb{Z}$ satisfies $v(x) + v(y) = v(x \vee y) + v(x \wedge y)$ for all x and y in L, then

$$v(x) = a + b\mathrm{rk}(x),$$

where rk is the rank function of L and a, b are integers.

3.6.3. Prove Theorem 3.1.7 using Möbius algebras.

3.6.4. *Residuated maps and homomorphisms of Möbius algebras.*[81]

Let P and Q be partially ordered sets. A function $\varphi : P \to Q$ is (*upper*) *residuated* if the inverse image of a principal filter is a (nonempty) principal filter.

(a) Show that φ is order-preserving.

The *adjoint* $\varphi^{\Delta} : Q \to P$ is the function defined by the following: if $y \in Q$, then $\varphi^{\Delta}(y)$ is the generator of the principal filter $\varphi^{-1}(F(y))$, where $F(y) = \{z\colon z \in Q,\ z \geq y\}$.

(b) Show that φ^{Δ} is order-preserving.

(c) Show that $\varphi : P \to Q$ and $\varphi^{\Delta} : Q \to P$ form a Galois coconnection.

(d) Prove *Greene's theorem.* Let $\varphi : P \to Q$ be a function between finite partially ordered sets. Then φ extends to an $\mathbb{A}$-algebra homomorphism $M(P, \mathbb{A}) \to M(Q, \mathbb{A})$ if and only if the inverse image of a principal filter in Q is a principal filter or empty.

3.7 Further Reading

The following is a selection of books or surveys on Dilworth's chain partition theorem, Sperner theory, and extremal set theory.

I. Anderson, *Combinatorics of Finite Sets*, Clarendon Press, Oxford, 1987.

K.P. Bogart, C. Greene, and J.P.S. Kung, The impact of the chain decomposition theorem on classical combinatorics, in K.P. Bogart, R. Freese, and J.P.S. Kung, eds., *The Dilworth Theorems*, Birkhäuser, Boston, 1990, pp. 19–29.

C. Greene and D.J. Kleitman, Proof techniques in the theory of finite sets, in G.-C. Rota, ed., *Topics in Combinatorics,* Mathematical Association of America, Washington, DC, 1978, pp. 22–79.

K. Engel, *Sperner Theory*, Cambridge University Press, Cambridge, 1997.

[81] Everett (1944).

There seems to be no book devoted to algebraic aspects of modular lattices. The book by von Neumann is a classic and full of ideas. Stern's book is a comprehensive survey of (nonatomic) semimodular lattices.

J. von Neumann, *Continuous Geometry*, Princeton University Press, Princeton, NJ, 1960.

M. Stern, *Semimodular Lattices. Theory and Applications*, Cambridge University Press, Cambridge, 1999.

4
Generating Functions and the Umbral Calculus

4.1 Generating Functions

There are many theories of generating functions, among them that proposed in *Foundations VI*. Many of these theories have received excellent expositions elsewhere. We will give an informal minimalist exposition.[1]

The idea of *ordinary* generating functions is due to Laplace. For example, Laplace used the following argument:[2]

> On forming the product of the binomials $(1+a)$, $(1+b)$, $\ldots$, $(1+n)$, we obtain, on subtracting 1 from the expansion of this product, the sum of the combinations of all these letters taken one at a time, two at a time, three at a time, &c., each combination having 1 as coefficient. To get the number of combinations of these n letters taken s at a time, notice that if one supposes that all these letters are the same, the preceding product becomes $(1+a)^n$, and so the number of combinations of n letters taken s at a time will be the coefficient of a^s in the expansion of the binomial. This number is then given by the well-known binomial formula.

In modern terms, we think of each element of a set A as a variable and define the *ordinary generating function* $\mathrm{gf}(A)$ *of the set* A to be the sum

$$\sum_{a:a\in A} a.$$

Then, the *product formula*

$$\mathrm{gf}(A \times B) = \mathrm{gf}(A)\mathrm{gf}(B),$$

where $A \times B$ is the Cartesian product of A and B, is certainly true, especially if we regard it as a *definition* of multiplication. Note that it is not necessary to assume that this multiplication is commutative.

[1] Comprehensive accounts may be found in Graham et al. (1988), Petkovšek et al. (1996), Stanley (1986), and Wilf (2006). A short insightful introduction is Pólya (1969).

[2] Laplace (1995, p. 14).

In the spirit of Laplace, we use the product formula to derive the binomial distribution. If a coin is tossed, then there are two outcomes, head H and tail T, giving the generating function $H+T$. Hence, if a coin is tossed n times, the generating function is $(H+T)^n$. For example,

$$(H+T)^3 = HHH + HHT + HTH + HTT + THH + THT \\ + TTH + TTT.$$

Thus, the generating function "generates" all possible outcomes. Setting $H = pt$ and $T = q$, where t is a variable, p the probability of a head, and q the probability of a tail, then we obtain the generating function for the probability distribution of the number X of heads:

$$\sum_{k=0}^{n} \Pr(X=k)t^k = (pt+q)^n = \sum_{k=0}^{n} \binom{n}{k} p^k q^{n-k} t^k.$$

The random variable X is the sum $X_1 + X_2 + \cdots + X_n$, where X_i is the random variable that equals 1 if the ith toss is a head and 0 otherwise. Since the distribution of the sum of two random variables is the convolution of their distribution, taking the generating function converts convolution to multiplication. For continuous random variables, this conversion is done by Laplace or Fourier transforms. In this sense, (ordinary) generating functions are Laplace transforms in disguise.

The best way to introduce exponential generating functions is to discuss the exponential formula.[3] Let $\mathcal{C}_n$ be a finite set of "irreducible" labeled structures or *atoms* which can be put on a nonempty finite set of size n. Let $\mathcal{C} = \bigcup_{n=1}^{\infty} \mathcal{C}_n$. Using atoms, we can assemble a molecule on a finite set S as follows: choose a partition of S and on each block B, put an atom from $\mathcal{C}_{|B|}$. Formally, we define a *molecule with c components* on the set S to be a set $\{(B_1, \alpha_1), (B_2, \alpha_2), \ldots, (B_c, \alpha_c)\}$ of c pairs, where $B_1, B_2, \ldots, B_c$ is a partition of S and α_i is an atom on B_i.

The *exponential generating function* $f(\mathcal{C}; t)$ of the set $\mathcal{C}$ of atoms is defined by

$$f(\mathcal{C}; t) = \sum_{n=1}^{\infty} |\mathcal{C}_n| \frac{t^n}{n!}.$$

[3] There have been many attempts to formalize or axiomatize the combinatorics underlying the exponential formula. See, for example, Foata and Schützenberger (1970), Henle (1975), and Stanley (1978).

4.1.1. Theorem

(a) Another product formula: The coefficient of $t^n/n!$ in the power series

$$\frac{1}{c!}f(\mathcal{C};t)^c$$

is the number of molecules on a set of size n with c components.

(b) The exponential formula: Let a_{nc} be the number of molecules with c components on a set of size n and $a_n(x)$ be the polynomials defined by $a_0(x) = 1$, and if $n \geq 1$,

$$a_n(x) = \sum_{c=1}^{n} a_{nc}x^c.$$

Then

$$\sum_{n=0}^{\infty} a_n(x)\frac{t^n}{n!} = \exp(xf(\mathcal{C};t)).$$

Proof. Let $n_1, n_2, \ldots, n_c$ be positive integers such that $n_1 + n_2 + \cdots + n_c = n$. The number of ways to choose an ordered partition $B_1, B_2, \ldots, B_c$ of $\{1, 2, \ldots, n\}$ such that $|B_i| = n_i$ is $n!/n_1!n_2!\cdots n_c!$. On the block B_i, there are $|\mathcal{C}_{n_i}|$ ways of putting an atom. Hence, the total number of ordered c-tuples $((B_1, \alpha_1), (B_2, \alpha_2), \ldots, (B_c, \alpha_c))$ is

$$\sum_{n_1, n_2, \ldots, n_c} \frac{n!}{n_1!n_2!\cdots n_c!}|\mathcal{C}_{n_1}||\mathcal{C}_{n_2}|\cdots|\mathcal{C}_{n_c}|.$$

Dividing by $c!$, we obtain the number of molecules with c components. The product formula now follows from the multinomial formula. The exponential formula also follows formally from

$$\begin{aligned}\exp(xf(\mathcal{C};t)) &= \sum_{c=0}^{\infty}\frac{x^c f(\mathcal{C};t)^c}{c!}\\ &= \sum_{n=0}^{\infty}\left(\sum_{c=1}^{n} a_{nc}x^c\right)\frac{t^n}{n!}.\end{aligned}$$ □

The simplest case of the exponential formula is when there is exactly one atom for each finite set. Then a molecule is a partition and we have the generating functions

$$\sum_{n=1}^{\infty} S(n,c)\frac{t^n}{n!} = \frac{(e^x-1)^c}{c!},$$

$$\sum_{n=0}^{\infty}\left(\sum_{c=1}^{n} S(n,c)x^c\right)\frac{t^n}{n!} = e^{x(e^t-1)},$$

$$\sum_{n=0}^{\infty} B_n\frac{t^n}{n!} = e^{(e^t-1)},$$

where $S(n,c)$, a *Stirling number of the second kind*, is the number of partitions of a set of size n with c blocks and B_n, a *Bell number*, is the number of (all) partitions of a set of size n.[4]

Another classical application is in graphical enumeration.[5] Let $\mathcal{C}_n$ be the set of connected graphs on a vertex set of size n. Then a molecule is a graph on a finite vertex set. Graphs on the vertex set $\{1, 2, \ldots, n\}$ are in bijection with subsets of the set of 2-subsets of $\{1, 2, \ldots, n\}$. Hence, by the exponential formula,

$$\sum_{n=0}^{\infty} 2^{\binom{n}{2}}\frac{t^n}{n!} = \exp\left(\sum_{n=1}^{\infty} c_n\frac{t^n}{n!}\right),$$

where c_n is the number of connected graphs on the vertex set $\{1, 2, \ldots, n\}$.

The last application is also classical. Let $\mathcal{C}_n$ be the set of cyclic permutations on a set of size n. Then $|\mathcal{C}_n| = (n-1)!$ and

$$f(\mathcal{C};t) = \sum_{n=0}^{\infty} t^n/n = -\log(1-t).$$

The exponential formula yields

$$1 + \sum_{n=1}^{\infty}\left(\sum_{k=1}^{n}(-1)^k s(n,k)x^k\right)\frac{t^n}{n!} = \exp(-x\log(1-t)) = (1-t)^{-x},$$

where $s(n,k)$ is the number of permutations on $\{1, 2, \ldots, n\}$ with exactly k cycles in its cycle decomposition. The numbers $s(n,k)$ are the *(unsigned) Stirling numbers of the first kind*. Since

$$(1-t)^{-x} = \sum_{n=0}^{\infty}(-1)^n\binom{-x}{n}t^n = \sum_{n=0}^{\infty} x_{(n)}\frac{t^n}{n!},$$

4 Bell (1934a, b). 5 See Harary and Palmer (1973).

where $x_{(n)}$ is the *falling factorial* $x(x-1)(x-2)\cdots(x-n+1)$, we have

$$x(x-1)(x-2)\cdots(x-n+1) = x_{(n)} = \sum_{k=1}^{n}(-1)^k s(n,k)x^k.$$

Exercises

4.1.1. Let S be a finite set and x_a be a set of variables, one for each element a of A. Show that

$$\prod_{a:\, a\in S} x_a = \sum_{B:\, B\subseteq S}\ \prod_{b:\, b\in B}(x_b - 1).$$

For example, if $S = \{a, b\}$,

$$x_a x_b = (x_a - 1)(x_b - 1) + (x_a - 1) + (x_b - 1) + 1.$$

4.1.2. *H. Potter's q-binomial theorem.*[6]

Let x and y be variables and q a parameter or "quantum" satisfying the commutation relations $yx = qxy$, $qx = xq$, and $qy = yq$. Then

$$(x+y)^n = \sum_{k=0}^{n}\binom{n}{k}_q x^{n-k}y^k,$$

where

$$\binom{n}{k}_q = \frac{n!_q}{k!_q\,(n-k)!_q}$$

and

$$n!_q = (1+q)(1+q+q^2)\cdots(1+q+q^2+\cdots+q^{n-1}).$$

4.1.3. *Formal power series and Δ-matrices.*

Let $\mathbb{F}$ be a field, $\mathbb{F}[[t]]$ be the algebra of (formal) power series in the variable t with coefficients in t (under multiplication of power series), and $T_\infty(\mathbb{F})$ be the algebra of countably infinite square upper triangular matrices with entries in the field $\mathbb{F}$. A matrix $(a_{ij})_{0\le i,j<\infty}$ is a Δ-*matrix* if row k is obtained by shifting row 0 to the right by k entries and filling out the first k entries by zeros; that is,

[6] Potter (1950) and Schützenberger (1953).

(a_{ij}) has the form

$$\begin{pmatrix} a_0 & a_1 & a_2 & a_3 & \dots \\ 0 & a_0 & a_1 & a_2 & \dots \\ 0 & 0 & a_0 & a_1 & \\ \vdots & \vdots & & \ddots & \ddots \end{pmatrix}.$$

If $a_0, a_1, a_2, \dots$ is a sequence, let $\Delta(a_0, a_1, a_2, \dots)$ be the Δ-matrix with zeroth row equal to $a_0, a_1, a_2, \dots$.

(a) Show that the function $\Delta : \mathbb{F}[[t]] \to T_\infty(\mathbb{F})$ sending $\sum_{i=0}^{\infty} a_i t^i$ to $\Delta(a_0, a_1, a_2, \dots)$ is an injective $\mathbb{F}$-algebra homomorphism with image equal to the set $T_\infty^\Delta(\mathbb{F})$ of Δ-matrices. In particular, $T_\infty^\Delta(\mathbb{F})$ is a subalgebra and $\mathbb{F}[[t]]$ is isomorphic to $T_\infty^\Delta(\mathbb{F})$.

(b) Show that if $f(t)$ has an multiplicative inverse, then $\Delta(f(t)^{-1}) = \Delta(f(t))^{-1}$. Hence, $\sum_{i=0}^{\infty} a_i t^i$ has a multiplicative inverse if and only if $a_0 \neq 0$. Show that if $\sum_{i=0}^{\infty} b_i t^i$ is the inverse of $\sum_{i=0}^{\infty} a_i t^i$, then

$$b_n = \frac{(-1)^n}{a_0^{n+1}} \det \begin{vmatrix} a_1 & a_2 & a_3 & \dots & a_{n-1} & a_n \\ a_0 & a_1 & a_2 & \dots & a_{n-2} & a_{n-1} \\ 0 & a_0 & a_1 & \dots & a_{n-3} & a_{n-2} \\ 0 & 0 & a_0 & \dots & a_{n-4} & a_{n-3} \\ \vdots & \vdots & & \ddots & & \\ 0 & 0 & 0 & \dots & a_0 & a_1 \end{vmatrix}.$$

(The determinant on the right is an $n \times n$ determinant.)

(c) If $f(t)$ and $g(t)$ are power series and $f(t) = \sum_{n=0}^{\infty} a_n t^n$, the composition $f(g(t))$ is the power series defined by

$$f(g(t)) = \sum_{n=0}^{\infty} a_n g(t)^n.$$

Show that $f(t)$ has a compositional inverse $h(t)$; that is, there exists a power series $h(t)$ such that $f(h(t)) = t$, if and only if the constant term $f(0)$ is zero and the coefficient $f'(0)$ of t is nonzero.

4.1.4. *Lambert series.*[7]

Let $\ell(z) = \sum_{m=1}^{\infty} l_i z^i$. A *generalized Lambert series based on* $\ell(z)$ is a power series of the form

$$\sum_{n=1}^{\infty} a_n \ell(z^n).$$

When $\ell(z) = z/(1-z)$, we retrieve the classical Lambert series:

$$\sum_{n=1}^{\infty} a_n \frac{z^n}{1-z^n}.$$

Show that

$$\sum_{n=1}^{\infty} a_n \ell(z^n) = \sum_{s=1}^{\infty} b_s z^s$$

if and only if for all s, $b_s = \sum_{n:n|s} a_n l_{s/n}$. Thus, for classical Lambert series, the three conditions are equivalent:

$$\sum_{n=1}^{\infty} a_n \frac{z^n}{1-z^n} = \sum_{s=1}^{\infty} b_s z^s,$$

$$b_s = \sum_{n:n|s} a_n \quad \text{for all } s,$$

$$a_n = \sum_{s:s|n} \mu\left(\frac{n}{s}\right) b_s \quad \text{for all } n.$$

In particular,

$$z = \sum_{n=1}^{\infty} \mu(n) \frac{z^n}{1-z^n}.$$

4.1.5. *Cyclotomic products.*[8]

A *cyclotomic product* is a formal product of the form

$$C(z) = \prod_{n=1}^{\infty} (1-z^n)^{-a_n}$$

(a) Show that

$$C(z) = \exp\left(\sum_{s=1}^{\infty} b_s z^s\right),$$

[7] Hardy and Wright (1960). [8] Metropolis and Rota (1983).

where

$$b_s = \frac{1}{s} \sum_{n:\, n|s} n a_n \quad \text{and} \quad a_n = \frac{1}{n} \sum_{s:\, s|n} \mu\left(\frac{n}{s}\right) s b_s.$$

In particular, show that

$$\exp(z) = \prod_{n=1}^{\infty} (1 - z^n)^{-\mu(n)/n}$$

and

$$\frac{1}{1-\alpha z} = \prod_{n=1}^{\infty} (1 - z^n)^{-M(\alpha;n)}, \quad \text{where } M(\alpha; n) = \frac{1}{n} \sum_{d:\, d\,|\,n} \mu\left(\frac{n}{d}\right) \alpha^d.$$

(b) Develop a combinatorial theory of cyclotomic products.

4.2 Elementary Umbral Calculus

The umbral or symbolic method began as a heuristic device for deriving formulas or identities. A typical example is the following *inverse relation.*

4.2.1. Proposition. Let $(a_n)_{0 \le n < \infty}$ and $(b_n)_{0 \le n < \infty}$ be sequences. Then

$$b_n = \sum_{i=0}^{n} \binom{n}{i} a_i \text{ for all } n \quad \Leftrightarrow \quad a_n = \sum_{i=0}^{n} (-1)^{n-i} \binom{n}{i} b_i \text{ for all } n.$$

The inverse relation is motivated and, if one is optimistic, is proved by the following argument. Raising subscripts, we obtain, for all n,

$$b^n = \sum_{i=0}^{n} \binom{n}{i} a^i = (a+1)^n.$$

Hence, for all n,

$$a^n = (b-1)^n = \sum_{i=0}^{n} (-1)^{n-i} \binom{n}{i} b^i.$$

Lowering exponents, we obtain the inverse relation.

Using linear functionals on polynomial algebras, we will put such calculations on a rigorous basis.[9] For our example, we work over the ring $\mathbb{F}[\alpha]$ of

[9] The framework proposed here is due to Rota. Rota published many papers on the subject. The first is Rota (1964) and the last Rota and Shen (2000). An earlier attempt to make the umbral method rigorous, by a set of axioms, can be found in Bell (1940).

polynomials over a field $\mathbb{F}$ in the variable α. The variable α is thought of as a "shadow variable" or *umbra*. We consider $\mathbb{F}[\alpha]$ as a vector space and define the linear functional $U : \mathbb{F}[\alpha] \to \mathbb{F}$ by

$$U(\alpha^n) = a_n,$$

and extended to all of $\mathbb{F}[\alpha]$ by linearity. Then

$$\begin{aligned} b_n &= \sum_{i=0}^{n} \binom{n}{i} U(\alpha^i) \\ &= U\left(\sum_{i=0}^{n} \binom{n}{i} \alpha^i\right) = U((\alpha+1)^n). \end{aligned}$$

On the other hand,

$$\begin{aligned} a_n = U(\alpha^n) &= U((\alpha+1)-1)^n) \\ &= U\left(\sum_{i=0}^{n} (-1)^{n-i} \binom{n}{i} (\alpha+1)^i\right) \\ &= \sum_{i=0}^{n} (-1)^{n-i} \binom{n}{i} U((\alpha+1)^i) = \sum_{i=0}^{n} (-1)^{n-i} \binom{n}{i} b_i. \end{aligned}$$

This gives a rigorous proof of Proposition 4.1.1.

A somewhat deeper application of the umbral calculus is a rederivation of the exponential generating function for the Bell numbers B_n.

4.2.2. Theorem. Let t be another variable. Then

$$\sum_{n=0}^{\infty} \frac{B_n t^n}{n!} = e^{e^t - 1}.$$

Proof. We begin with a combinatorial observation. Let X be a finite set of size x. If $f : S \to X$ is a function, then those inverse images $f^{-1}(x)$ that are nonempty give a partition of S. This partition is the coimage of f defined in Section 1.4. If π is a partition of S into $c(\pi)$ blocks, then the number of functions with coimage equal to π equals $x_{(c(\pi))}$, where $x_{(k)}$ is a falling factorial. Since the total number of functions $S \to X$ is x^n, we have the combinatorial version of *Stirling's identity*

$$\sum_{\pi} x_{(c(\pi))} = x^n,$$

the sum ranging over all the partitions π of S.

Let $\mathbb{Q}[\beta]$ be the ring of polynomials with rational coefficients. Then as the set $\{\beta_{(k)}: 0 \le k < \infty\}$ is a basis for $\mathbb{Q}[\beta]$, there is a linear functional

$L : \mathbb{Q}[\beta] \to \mathbb{Q}$ such that

$$L(1) = 1, \quad L(\beta_{(i)}) = 1 \text{ for } i \geq 1.$$

By Stirling's identity,

$$B_n = \sum_{\pi} 1 = \sum_{\pi} L(\beta_{(c(\pi))}) = L(\beta^n).$$

4.2.3. Lemma.

$$B_{n+1} = \sum_{i=0}^{n} \binom{n}{i} B_{n-i},$$

or $L(\beta^{n+1}) = L((\beta+1)^n)$.

Proof. We have, for every nonnegative integer k, $\beta(\beta-1)_{(k)} = \beta_{(k)}$ and $L(\beta(\beta-1)_{(k)}) = L(\beta_{(k)})$. Since $\{\beta_{(k)}: 0 \leq k < \infty\}$ is a basis, we also have, by linearity,

$$L(\beta p(\beta-1)) = L(p(\beta))$$

for every polynomial $p(\beta)$. In particular,

$$L(\beta\beta^n) = L(\beta^{n+1}) = L((\beta+1)^n).$$

This is the required formula in umbral form. □

Next, write

$$\sum_{n=0}^{\infty} \frac{g_n t^n}{n!} = e^{e^t-1}$$

and let $M : \mathbb{Q}[\beta] \to \mathbb{Q}$ be the linear functional defined by $M(\beta^n) = g_n$. We will show that the linear functionals L and M are identical. We begin by observing that

$$e^{e^t-1} = \sum_{n=0}^{\infty} \frac{M(\beta^n)t^n}{n!} = M(e^{\beta t}).$$

Differentiating formally relative to t, we obtain

$$e^t e^{e^t-1} = M(\beta e^{\beta t});$$

that is,

$$e^t M(e^{\beta t}) = M(e^{(\beta+1)t}) = M(\beta e^{\beta t}).$$

Expanding as Taylor series,

$$\sum_{n=0}^{\infty} \frac{M((\beta+1)^n)t^n}{n!} = \sum_{n=0}^{\infty} \frac{M(\beta^{n+1})t^n}{n!}.$$

We conclude that $M((\beta+1)^n) = M(\beta^{n+1})$ and by linearity, $M(p(\beta+1)) = M(\beta p(\beta))$. Hence, $M = L$ and $g_n = B_n$. □

4.2.4. Dobinski's formula.

$$\begin{aligned} B_{n+1} &= \frac{1}{e}\sum_{j=0}^{\infty} \frac{(j+1)^n}{j!} \\ &= \frac{1}{e}\left(1^n + \frac{2^n}{1!} + \frac{3^n}{2!} + \frac{4^n}{3!} + \cdots\right). \end{aligned}$$

Proof. If n and k are nonnegative integers, then $k_{(n)} = 0$ if $k < n$ and $k_{(n)}/k! = 1/(k-n)!$. Thus,

$$L(\beta_{(n)}) = 1 = \frac{1}{e}\sum_{k=0}^{\infty} \frac{k_{(n)}}{k!},$$

and hence, since $\{\beta_{(n)}\}$ is a basis,

$$L(p(\beta)) = \frac{1}{e}\sum_{k=0}^{\infty} \frac{p(k)}{k!}.$$

Dobinski's formula now follows by putting $p(\beta) = \beta^{n+1}$. □

Exercises

4.2.1. Give a combinatorial proof of Lemma 4.2.3.

4.2.2. Find an umbral calculus proof of the exponential formula.

4.2.3. (Research problem) A random variable X defines a sequence $E(X^n)$, its sequence of moments, while an umbra α represents a sequence $U(\alpha^n)$. Explain or develop this analogy.

4.3 Polynomial Sequences of Binomial Type

In the next three sections, we give a short introduction to polynomial sequences of binomial type. We will use the finite operator calculus developed in

Foundations III and *VIII*. Let $\mathbb{F}$ be a field of characteristic zero and $\mathbb{F}[x]$ be the $\mathbb{F}$-algebra of polynomials with coefficients in $\mathbb{F}$ in the variable x.

A *polynomial sequence* $(p_n(x))_{n=0}^{\infty}$ in $\mathbb{F}[x]$ is a sequence of polynomials such that $p_n(x)$ has degree (exactly) n. A polynomial sequence $(p_n(x))$ has *binomial type* if it satisfies the *(generalized) binomial identities*: for all n,

$$p_n(x+y) = \sum_{k=0}^{n} \binom{n}{k} p_k(x) p_{n-k}(y).$$

The binomial identities are assumed to hold as polynomial identities in the variables x and y.

An *operator* on polynomials is a linear transformation defined from $\mathbb{F}[x]$ to $\mathbb{F}[x]$. Since a polynomial sequence $(p_n(x))$ gives a basis of $\mathbb{F}[x]$, an operator Q is determined by the (labeled) values $Qp_n(x)$. The *identity operator* I sends every polynomial to itself and if T is an operator, $T^0 = I$ by convention. If a is an element in $\mathbb{F}$, then E^a is the *shift operator* sending a polynomial $p(x)$ to $p(x+a)$. We write E^1 simply as E. We will also use the linear functional $\epsilon(a) : \mathbb{F}[x] \to \mathbb{F},\ p(x) \mapsto p(a)$ that *evaluates* $p(x)$ at a, as well as the notation $[p(x)]_{x=a} = p(a)$.

4.3.1. Boole's formula.[10]

$$E^a = e^{aD} = \sum_{k=0}^{\infty} \frac{a^k}{k!} D^k.$$

Proof. Since x^n is a basis for $\mathbb{F}[x]$, Boole's formula is a consequence of the equations: for all nonnegative integers n,

$$E^a\, x^n = (x+a)^n = \sum_{k=0}^{n} \frac{a^k}{k!} n_{(k)} x^{n-k} = e^{aD} x^n. \qquad \square$$

The proof of Boole's formula used the following fact: if $(a_n)_{n=0}^{\infty}$ is a sequence of elements in $\mathbb{F}$ and T is any operator such that for all polynomials $p(x)$, $Tp(x)$ has degree strictly less than the degree of $p(x)$, then the formal sum

$$\sum_{n=0}^{\infty} a_n T^n$$

is a well-defined operator. The reason is that when the infinite sum is applied to a given polynomial, only a finite number of terms are nonzero.

[10] Boole (1872, p. 18). Boole's formula was known much earlier.

An operator Q is *shift-invariant* if $QE^a = E^a Q$ for any element a in $\mathbb{F}$. The most familiar shift-invariant operator is the *differentiation operator* D, defined by $Dx^n = nx^{n-1}$. Operators of the form $\sum_{n=0}^{\infty} a_n D^n$ are also shift-invariant.

An operator $Q : \mathbb{F}[x] \to \mathbb{F}[x]$ is a *delta operator* if Q is shift-invariant and $Qx = c$, where c is a nonzero constant.

4.3.2. Lemma. Let Q be a delta operator and $p(x)$ a polynomial of degree n. Then $Qp(x)$ is a polynomial of degree (exactly) $n-1$.

Proof. Since $\{x^n : 0 \le n < \infty\}$ is a basis for $\mathbb{F}[x]$, it suffices to prove the lemma for x^n. We begin with the constant polynomial 1. By linearity, $QEx = Q(x+1) = Qx + Q1 = c + Q1$. By shift-invariance, $QEx = EQx = Ec = c$. Hence, $c + Q1 = c$ or $Q1 = 0$.

Next suppose that $n \ge 1$. Let $r(x) = Qx^n$. By the binomial theorem and shift-invariance,

$$\begin{aligned}\sum_{k=0}^{n} \binom{n}{k} a^k Qx^{n-k} &= Q(x+a)^n \\ &= QE^a x^n \\ &= E^a Qx^n = r(x+a).\end{aligned}$$

Setting $x = 0$, we have

$$r(a) = \sum_{k=0}^{n} \binom{n}{k} a^k [Qx^{n-k}]_{x=0}.$$

The coefficient of a^n is $[Qx^0]_{x=0}$. Since $x^0 = 1$, the coefficient of a^n is zero by the first part. Further, the coefficient of a^{n-1} is $n[Qx]_{x=0}$. This equals nc and is nonzero. Hence, $r(a)$ is a polynomial in a of degree exactly $n-1$. □

A polynomial sequence $(p_n(x))$ is a *basic sequence* for the delta operator Q if $p_0(x) = 1$, $p_n(0) = 0$ whenever $n \ge 1$, and $Qp_n(x) = p_{n-1}(x)$.

Polynomial sequences of binomial type can be described in many different equivalent ways.

4.3.3. Theorem. Let $(p_n(x))$ be a polynomial sequence and $Q : \mathbb{F}[x] \to \mathbb{F}[x]$ be the operator defined by $Qp_n(x) = np_{n-1}(x)$. Then the following are equivalent:

BT1. The sequence $(p_n(x))$ is of binomial type.
BT2. The operator Q is a delta operator.

BT3. The operator Q has a formal expansion:

$$Q = q(D) = \sum_{n=1}^{\infty} a_n \frac{D^n}{n!},$$

where D is differentiation and $q(t)$ is a power series with zero constant term and $a_1 \neq 0$.

BT4. The following two equivalent formal power series relations hold:

$$e^{xt} = \sum_{n=0}^{\infty} p_n(x) \frac{q(t)^n}{n!}, \qquad \sum_{n=0}^{\infty} p_n(x) \frac{t^n}{n!} = \exp(x f(t)),$$

where $f(t)$ and $q(t)$ are power series with zero constant term that are compositional inverses of each other in the sense that $f(q(t)) = q(f(t)) = t$.

The power series $f(t)$ is the *indicator* of the polynomial sequence $(p_n(x))$.

We begin the proof of Theorem 4.3.3 by showing that BT1 is equivalent to BT2. Suppose that $(p_n(x))$ is the basic sequence for the delta operator Q. Then

$$Q^k p_n(x) = n_{(k)} p_{n-k}(x),$$

where $n_{(k)} = n(n-1)\cdots(n-k+1)$. Hence, $\epsilon(0)Q^n p_n(x) = n!$ and $\epsilon(0)$ $Q^k p_n(x) = 0$ if $k < n$. Thus,

$$p_n(x) = \sum_{k=0}^{n} \frac{p_k(x)}{k!} \epsilon(0) Q^k p_n(x).$$

Since $\{p_n(x)\}$ is a basis of $\mathbb{F}[x]$ and $\epsilon(0)Q^k$ is a linear functional,

$$p(x) = \sum_{k=0}^{n} \frac{p_k(x)}{k!} \epsilon(0) Q^k p(x)$$

for every polynomial $p(x)$. In particular, if $p(x) = p_n(x+y)$, then

$$p_n(x+y) = \sum_{k=0}^{n} \frac{p_k(x)}{k!} \epsilon(0) Q^k p(x).$$

Since

$$\begin{aligned} \epsilon(0)Q^k p(x+y) &= \epsilon(0)Q^k E^y p(x) \\ &= \epsilon(0)E^y Q^k p(x) \\ &= \epsilon(y) n_{(k)} p_{n-k}(x) \\ &= n_{(k)} p_{n-k}(y), \end{aligned}$$

we conclude that

$$p_n(x+y) = \sum_{k=0}^{n} \frac{n_{(k)}}{k!} p_k(x) p_{n-k}(y).$$

Now suppose that the polynomial sequence $(p_n(x))$ has binomial type. Setting $y = 0$ in the generalized binomial identities, we obtain

$$p_n(x) = p_n(x)p_0(0) + np_{n-1}(x)p_1(0) + \binom{n}{2} p_{n-2}(x)p_2(0) + \cdots .$$

Since we are hypothesizing that $p_k(x)$ has degree exactly k, it follows that $p_0(x) = p_0(0) = 1$ and $p_k(0) = 0$ if $k \geq 1$.

There is only one choice for the delta operator: Q must send $p_0(x)$ to 0 and $p_n(x)$ to $np_{n-1}(x)$ if $n \geq 1$. Since $p_1(x) = x$, $Qx = Qp_1(x) = p_0(x) = 1$, it remains to show that Q is shift-invariant. To do this, write the binomial identity in the form

$$p_n(x+y) = \sum_{k=0}^{n} \frac{p_k(x)}{k!} Q^k p_n(y).$$

Since $\{p_n(x)\}$ is a basis, we obtain, by linearity,

$$p(x+y) = \sum_{k=0}^{n} \frac{p_k(x)}{k!} Q^k p(y).$$

Replacing p by Qp and exchanging x and y, we have

$$(Qp)(x+y) = \sum_{k=0}^{n} \frac{p_k(y)}{k!} Q^{k+1} p(x).$$

Further,

$$\begin{aligned} E^y Qp(x) &= (Qp)(x+y) \\ &= Q\left[\sum_{k=0}^{n} \frac{p_k(y)}{k!} Q^k p(x)\right] \\ &= Qp(x+y) = QE^y p(x). \end{aligned}$$

We conclude that $E^y Q = QE^y$; that is, Q is shift-invariant. We have now proved that BT1 and BT2 are equivalent.

To prove BT2 implies BT3, we need the following result.

4.3.4. The first expansion theorem. Let T be a shift-invariant operator and Q be a delta operator with basic sequence $(p_n(x))$. Then

$$T = \sum_{k=0}^{\infty} \frac{[Tp_k(x)]_{x=0}}{k!} Q^k.$$

Proof. Since the sequence $(p_n(x))$ has binomial type,

$$p_n(x+y) = \sum_{k=0}^{n} \frac{p_k(y)}{k!} Q^k p_n(x).$$

Regard both sides as polynomials in y and apply T to obtain

$$Tp_n(x+y) = \sum_{k=0}^{n} \frac{Tp_k(y)}{k!} Q^k p_n(x).$$

Since $\{p_n(x)\}$ is a basis, we have, by linearity,

$$Tp(x+y) = \sum_{k=0}^{n} \frac{Tp_k(y)}{k!} Q^k p(x).$$

Setting $y = 0$, we conclude that

$$Tp(x) = \sum_{k=0}^{n} \frac{[Tp_k(y)]_{y=0}}{k!} Q^k p(x). \qquad \square$$

Now assume BT2. Then since Q is shift-invariant, we can expand Q in terms of the differentiation operator D with basic sequence x^n. We conclude that $Q = q(D)$, where

$$q(t) = \sum_{n=0}^{\infty} [Qx^n]_{x=0} \frac{t^n}{n!}.$$

Since $Q1 = 0$ and $Qx = c$, where c is a nonzero constant, $q(t)$ is the form described in BT3.

To prove BT3 implies BT4, we need an isomorphism theorem. The operators on the polynomial algebra $\mathbb{F}[x]$ form an $\mathbb{F}$-algebra under composition. The shift-invariant operators form a commutative subalgebra $\mathcal{S}$ of this algebra. This is easy to prove formally from the definition and also follows from the next result. Let $\mathbb{F}[[t]]$ be the $\mathbb{F}$-algebra of formal power series with coefficients in $\mathbb{F}$ in the variable t.

4.3.5. The isomorphism theorem. Let Q be a delta operator with basic sequence $(q_n(x))$. Then the function $\mathcal{S} \to \mathbb{F}[[t]]$ given by

$$T \mapsto \sum_{k=0}^{\infty} [Tq_k(x)]_{x=0} \frac{t^k}{k!}$$

is an $\mathbb{F}$-algebra isomorphism from $\mathcal{S}$ onto $\mathbb{F}[[t]]$.

Proof. The function is $\mathbb{F}$-linear because $(T+S)q_k(x) = Tq_k(x) + Sq_k(x)$ by definition. It is injective by the first expansion theorem and surjective because operators of the form $\sum_{n=0}^{\infty} a_n Q^n$ are shift-invariant.

It remains to check that it sends composition to multiplication. Let S and T be shift-invariant operators and

$$f(t) = \sum_{k=0}^{\infty} [Sq_k(x)]_{x=0} \frac{t^k}{k!}, \qquad g(t) = \sum_{n=0}^{\infty} [Tq_n(x)]_{x=0} \frac{t^n}{n!}.$$

We will show that

$$f(t)g(t) = \sum_{k=0}^{\infty} [TSq_k(x)]_{x=0} \frac{t^k}{k!}$$

by looking at the coefficients. Specifically, we use the fact that

$$[TSq_r(x)]_{x=0} = \left[\sum_{k,n=1}^{\infty} \frac{[Tq_k(x)]_{x=0}[Sq_n(x)]_{x=0}}{k!n!} Q^{k+n} q_r(x) \right]_{x=0}.$$

Since $q_n(0) = 0$ if $n > 0$ and $q_0(0) = 1$, a summand is nonzero only if $n = r - k$. Further, if $r = k + n$, then $Q^{k+n} q_r(x) = r! = [Q^{k+n} q_r(x)]_{x=0}$. Hence,

$$\begin{aligned} [TSq_r(x)]_{x=0} &= \sum_{k=1}^{\infty} \frac{[Tq_k(x)]_{x=0}[Sq_{r-k}(x)]_{x=0}}{k!(r-k)!} r! \\ &= \sum_{k=1}^{\infty} \binom{r}{k} [Tq_k(x)]_{x=0}[Sq_{r-k}(x)]_{x=0}. \end{aligned}$$

□

4.3.6. Corollary. A shift-invariant operator T has a compositional inverse if and only if $T1 \neq 0$.

4.3.7. Proposition. Let Q be a delta operator with basic sequence $(q_n(x))$. If $Q = q(D)$, then

$$\sum_{n=0}^{\infty} q_n(x) \frac{t^n}{n!} = \exp(xf(t)),$$

where $f(t)$ is the compositional inverse of $q(t)$.

Proof. Expand the shift operator E^a in terms of Q, obtaining

$$E^a = \sum_{n=0}^{\infty} [q_n(x+a)]_{x=0} \frac{Q^n}{n!} = \sum_{n=0}^{\infty} q_n(a) \frac{Q^n}{n!}.$$

By the isomorphism theorem and Boole's formula (4.3.1) that $E^a = e^{aD}$, we have the power series identity

$$e^{au} = \sum_{n=0}^{\infty} q_n(a) \frac{q(u)^n}{n!}.$$

Making the changes of variables $a = x$, $t = q(u)$, and $u = q^{-1}(t) = f(t)$, we obtain the required formula. □

Finally, we show that BT4 implies BT1. This follows from the formal power series calculation

$$\begin{aligned}
\sum_{n=0}^{\infty} p_n(x+y) \frac{t^n}{n!} &= \exp((x+y)f(t)) \\
&= \exp(xf(t)) \exp(yf(t)) \\
&= \sum_{n=0}^{\infty} \left(\sum_{k=0}^{n} \binom{n}{k} p_k(x) p_{n-k}(x) \right) \frac{t^n}{n!}.
\end{aligned}$$

This completes the proof of Theorem 4.3.2.

We conclude the theoretical part of this section with a discussion of closed forms. A "closed form" is a formula that appears to be simple (and may in fact be simple). To obtain closed forms, we need a lemma, a definition, and then another lemma.

4.3.8. Lemma. An operator Q is a delta operator if and only if $Q = DP$, where P is an invertible shift-invariant operator.

Proof. By BT3, $Q = q(D)$. Writing $q(t) = tf(t)$ and setting $P = f(D)$, we obtain $Q = DP$, where P, being a power series in D with nonzero constant term, is shift-invariant and invertible. Conversely, if $Q = DP$, then Q is a delta operator by BT3. □

If $T : \mathbb{F}[x] \to \mathbb{F}[x]$ is an operator, then its *Pincherle derivative*[11] T' is the operator

$$T\mathbf{x} - \mathbf{x}T,$$

where $\mathbf{x}$ is the *multiplication-by-x operator* $p(x) \mapsto xp(x)$. The Pincherle derivative is a derivation; that is, $(TS)' = T'S + TS'$.

4.3.9. Lemma. If Q is a delta operator with basic sequence $(p_n(x))$, $Q = q(D)$, and Q' is the Pincherle derivative of Q, then $Q' = q'(D)$, where $q'(t)$ is the derivative of the formal power series $q(t)$. In particular, Q' is an invertible shift-invariant operator.

Proof. Use the fact that

$$[(Q\mathbf{x} - \mathbf{x}Q)x^n]_{x=0} = [Qx^{n+1} - xQx^n]_{x=0} = [Qx^{n+1}]_{x=0}. \qquad \square$$

4.3.10. Proposition. Let Q be a delta operator, $Q = DP$, $(p_n(x))$ be the basic sequence of Q, and $n \geq 1$. Then,

(a) $p_n(x) = Q'P^{-(n+1)}x^n$,
(b) $p_n(x) = P^{-n}x^n - (P^{-n})'x^{n-1}$,
(c) $p_n(x) = xP^{-n}x^{n-1}$,
(d) $p_n(x) = x(Q')^{-1}p_{n-1}(x)$.

Proof. We begin by showing that the right sides of the first three formulas are equal. We begin with

$$\begin{aligned} Q'P^{-(n+1)} &= (DP)'P^{-(n+1)} = (D'P + DP')P^{-(n+1)} \\ &= P^{-n} + P'P^{-(n+1)}D = P^{-n} - \frac{1}{n}(P^{-n})'D, \end{aligned}$$

where we have used $D' = I$ and the chain rule for derivations. Applied to x^n, we obtain

$$Q'P^{-(n+1)}x^n = P^{-n}x^n - \frac{1}{n}(P^{-n})'Dx^n = P^{-n}x^n - (P^{-n})'x^{n-1}.$$

We also have

$$P^{-n}x^n - (P^{-n})'Dx^{n-1} = P^{-n}x^n - (P^{-n}\mathbf{x} - \mathbf{x}P^{-n})x^{n-1} = xP^{-n}x^{n-1}.$$

This shows that the first three formulas are equivalent.

[11] Pincherle (1901, 1933). The work of Pincherle was ahead of its time. However, as Rota put it in *Foundations VIII*, "although Pincherle was fully aware of the abstract possibilities of the concept of operator, he was ignorant of the nitty-gritty of numerical analysis, where he would have found fertile ground for his ideas."

Next, let $\tilde{p}_n(x) = Q'P^{-(n+1)}x^n$. First, as $Q = DP$ and shift-invariant operators commute,

$$Q\tilde{p}_n(x) = DPQ'P^{-(n+1)}x^n = Q'P^{-n}Dx^n = n(Q'P^{-n})x^{n-1} = n\tilde{p}_{n-1}(x).$$

Second, by the third formula, $\tilde{p}_n(x) = xP^{-n}x^{n-1}$, and hence if $n \geq 1$, $\tilde{p}_n(0) = 0$. Thus, $\tilde{p}_n(x)$ is the basic sequence for Q, and hence $\tilde{p}_n(x) = p_n(x)$. We have now proved the first three formulas.

To prove the fourth formula, we invert the first formula, obtaining

$$x^n = (Q')^{-1}P^{n+1}p_n(x).$$

Substituting this into the third formula, we obtain

$$p_n(x) = xP^{-n}(Q')^{-1}P^n p_{n-1}(x) = x(Q')^{-1}p_{n-1}(x).$$

This proves the fourth formula. □

We will show how the theory developed so far works with two classical polynomial sequences. We begin with the delta operator E^aD. By Lemma 4.3.10(c), the degree-n polynomial $A_n(x;a)$ can be calculated as follows:

$$A_n(x;a) = xE^{-na}x^{n-1} = x(x-na)^{n-1}.$$

The polynomials $A_n(x;a)$ are the *Abel polynomials*. The generalized binomial identity yields *Abel's binomial identity:*

$$(x+y)(x+y-na)^{n-1} = \sum_{k=0}^{n} \binom{n}{k} x(x-ka)^{k-1} y(y-(n-k)a)^{n-k-1}.$$

BT4 yields another identity of Abel:

$$e^{xt} = \sum_{k=0}^{\infty} \frac{x(x-ka)^{k-1}}{k!} t^k e^{kat}.$$

Next, we consider the *Laguerre operator* $L : \mathbb{F}[x] \to \mathbb{F}[x]$ defined by

$$Lp(x) = -\int_0^{\infty} e^{-u} p'(x+u)du.$$

The Laguerre operator seems to be analytic, but it can be rewritten algebraically using integration by parts. To see this, note that

$$\int_0^{\infty} e^{-u} f(x+u)du = [e^{-u}f(x+u)]_{u=0}^{u=\infty} + \int_0^{\infty} e^{-u} f'(x+u)du.$$

Hence, if f is a polynomial, we can iterate the integration-by-parts formula to obtain

$$-\int_0^\infty e^{-u} f(x+u)du = -[f(x) + Df(x) + D^2 f(x) + \cdots].$$

Thus,

$$L = \frac{D}{D-I}.$$

Hence, by Proposition 4.3.10(c), the degree-n polynomial $L_n(x)$ in the basic sequence of L can be calculated as follows:

$$\begin{aligned} L_n(x) = x(D-I)^n x^{n-1} &= \sum_{k=1}^{n} (-1)^{n-k} \binom{n}{k} (n-1)(n-2)\cdots(n-k)x^{n-k}, \\ &= \sum_{k=1}^{n} \frac{n!}{k!} \binom{n-1}{k-1} (-x)^k, \end{aligned}$$

where, in the last step, we change the index of summation from k to $n-k$. It is easy to check formally that $e^x De^{-x} = D - I$; thus, we have *Rodrigues' formula*,[12]

$$L_n(x) = xe^x D^n e^{-x} x^{n-1}.$$

By BT4 and the fact that $q(t) = t/(t-1)$ is its own compositional inverse, we have the following identities:

$$\exp(xt) = \sum_{n=0}^{\infty} L_n(x) \frac{t^n}{n!(t-1)^n}, \quad \exp\left(\frac{xt}{t-1}\right) = \sum_{n=0}^{\infty} L_n(x) \frac{t^n}{n!}.$$

The Abel and Laguerre polynomials have similar combinatorial interpretations. Let S and X be disjoint finite sets, with $|S| = n$ and $|X| = x$. A *reluctant function* $f : S \Rightarrow X$ from S to X is a function $f : S \to S \cup X$ satisfying the following property: for each a in S, there is a positive integer k (depending on a) such that the image $f^k(a)$, when f is applied k times, is in X. Intuitively, a reluctant function drags every element a in S to X. The *graph* of a reluctant function $f : S \Rightarrow X$ is the directed graph on the vertex set $S \cup X$ with a directed edge (a, b) if $f(a) = b$. The condition on reluctant functions implies that the graph of a reluctant function is a disjoint union or *forest* of x directed trees, with all edges in each tree directed to the unique vertex, the *root*, in X. Conversely, a forest of rooted trees, with a total of $|S|$ nonroot vertices labeled by S and roots labeled by a subset of X, determines a reluctant function.

[12] Rodrigues derived the analogous formula for Legendre polynomials in his 1816 thesis. See Altmann and Ortiz (2005).

4.3.11. Theorem. Let t_{nk} be the number of forests of rooted labeled trees with n vertices and k connected components. Then

$$\sum_{k=1}^{n} t_{nk}x^k = x(x+n)^{n-1}.$$

In particular, the number t_{n1} of rooted labeled trees on n vertices is n^{n-1}.

To prove this, we use the following lemma.

4.3.12. Lemma. Let P be an invertible shift-invariant operator and $(p_n(x))$ a basic sequence satisfying

$$[x^{-1}p_n(x)]_{x=0} = n[P^{-1}p_{n-1}(x)]_{x=0}$$

for $n \geq 1$. Then $(p_n(x))$ is the basic sequence for the delta operator DP.

Proof. Let Q be the delta operator for $(p_n(x))$. If $p(0) = 0$, $[x^{-1}p(x)]_{x=0} = [Dp(x)]_{x=0}$, and hence,

$$[Dp(x)]_{x=0} = [P^{-1}Qp(x)]_{x=0}. \tag{Z}$$

In addition, when $p(x)$ is a constant, both sides of (Z) equal zero. Hence, by linearity, (Z) holds for all polynomials $p(x)$. Setting $p(x) = q(x+a)$ and using shift-invariance, we have

$$Dq(a) = [P^{-1}QE^a q(x)]_{x=0} = [E^a P^{-1}Qq(x)]_{x=0} = P^{-1}Qq(a).$$

This relation holds for all polynomials $q(x)$ and all constants a. We conclude that $D = P^{-1}Q$. □

Returning to the proof of Theorem 4.3.11, let $\tilde{A}_n(x) = \sum_{k=1}^{n} t_{nk}x^k$. Then,

$$t_{n1} = n\sum_{k=1}^{n} t_{n-1,k} = n\tilde{A}_{n-1}(1).$$

To see this, observe that we can construct a rooted tree on the vertex set $\{1, 2, \ldots, n\}$ by taking a forest of rooted trees on the vertex set $\{1, 2, \ldots, n-1\}$ and joining all the roots to a new root by edges. The new root can be labeled in n ways by the new label n or by transferring one of the $n-1$ old labels from an old vertex to the new root and relabeling that old vertex by the new label n. This construction is reversible. Hence,

$$[x^{-1}\tilde{A}_n(x)]_{x=0} = t_{n1} = n\tilde{A}_{n-1}(1) = n[E\tilde{A}_{n-1}(x)]_{x=0}.$$

By Lemma 4.3.12, we conclude that $\tilde{A}_n(x)$ is the Abel polynomial $A_n(x;-1)$. This completes the proof of Theorem 4.3.11.

Reluctant functions also offer a counting interpretation for Laguerre polynomials. If $f : S \Rightarrow X$ is a reluctant function, let the *(rooted) forest* of f be the rooted forest obtained when the vertices in X (and edges incident on them) are deleted from the graph of *f*. Let $\mathcal{L}(S, X)$ be the set of reluctant functions $f : S \Rightarrow X$ such that the forest of f is a disjoint union of directed paths. If $|S| = n$ and $|X| = x$, then

$$|\mathcal{L}(S, X)| = L_n(-x), \tag{LG}$$

where $L_n(x)$ is a Laguerre polynomial. For a proof, see Exercise 4.3.4. We end with the remark that reluctant functions were the motivation for species.[13]

Exercises

4.3.1. Show that if $p(x)$ is a polynomial with nonzero constant term and $g(x)$ is any polynomial, then there exists a unique polynomial $y(x)$ such that

$$p(D)y = g(x).$$

4.3.2. Find an explicit formula for the coefficients a_n in the power series relation[14]

$$\sin xt = \sum_{n=1}^{\infty} \frac{a_n}{n!}[t(1-t)]^n.$$

4.3.3. *Rising and falling factorials.*

(a) Show that the degree-n term in the basic sequence of the forward difference operator Δ, defined equivalently by $\Delta = E - I$ or $\Delta : p(x) \mapsto p(x+1) - p(x)$, is the falling factorial $x_{(n)}$, where $x_{(n)} = x(x-1)(x-2)\cdots(x-n+1)$. Hence, deduce

$$(x+y)_{(n)} = \sum_{k=0}^{n} \binom{n}{k} x_{(k)} y_{(n-k)}.$$

(b) Show that the degree-n term in the basic sequence of the backward difference operator ∇, defined equivalently by $\nabla = I - E^{-1}$ or $\nabla : p(x) \mapsto p(x) - p(x-1)$, is the rising factorial $x^{(n)}$, where $x^{(n)} = x(x+1)(x+2)\cdots(x+n-1)$.

[13] For species, see Bergeron et al. (1998).
[14] This exercise is attributed to I. Schur in Carlitz (1966).

4.3.4. Let $(p_n(x))$ be a sequence of binomial type and when $n \geq 1$, let

$$p_n(x) = \sum_{k=1}^{n} c_{nk} x^k.$$

(a) Show that

$$\sum_{n=0}^{\infty} p_n(x) \frac{t^n}{n!} = \exp\left(x \sum_{j=1}^{\infty} c_{j1} t^j / j! \right).$$

In particular, the sequence $(p_n(x))$ is determined by the sequence (c_{n1}), where $c_{n1} = p_n'(0)$.

(b) Prove Equation (LG).

(c) (Research problem) Study polynomial sequences enumerating other kinds of reluctant functions, for example, the reluctant functions whose rooted forests are binary trees or plane binary trees.

4.3.5. *Euler's summation of alternating series.*

Prove the formal identity:

$$\sum_{n=0}^{\infty} (-1)^n f(n) = \sum_{n=0}^{\infty} \frac{(-1)^n}{2^{n+1}} [\Delta^n f(x)]_{x=0}.$$

4.3.6. *Abstract Poisson processes.*[15]

Let Σ be a field of subsets on the set S. A subset R of S is *rare* if the intersection $R \cap A$ is finite for every subset A in Σ. Let $\mathcal{R}$ be a collection of rare subsets of $\mathbb{R}$. If $A \in \Sigma$ and k is a nonnegative integer, let

$$[A|k] = \{R\colon R \in \mathcal{R},\ |R \cap A| = k\}.$$

The *Poisson algebra* $\mathcal{P}$ of the probability space S is the field of subsets consisting of finite unions and intersections of all subsets in $2^{\mathcal{R}}$ of the form $[A|k]$ or $[A|k]^c$, where $A \in \Sigma$ and $0 \leq k < \infty$.

(a) Show that if A and B are disjoint subsets in Σ,

$$[A \cup B|n] = \bigcup_{k=0}^{n} [A|k] \cap [B|n-k].$$

A *Poisson valuation* π on the Poisson algebra is a valuation $P \to \mathbb{R}$ that is *normalized,* $\pi([\emptyset|0]) = 1$, and satisfies the *independence property*

$$\pi([A|k] \cap [B|l]) = \pi([A|k])\pi([B|l])$$

for disjoint subsets A and B in Σ.

[15] Kung (1981) and G.-C. Rota (*Twelve problems*).

(b) Let $\nu : \Sigma \to \mathbb{R}$ be a real valuation on Σ and let I be the image $\{r: r = \nu(A)$ for some $A \in \Sigma\}$. Suppose that the image I of ν is infinite. A Poisson valuation is *homogeneous (relative to* ν*)* if $\pi([A|n])$ depends only on n and $\nu(A)$. From a homogeneous Poisson valuation, we obtain a sequence $\pi_n : I \to \mathbb{R}$ of functions defined by $\pi_n(x) = \pi([A|n])$, where A is a subset in Σ such that $x = \nu(A)$. The homogeneous Poisson measure π is *equicontinuous* if for all x_0 in I and $\epsilon > 0$, there exists a positive real number δ such that for all n,

$$|x - x_0| < \delta \text{ and } x \in I \text{ imply } |\pi_n(x) - \pi_n(x_0)| < \epsilon.$$

Show that if π is an equicontinuous homogeneous Poisson measure, then

$$\pi([A|n]) = \frac{e^{-\lambda\nu(A)} p_n(\lambda\nu(A))}{n!},$$

where λ is a real number and $(p_n(x))$ is a sequence of polynomials of binomial type.

(c) If π is a Poisson valuation, let

$$v(A) = \sum_{k=0}^{\infty} \pi([A|k])t^k.$$

Show that π is a Poisson valuation if and only if it is *consistent,* that is, $v(A)$ is a polynomial of degree at most $|A|$ when A is finite, and *multiplicative*, that is, for subsets A and B in Σ,

$$v(A)v(B) = v(A \cup B)v(A \cap B).$$

This result gives a pointless description of inhomogeneous Poisson processes.

4.3.7. *Coalgebras.*[16]

We give an informal introduction to coalgebras in enumerative combinatorics. Much of algebra is about multiplications on a set A, that is, binary operations $A \times A \to A$ that give rules for *combining* two elements. The opposite of multiplication is comultiplication, which gives a rule for *breaking up* an element. Comultiplication is by nature many valued; one way to express this is by formal sums in tensor products. We will illustrate this with three examples. Some familiarity with tensor products is required for this exercise (see Section 2.8 for an introduction to tensor algebra).

The first example is the *polynomial coalgebra.* Let $\mathbb{F}[x]$ be the algebra of polynomials over a field $\mathbb{F}$ in the variable x. Then $\mathbb{F}[x]$ is a coalgebra by the

[16] Barnabei et al. (1980), Joni and Rota (1979), and Rota (1978). For more work on coalgebras, see, for example, Haiman and Schmitt (1989) and Schmitt (1995).

comultiplication $\Delta : \mathbb{F}[x] \to \mathbb{F}[x] \otimes \mathbb{F}[x]$ defined by

$$\Delta x^n = \sum_{i=0}^{n} x^i \otimes x^{n-i}$$

and extended by linearity. Intuitively, Δx^n is the formal sum over all possible ways of breaking up a sequence of n x's into two segments: the initial segment with k x's and the final segment with $n-k$ x's.

Another example is the *Boolean coalgebra*. Let S be a finite set, A be a set of symbols, one for each subset A of S, and $\mathbb{F}(S)$ be the vector space of formal $\mathbb{F}$-linear combinations $\sum_{A:\, A \in S} c_A A$. The comultiplication $\Delta : \mathbb{F}(S) \to \mathbb{F}(S) \otimes \mathbb{F}(S)$ is defined by

$$\Delta A = \sum_{B:\, B \subseteq A} B \otimes (A \backslash B)$$

and extended by linearity. Intuitively, ΔA is the formal sum of all possible ways of breaking up A into an ordered pair of subsets.

Many combinatorial identities say that a counting process "commutes" with comultiplication. The classic example is an identity of Tutte.[17] Let $P(\Gamma; x)$ be the chromatic polynomial of the graph Γ (see Exercise 3.5.2). If A is a subset of the vertex set $V(\Gamma)$ of Γ, then $\Gamma|A$ is the graph on the vertex set A with all the edges of Γ with both endpoints in A. Then

$$P(\Gamma; x+y) = \sum_{A:\, A \subseteq V(\Gamma)} P(\Gamma|A; x) P(\Gamma|(V \backslash A); y).$$

If we replace a subset A by $x^{|A|}$ in the Boolean coalgebra, we obtain the *binomial coalgebra* with comultiplication

$$\Delta x^n = \sum_{k=0}^{n} \binom{n}{k} x^k \otimes x^{n-k}.$$

The relation of this comultiplication with binomial identities is clear. An alternate theory of polynomial sequences of binomial type can be constructed starting with a bilinear pairing between the algebra of formal power series and the binomial coalgebra.[18]

With the examples in mind, we can now define a combinatorial coalgebra. A *system of section coefficients* on a set $\mathcal{P}$ of *pieces* is a system of numbers

$$\left[\begin{matrix} a \\ a_1, a_2, \ldots, a_m \end{matrix} \right],$$

where $a, a_1, a_2, \ldots, a_m$ are pieces satisfying the following axioms:

[17] Tutte (1967). [18] Roman and Rota (1978).

Consistency: For $a,\ a_1 \in \mathcal{P}$,

$$\begin{bmatrix} a \\ a_1 \end{bmatrix} = \begin{cases} 1 & \text{if } a = a_1 \\ 0 & \text{otherwise.} \end{cases}$$

Finiteness: For each piece a,

$$\begin{bmatrix} a \\ a_1, a_2, \ldots, a_m \end{bmatrix} \neq 0$$

for only finitely many sequences $a_1, a_2, \ldots, a_m$.

Coassociativity: For positive integer m and k such that $k \leq m$,

$$\begin{aligned} \begin{bmatrix} a \\ a_1, a_2, \ldots, a_m \end{bmatrix} &= \sum_{p:\, p \in \mathcal{P}} \begin{bmatrix} a \\ a_1, a_2, \ldots, a_k, p \end{bmatrix} \begin{bmatrix} p \\ a_{k+1}, a_{k+2}, \ldots, a_m \end{bmatrix} \\ &= \sum_{p:\, p \in \mathcal{P}} \begin{bmatrix} p \\ a_1, a_2, \ldots, a_k \end{bmatrix} \begin{bmatrix} a \\ p, a_{k+1}, a_{k+2}, \ldots, a_m \end{bmatrix}. \end{aligned}$$

Section coefficients define a coalgebra on the vector space $\mathcal{C}$ of (finite) formal linear combinations of pieces by the comultiplication $\Delta : \mathcal{C} \to \mathcal{C} \otimes \mathcal{C}$ given by

$$\Delta a = \sum_{(a_1, a_2)} \begin{bmatrix} a \\ a_1, a_2 \end{bmatrix} a_1 \otimes a_2.$$

When Δ is iterated, we need to choose one component of the tensor product to apply Δ. Specifically, we can do it in the two possible ways:

$$\mathcal{C} \xrightarrow{\Delta} \mathcal{C} \otimes \mathcal{C} \xrightarrow{\Delta \otimes I} (\mathcal{C} \otimes \mathcal{C}) \otimes \mathcal{C} \quad \text{and} \quad \mathcal{C} \xrightarrow{\Delta} \mathcal{C} \otimes \mathcal{C} \xrightarrow{I \otimes \Delta} \mathcal{C} \otimes (\mathcal{C} \otimes \mathcal{C}).$$

Coassociativity says that the two ways yield the same answer. In general, coassociativity says that

$$\Delta^{m-1} a = \sum_{(a_1, a_2, \ldots, a_m)} \begin{bmatrix} a \\ a_1, a_2, \ldots, a_m \end{bmatrix} a_1 \otimes a_2 \otimes \cdots \otimes a_m,$$

an expression independent of the choices made when iterating Δ.

4.4 Sheffer Sequences

A polynomial sequence $(s_n(x))$ is a *Sheffer sequence*[19] for the delta operator Q if $s_0(x) = c$, where c is a nonzero constant, and for $n \geq 1$, $Qs_n(x) = ns_{n-1}(x)$. Like sequences of binomial type, Sheffer sequences can be described in several equivalent ways.

4.4.1. Theorem. Let Q be a delta operator with basic sequence $(p_n(x))$ and indicator function $q(t)$ (so that $Q = q(D)$). Let $(s_n(x))$ be a polynomial sequence. Then the following statements are equivalent:

SS1. The sequence $(s_n(x))$ is a Sheffer sequence for Q.

SS2. There exists a (unique) invertible shift-invariant operator S such that

$$p_n(x) = Ss_n(x) \ \text{ or } \ s_n(x) = S^{-1}p_n(x).$$

SS3. The following two equivalent power series relations hold:

$$\frac{e^{xt}}{s(t)} = \sum_{n=0}^{\infty} s_n(x)\frac{q(t)^n}{n!} \quad \text{and} \quad \sum_{n=0}^{\infty} s_n(x)\frac{t^n}{n!} = \frac{\exp(xf(t))}{s(f(t))},$$

where $f(t)$ and $q(t)$ are power series with zero constant term that are compositional inverses of each other.

SS4. The binomial identities hold: for all nonnegative integers n,

$$s_n(x+y) = \sum_{k=0}^{n} \binom{n}{k} s_k(x)p_{n-k}(y)$$

as a polynomial identity in the variables x and y.

We first prove that SS2 implies SS1. Suppose that $s_n(x) = S^{-1}p_n(x)$. Then by Corollary 4.3.6, $s_0(x) = S^{-1}1 \neq 0$. If $n \geq 1$, then as shift-invariant operators commute,

$$Qs_n(x) = QS^{-1}p_n(x) = S^{-1}Qp_n(x) = S^{-1}np_{n-1}(x) = ns_{n-1}(x).$$

This verifies SS1.

Next, we prove that SS1 implies SS2. Suppose $(s_n(x))$ is a Sheffer sequence for the delta operator Q. Define the operator S by $s_n(x) \mapsto p_n(x)$ and extending S by linearity. Since the polynomials $s_n(x)$ and $p_n(x)$ have the same degree n and $s_0(x) \neq 0$ and $p_0(x) = 1$, the operator S is invertible. It remains to show that S is shift-invariant. Note first that S commutes with Q. Indeed,

$$SQs_n(x) = nSs_{n-1}(x) = np_{n-1}(x) = Qp_n(x) = QSs_n(x).$$

[19] Sheffer (1939).

Since $\{s_n(x)\}$ is a basis, $SQ = QS$, whence $SQ^n = Q^n S$. Now by the first expansion theorem (4.3.3), we can write E^a as a power series in Q. From this, we conclude that Q commutes with E^a.

We have now proved that SS1 and SS2 are equivalent. We say that a polynomial sequence $(s_n(x))$ is (Q, S)-*Sheffer* if the sequence $(s_n(x))$, the delta operator Q, and the invertible shift-invariant operator S are related as described in SS1 and SS2. Next, we prove that SS2 implies SS3. By the first expansion theorem (4.3.4),

$$E^x = \sum_{n=0}^{\infty} \frac{p_n(x)}{n!} Q^n.$$

Hence,

$$S^{-1} E^x = \sum_{n=0}^{\infty} \frac{S^{-1} p_n(x)}{n!} Q^n = \sum_{n=0}^{\infty} \frac{s_n(x)}{n!} Q^n.$$

By Boole's formula, $E^x = \exp(xD)$. Hence, by the isomorphism theorem (4.3.5),

$$\frac{e^{xt}}{s(t)} = \sum_{n=0}^{\infty} \frac{s_n(x)}{n!} q(t)^n.$$

To prove SS3 implies SS4, we can use the generating function argument used to prove that BT4 implies BT1. Alternatively, we can use the following argument: since $p_n(x)$ is of binomial type, we have

$$p_n(x+y) = \sum_{k=0}^{n} \binom{n}{k} p_k(x) p_{n-k}(y).$$

Let S be the operator defined by $S(s_n(x)) = p_n(x)$. Applying S^{-1}, with x as the variable, we obtain

$$S^{-1} p_n(x+y) = \sum_{k=0}^{n} \binom{n}{k} s_k(x) p_{n-k}(y).$$

Since

$$S^{-1} p_n(x+y) = S^{-1} E_n^y(x) = E^y S^{-1} p_n(x) = E^y s_n(x) = s_n(x+y),$$

the binomial identity follows.

Setting $x = 0$ in SS4, we obtain

$$s_n(y) = \sum_{k=0}^{n} \binom{n}{k} s_k(0) p_{n-k}(y).$$

Thus, a Sheffer sequence $(s_n(x))$ is determined by the sequence of constant terms $(s_n(0))$ and the basic sequence $(p_n(x))$.

We will now close the circle of implication in the proof of Theorem 4.4.1 by showing that SS4 implies SS1. Exchanging the variables x and y and setting $y = 0$ in the binomial identity, we obtain

$$s_n(x) = \sum_{k=0}^{n} \binom{n}{k} s_k(0) p_{n-k}(x).$$

Then

$$\begin{aligned} Qs_n(x) &= \sum_{k=0}^{n} \binom{n}{k} s_k(0) Qp_{n-k}(x) \\ &= n \sum_{k=0}^{n} \binom{n-1}{k} s_k(0) p_{(n-1)-k}(x) = ns_{n-1}(x). \end{aligned}$$

Thus, SS1 holds and the proof of Theorem 4.4.1 is complete.

We end with the Sheffer sequence analog of the first expansion theorem (4.3.4).

4.4.2. Second expansion theorem. Let Q be a delta operator with basic sequence $(p_n(x))$, S be an invertible shift-invariant operator, and $(s_n(x))$ be a (Q, T)-Sheffer. If T is any shift-invariant operator and $p(x)$ is any polynomial, the following identity holds, as a polynomial identity in the variables x and y:

$$Tp(x + y) = \sum_{n=0}^{\infty} \frac{s_n(y)}{n!} Q^n STp(x).$$

In particular,

$$S^{-1} = \sum_{n=0}^{\infty} \frac{s_n(0)}{n!} Q^n.$$

Conversely, let S be an invertible shift-invariant operator, Q a delta operator, and $(s_n(x))$ a polynomial sequence. If

$$E^a f(x) = \sum_{n=0}^{\infty} \frac{s_n(a)}{n!} Q^n Sf(x) \tag{C}$$

for all polynomials $f(x)$ and constants a, then $(s_n(x))$ is (Q, S)-Sheffer.

Proof. By Theorem 4.3.4,

$$E^a = \sum_{n=0}^{\infty} \frac{p_n(a)}{n!} Q^n.$$

Applied to the polynomial $p(y)$, we obtain

$$p(x+y) = E^x p(y) = \sum_{n=0}^{\infty} \frac{p_n(x)}{n!} Q^n p(y).$$

Next, we apply S^{-1} as an operator on polynomials in the variable x and obtain

$$(S^{-1}p)(x+y) = \sum_{n=0}^{\infty} \frac{S^{-1}p_n(x)}{n!} Q^n p(y) = \sum_{n=0}^{\infty} \frac{s_n(x)}{n!} Q^n p(y).$$

This identity holds as a polynomial in the variable y. Hence, we may interchange x and y to get

$$(S^{-1}p)(x+y) = \sum_{n=0}^{\infty} \frac{s_n(y)}{n!} Q^n p(x).$$

To finish, regard x as the variable and apply TS and use the fact that shift-invariant operators commute to conclude that

$$Tp(x+y) = \sum_{n=0}^{\infty} \frac{s_n(y)}{n!} Q^n STp(x).$$

The special case now follows by setting $T = S^{-1}$ and $y = 0$.

To prove the converse, regard x as the variable and a as a parameter in Equation (C) and apply S^{-1} to both sides. Then

$$S^{-1}E^a f(x) = \sum_{n=0}^{\infty} \frac{s_n(a)}{n!} Q^n f(x)$$

for all constants a. Since S^{-1} is shift-invariant,

$$S^{-1}E^a f(x) = S^{-1} f(x+a) = E^a S^{-1} f(x).$$

Hence, we can exchange the variables x and a to obtain

$$S^{-1}E^a f(x) = \sum_{n=0}^{\infty} \frac{s_n(x)}{n!} Q^n f(a).$$

We next apply S on both sides and exchange x and a again, obtaining

$$f(x+a) = E^a f(x) = \sum_{n=0}^{\infty} \frac{(Ss_n)(a)}{n!} Q^n f(x).$$

Setting $f(x) = p_m(x)$, this equation becomes

$$p_m(x+a) = \sum_{n=0}^{\infty} \binom{m}{n} (Ss_n)(a) p_{m-n}(x).$$

Finally, set $x = 0$ and observe that $p_{m-n}(0) = 0$ except when $m = n$. Hence, $p_m(a) = (Ss_m)(a)$ for all constants a, and hence, $p_m(x) = Ss_m(x)$. By SS2, $(s_n(x))$ is (Q, S)-Sheffer. □

The simplest Sheffer sequences $(s_n(x))$ are those with delta operator D. These have generating function $e^{xt}/s(t)$. Hence, if

$$\frac{1}{s(t)} = \sum_{n=0}^{\infty} a_n \frac{t^n}{n!},$$

then

$$s_n(x) = \sum_{k=0}^{n} \binom{n}{k} a_{n-k} x^k.$$

Such Sheffer sequences were studied by Appell.[20] Two natural examples, Hermite and Bernoulli polynomials, are discussed in the exercises.

Exercises

4.4.1. *Recurrences.*

Let $(s_n(x))$ be a polynomial sequence with $s_0(x) = 1$. If $(s_n(x))$ is a Sheffer sequence, then for every delta operator A, there exists a sequence of constants a_n such that for all nonnegative integers n,

$$As_n(x) = \sum_{k=0}^{n} \binom{n}{k} a_k s_{n-k}(x).$$

Conversely, if the recurrence holds for some delta operator A and sequence (a_n), then $(s_n(x))$ is a Sheffer sequence for a delta operator Q (where Q may be different from A).

4.4.2. *Hermite polynomials of variance v.*

Let W_v be the *Weierstrass operator* defined by

$$W_v p(x) = \frac{1}{\sqrt{2\pi v}} \int_{-\infty}^{\infty} e^{-u^2/2v} p(x+u)du.$$

Define the *Hermite polynomials* $H_n^{(v)}(x)$ *of variance* v by

$$H_n^{(v)}(x) = W_v^{-1} x^n.$$

[20] Appell (1880).

The Hermite polynomials of variance v are (D, W_v)-Sheffer. We write $H_n(x) = H_n^{(1)}(x)$.

(a) Show that

$$\begin{aligned} W_v &= \sum_{m=0}^{\infty} \frac{v^m (2m-1)!!}{(2m)!} D^{2m} \\ &= \sum_{m=0}^{\infty} \frac{v^m}{2^m m!} D^{2m} \\ &= \exp(vD^2/2), \end{aligned}$$

where $(2m-1)!! = (2m-1)(2m-3)\cdots 5 \cdot 3 \cdot 1$ if $m \geq 1$, and $0!! = 1$.

(b) Derive formulas for Hermite polynomials from the theory of Sheffer sequences. In particular, derive

$$\begin{aligned} H_n^{(v)}(x) &= \sum_{k=0}^{\lfloor n/2 \rfloor} \binom{n}{2k} (-v)^k (2k-1)!!\, x^{n-2k}, \\ e^{-vt^2/2} e^{xt} &= \sum_{n=0}^{\infty} H_n^{(v)}(x) \frac{t^n}{n!}, \\ H_n^{(v)}(x) &= v^{n/2} H_n(x/\sqrt{v}). \end{aligned}$$

(c) Prove identities for Hermite polynomials from the literature. For example, show that for $k \geq j$,

$$H_j(x) H_k(x) = \sum_n \frac{j!k!}{\left(\frac{n+j-k}{2}\right)! \left(\frac{n+k-j}{2}\right)! \left(\frac{k+j-n}{2}\right)!} H_n(x),$$

where the sum ranges over integers n such that $k - j \leq n \leq k + j$ and $n \equiv k + j \bmod 2$.

(d) (Research problem) Using the theory of Sheffer sequences, prove *Mehler's formula:*[21]

$$1 + \sum_{n=1}^{\infty} H_n(x) H_n(y) \frac{t^n}{n!} = \frac{1}{\sqrt{1-4t^2}} \exp\left(\frac{4xyt - 4(x^2+y^2)t^2}{1-4t^2} \right).$$

Are there similar formulas for other Sheffer sequences?

[21] For a combinatorial proof, see Foata (1978).

4.4.3. *Bernoulli polynomials.*

Let J^r be the operator defined by

$$J^r p(x) = \int_x^{x+r} p(u)du$$

and $J = J^1$. Then $DJ = \Delta$ and

$$J^r = \left(\frac{\Delta}{D}\right)^r = \left(\frac{e^D - I}{D}\right)^r.$$

The *Bernoulli polynomials* $B_n^{(r)}(x)$ are defined by $B_n^{(r)}(x) = J^{-r}x^n$. For $r = 1$, we have the generating function

$$\sum_{n=0}^{\infty} B_n(x)\frac{t^n}{n!} = \frac{te^{xt}}{e^t - 1}.$$

The *Bernoulli number* B_n is the constant term $B_n(0)$.

Show the formal *Euler–MacLaurin summation formula.* If $f(x)$ is a polynomial, then

$$\sum_{j=a}^{b} f(j) = \int_a^b f(x)dx + \sum_{k=1}^{\infty} \frac{B_k[D^{k-1}f(b) - D^{k-1}f(a)]}{k!}.$$

4.5 Umbral Composition and Connection Matrices

An operator $U : \mathbb{F}[x] \to \mathbb{F}[x]$ is an *umbral operator* if there are two basic sequences $(p_n(x))$ and $(q_n(x))$ such that for all n, $Up_n(x) = q_n(x)$. An umbral operator is invertible, but it is usually not shift-invariant. If $(a_n(x))$ and $(b_n(x))$ are polynomial sequences with

$$a_n(x) = \sum_{k=0}^{n} a_{nk}x^k,$$

then the *umbral composition* is the sequence $(c_n(x))$ defined by

$$c_n(x) = \sum_{k=0}^{n} a_{nk}b_k(x).$$

Put another way, $c_n(x) = Ua_n(x)$, where U is the umbral operator defined by $U : x^n \mapsto b_n(x)$. We will use the following notation: $c_n(x) = a_n(\underline{b}(x))$ if $c_n(x) = Ua_n(x)$. In particular, $b_n(x) = \underline{b}(x)^n$. In this notation, the two binomial identities can be written

$$\underline{p}(x+y)^n = [\underline{p}(x) + \underline{p}(y)]^n \quad \text{and} \quad \underline{s}(x+y)^n = [\underline{s}(x) + \underline{p}(y)]^n.$$

4.5.1. The automorphism theorem. Let U be an umbral operator.

(a) The function $S \mapsto USU^{-1}$ is an automorphism of the algebra $\mathcal{S}$ of shift-invariant operators. If Q is a delta operator, then UQU^{-1} is a delta operator. If $S = s(Q)$, where $s(t)$ is a formal power series, then $USU^{-1} = s(UQU^{-1})$.

(b) If $(r_n(x))$ is a basic sequence, then $(Ur_n(x))$ is a basic sequence.

(c) If $(s_n(x))$ is a Sheffer sequence, then $(Us_n(x))$ is a Sheffer sequence.

Proof. Let $(p_n(x))$ and $(q_n(x))$ be basic sequences for the delta operators P and Q and U be the umbral operator defined by $Up_n(x) = q_n(x)$. To prove (a), observe that

$$UPp_n(x) = Unp_{n-1}(x) = nUp_{n-1}(x) = nq_{n-1}(x) = Qq_n(x) = QUp_n(x).$$

As $\{p_n(x)\}$ is a basis, $UPp(x) = QUp(x)$ for all polynomials, and hence, $UP = QU$ as operators on $\mathbb{F}[x]$. In particular, $UPU^{-1} = Q$ and $UP^nU^{-1} = Q^n$. Let S be a shift-invariant operator. Expand S as a power series in P, obtaining

$$S = \sum_{n=0}^{\infty} \frac{a_n}{n!} P^n.$$

Then

$$USU^{-1} = \sum_{n=0}^{\infty} \frac{a_n}{n!} UP^nU^{-1} = \sum_{n=0}^{\infty} \frac{a_n}{n!} Q^n.$$

As Q is shift-invariant, USU^{-1} is shift-invariant. By the first expansion theorem (4.3.4), every shift-invariant operator can be expanded as a power series in Q. Hence, $S \mapsto USU^{-1}$ is surjective. Since $S \mapsto U^{-1}SU$ is the inverse of $S \mapsto USU^{-1}$, the function is an automorphism of the algebra $\mathcal{S}$.

Since S is a delta operator if and only if $a_0 = 0$ and $a_1 \neq 0$, TST^{-1} is a delta operator if and only if S is a delta operator. The final assertion in (a) follows from the isomorphism theorem (4.3.5).

To prove (b), let $(r_n(x))$ be the basic sequence for the delta operator R. Then TRT^{-1} is a delta operator and it is routine to show that $(Tr_n(x))$ is its basic sequence. The proof of (c) is similar. □

Parts (b) and (c) say that the umbral composition of two basic sequences is basic, and the umbral composition of two Sheffer sequences is Sheffer. The next result shows how this happens when the delta and shift-invariant operators are expressed as power series in D.

4.5.2. The umbral composition lemma. Let $(p_n(x))$ and $(q_n(x))$ be basic sequences with delta operators $p(D)$ and $q(D)$. Let $(s_n(x))$ be $(q(D), s(D))$-Sheffer and $(t_n(x))$ be $(p(D), t(D))$-Sheffer.

(a) The umbral composition $(q_n(\underline{p}(x))$ is basic with delta operator $q(p(D))$.

(b) The umbral composition $(s_n(\underline{t}(x)))$ is $(q(p(D)), t(D)s(p(D))$-Sheffer.

Proof. Let U be the umbral operator defined by $Ux^n = p_n(x)$. If $q_n(x) = \sum_{k=1}^{n} b_{nk}x^k$, then

$$Uq_n(x) = \sum_{k=1}^{n} b_{nk}Ux^k = \sum_{k=1}^{n} b_{nk}\,p_n(x) = q_n(\underline{p}(x)).$$

By Theorem 4.5.1, $q_n(\underline{p}(x))$ is basic with delta operator $Tq(D)T^{-1}$. By the automorphism theorem (4.5.1), $Tq(D)T^{-1} = q(p(D))$. This proves (a). The proof of (b) is similar. □

Let $(p_n(x))$ and $(q_n(x))$ be two polynomial sequences. The sequence $(q_n(x))$ is *inverse* to the polynomial sequence $(p_n(x))$ if the umbral composition $(p_n(\underline{q}(x))$ is the sequence x^n. Suppose that $p_0(x) = c$, where c is a nonzero constant, so that $(p_n(x))$ gives a basis for $\mathbb{F}[x]$. The *connection matrix* $(c_{nk})_{0\leq n,k<\infty}$ *connecting* $(p_n(x))$ *to* $(q_n(x))$ is the matrix whose entries are defined by

$$q_n(x) = \sum_{k=0}^{n} c_{nk}\,p_k(x).$$

Lemma 4.5.2 gives recipes for calculating inverse sequences and connection matrices.

4.5.3. Corollary. Let $(p_n(x))$ and $(q_n(x))$ be basic with delta operators $p(D)$ and $q(D)$.

(a) The sequence $(q_n(x))$ is inverse to $(p_n(x))$ if and only if $p(q(t)) = t$; that is, the power series $q(t)$ is the compositional inverse of $p(t)$.

(b) Let (c_{nk}) be the matrix connecting $(p_n(x))$ and $(q_n(x))$, and $(r_n(x))$ be the polynomial sequence defined by

$$r_n(x) = \sum_{k=0}^{n} c_{nk}x^k.$$

Then $(r_n(x))$ is a basic sequence with delta operator $q(p^{-1}(D))$.

Proof. Part (a) is immediate from Lemma 4.5.2 (a). To prove (b), consider the umbral operator U defined by $Ux^n = p_n(x)$. Then $q_n(x) = Ur_n(x) = r_n(\underline{p}(x))$.

By Theorem 4.5.1, $(r_n(x))$ is basic, and if $r(D)$ is the delta operator of $(r_n(x))$, then $q(t) = r(p(t))$ or $r(t) = q(p^{-1}(t))$. □

We end this section with some examples. Since the power series $t/(1-t)$ is its own compositional inverse, the Laguerre polynomials are self-inverse. Explicitly,

$$\sum_{k=1}^{n}(-1)^k \frac{n!}{k!}\binom{n-1}{k-1} L_k(x) = x^n.$$

The sequence $(\varphi_n(x))$ of *exponential polynomials*[22] is the basic sequence for the delta operator $\log(I+D)$. Since $\exp(t)-1$ is the compositional inverse of $\log(1+t)$ and $\Delta = e^D - I$, the exponential polynomials are inverse to the falling factorials; that is, in umbral notation,

$$\underline{\varphi}(\underline{\varphi}-1)(\underline{\varphi}-2)\cdots(\underline{\varphi}-n+1) = x^n$$

and, explicitly,

$$\sum_{k=1}^{n}(-1)^k s(n,k)\varphi_k(x) = x^n,$$

where $s(n,k)$ are (unsigned) Stirling numbers of the first kind. By BT4, we have

$$\sum_{n=0}^{\infty} \varphi_n(x)\frac{t^n}{n!} = \exp(x(e^t-1)).$$

Thus, by the exponential formula (4.1.1),

$$\varphi_n(x) = \sum_{k=1}^{n} S(n,k)x^k,$$

where $S(n,k)$ is a Stirling number of the second kind. The inverse relation implies the matrix relation

$$\sum_{k=1}^{n}(-1)^k s(n,k)S(k,m) = \delta_{nm},$$

where $\delta_{nm} = 0$ if $m \neq n$ and 1 if $m = n$.

[22] Touchard (1956).

Consider now the basic sequence $(\varphi_n(-x))$. This has delta operator $\log(I - D)$. Let $f(t) = 1 - e^{-t}$. Then

$$f(\log(1-t)) = 1 - \frac{1}{1-t} = \frac{t}{1-t}.$$

Since the rising factorials $x(x+1)(x+2)\cdots(x+n-1)$ form the basic sequence for the backward difference operator $I - E^{-1}$, we have the following expansions for the Laguerre polynomials:

$$L_n(x) = \underline{\varphi}(\underline{\varphi}+1)(\underline{\varphi}+2)\cdots(\underline{\varphi}+n-1) = \sum_{k=1}^{n} s(n,k)\varphi_k(x).$$

Exercises

4.5.1. *The Gould or difference-Abel polynomials.*[23]

Consider the delta operator $E^{-b}\Delta$ and let $(G_n(x;b))$ be its basic sequence. Show that

$$\begin{aligned} G_n(x;b) &= x(x+bn-1)(x+bn-2)\cdots(x+bn-n+1) \\ &= x(x+bn-1)_{(n-1)} = n!\frac{x}{x+bn}\binom{x+bn}{n}. \end{aligned}$$

The polynomials $G_n(x;b)$ are the *Gould polynomials*. Hence, derive the convolution

$$\begin{aligned} &\frac{x+y}{x+y+bn}\binom{x+y+bn}{n} \\ &\quad = \sum_{k=0}^{n} \frac{x}{x+bk}\binom{x+bk}{k}\frac{y}{y+b(n-k)}\binom{y+b(n-k)}{n-k}. \end{aligned}$$

4.5.2. A polynomial sequence $(p_n(x))$ is *permutable* if for all nonnegative integers k and n,[24]

$$p_n(k) = p_k(n).$$

(a) Show that $(p_n(x))$ is permutable if and only if there is a sequence (λ_j) of nonzero constants such that

$$p_n(x) = \sum_{j=0}^{n} \binom{n}{j}\lambda_j x_{(j)}.$$

[23] Gould (1961). [24] *Foundations VIII*, p. 746.

(b) Show that a polynomial sequence $(s_n(x))$ is permutable and Sheffer if and only if it is $(\Delta, (I - a\Delta)^{-1})$-Sheffer; that is,

$$s_n(x) = a^n x_{(n)} + \binom{n}{1} a^{n-1} x_{(n-1)} + \binom{n}{2} a^{n-2} x_{(n-2)} + \cdots + \binom{n}{n-1} a^{n-1} x_{(1)} + 1.$$

4.5.3. Prove the following *duplication formula* for Laguerre polynomials:

$$L_n(ax) = \sum_{k=1}^{n} \frac{n!}{k!} \binom{n-1}{k-1} (1-a)^{n-k} a^k L_k(x).$$

4.5.4. *Sequences of higher type.*[25]

A sequence $(h_n(x))$ of polynomials has *Sheffer type k* if

$$\sum_{n=0}^{\infty} h_n(x) \frac{t^n}{n!} = g(t) \exp(x f_1(t) + x^2 f_2(t) + \cdots + x^{k+1} f_{k+1}(t)),$$

where $g(t)$ is a power series with $g(0) \neq 0$ and $f_j(t)$ is a power series such that the coefficients of $1, x, x^2, \ldots, x^{j-1}$ are zero and the coefficient of x^j is nonzero.

(Research problem) Study sequences of polynomials of higher type.

4.5.5. (Research problem posed by Sheffer) Characterize the Sheffer sequences that are sequences of orthogonal polynomials.

4.5.6. *Sequences of biorthogonal polynomials.*[26]

Let $\phi_s(D)$, $s = 0, 1, 2, \ldots$, be a sequence of linear operators on $\mathbb{F}[x]$ of the form

$$\phi_s(D) = D^s \sum_{r=0}^{\infty} b_{sr} D^r,$$

where the coefficients $b_{s,0}$ are assumed to be nonzero.

(a) Prove that there exists a unique polynomial sequence $(p_n(x))$ such that

$$\epsilon(0)\phi_s(D) q_n(x) = n!\delta_{sn},$$

[25] Sheffer (1939). Our definition is not exactly the same as Sheffer's.
[26] Boas and Buck (1964) and Kung and Yan (2003).

where δ_{sn} equals 1 if $s = n$ and 0 if $s \neq n$. In particular, $q_n(x)$ has the following explicit *determinant formula*:

$$p_n(x) = \frac{n!}{b_{00}b_{10}\cdots b_{n0}} \begin{vmatrix} b_{00} & b_{01} & b_{02} & \dots & b_{0,n-1} & b_{0,n} \\ 0 & b_{10} & b_{11} & \dots & b_{1,n-2} & b_{1,n-1} \\ 0 & 0 & b_{20} & \dots & b_{2,n-3} & b_{2,n-2} \\ \vdots & \vdots & \vdots & \ddots & \vdots & \vdots \\ 0 & 0 & 0 & \dots & b_{n-1,0} & b_{n-1,1} \\ 1 & x & x^2/2! & \dots & x^{n-1}/(n-1)! & x^n/n! \end{vmatrix}.$$

The polynomial sequence $(p_n(x))$ is said to be *biorthogonal* to the sequence $\phi_s(D)$ of operators or the sequence $\epsilon(0)\phi_s(D)$ of linear functionals.

(b) Prove the following formulas:

The *expansion formula*. Let $p(x)$ be a degree-n polynomial. Then

$$p(x) = \sum_{k=0}^{n} \frac{\epsilon(0)\phi_k(D)p(x)}{k!} p_k(x).$$

The *interpolation formula*. Given a sequence $d_0, d_1, \dots, d_n$ of numbers, then $p(x) = \sum_{k=0}^{n} d_k p_k(x)/k!$ is the unique degree-n polynomial satisfying

$$\epsilon(0)\phi_k(D)p(x) = d_k, \quad 0 \leq k \leq n.$$

The *Appell relation*.

$$e^{xt} = \sum_{n=0}^{\infty} \frac{p_n(x)\phi_n(t)}{n!}, \quad \text{where } \phi_n(t) = t^s \sum_{r=0}^{\infty} b_{sr}t^r.$$

(c) Show that a polynomial sequence $(p_n(x))$ has binomial type if and only if it is biorthogonal to an operator sequence of the form

$$\phi_s(D) = [f(D)]^s,$$

where $f(t)$ is a formal power series with $f(0) = 0$ and $f'(0) \neq 0$.

(d) (Research problem) Let ∇ be the backward difference operator. Develop a theory of biorthogonality with operators that are power series in ∇.

4.5.7. *Gončarov polynomials.*

Let $(a_0, a_1, a_2, \dots)$, be a sequence of numbers. The sequence of *Gončarov polynomials* $(g_n(x; a_0, a_1, \dots, a_{n-1}))_{n=0}^{\infty}$ is the sequence of polynomials biorthogonal to the operators

$$\phi_s(D) = E^{a_s} D^s = D^s \sum_{r=0}^{\infty} \frac{a_s^r D^r}{r!}.$$

When all the terms a_i are equal to a number a, then $g_n(x; a, a, \ldots, a)$ is the Abel polynomial $x(x - na)^{n-1}$.

(a) Prove that the Gončarov polynomials are determined by the *differential relations*

$$Dg_n(x; a_0, a_1, \ldots, a_{n-1}) = ng_{n-1}(x; a_1, a_2, \ldots, a_{n-1})$$

and the initial conditions

$$g_n(a_0; a_0, a_1, \ldots, a_{n-1}) = \delta_{0,n}.$$

(b) Show that

$$g_n(x + y; a_0, a_1, \ldots, a_{n-1}) = \sum_{k=0}^{n} \binom{n}{k} g_{n-k}(y; a_k, a_{k+1}, \ldots, a_{n-1})x^k.$$

(c) If $\underline{x}$ is the finite length-n sequence $x_1, x_2, \ldots, x_n$, the sequence of *order statistics* is the sequence $x_{(1)}, x_{(2)}, \ldots, x_{(n)}$ obtained from $\underline{x}$ by rearranging it in nondecreasing order. Let $\underline{u}$ be a given infinite sequence $u_1, u_2, u_3, \ldots$ of nondecreasing positive integers. A *length-n $\underline{u}$-parking function* is a sequence $x_1, x_2, \ldots, x_n$ such that its order statistics satisfy $x_{(i)} \le u_i$, $1 \le i \le n$. Let $P_n(u_1, u_2, \ldots, u_n)$ be the number of length-n $\underline{u}$-parking functions. Show that

$$P_n(u_1, u_2, \ldots, u_n) = (-1)^n g_n(0; u_1, u_2, \ldots, u_n).$$

In particular, show that

$$P_n(a, a + b, a + 2b, \ldots, a + (n - 1)b) = a(a + nb)^{n-1}.$$

(d) Show that $P_n(u_1, u_2, \ldots, u_n) = n! \det \mathcal{D}$, where $\mathcal{D}$ is the matrix with ij-entry equal to

$$\frac{u_i^{j-i+1}}{(j - i + 1)!}$$

if $j - i + 1 \ge 0$ and 0 otherwise.[27]

4.6 The Riemann Zeta Function

The *Riemann zeta function* is the Dirichlet series $\sum_{n=1}^{\infty} n^{-s}$. It is the obscure object of desire in analytic number theory. In this section, we sketch a combinatorial interpretation of the zeta function.[28]

[27] This is the discrete analog of a formula in Steck (1969). [28] Rota (2003).

Let C_N be the cyclic group of order N. A *character* χ of C_N is a homomorphism from C_N to the complex numbers $\mathbb{C}$. The kernel $\ker \chi$ of a character χ is a subgroup of C_N. If $(\chi_1, \chi_2, \ldots, \chi_s)$ is an s-tuple of characters, then we define their *joint kernel* to be the intersection $\ker \chi_1 \cap \ker \chi_2 \cap \cdots \cap \ker \chi_s$.

4.6.1. Lemma. Let n divides N, C_n be the subgroup of order n in C_N, and $\Pr_s(C_n)$ be the probability that the joint kernel of an s-tuple $\chi_1, \chi_2, \ldots, \chi_s$ of characters, chosen independently and at random, equals C_n. Then

$$\Pr_s(C_n) = \sum_{d:\, n|d \text{ and } d|N} \mu(d/n) d^{-s},$$

where μ is the number-theoretic Möbius function.

Proof. The probability that the kernel of a random character χ contains the subgroup C_n equals $1/n$, because there are N characters of the group C_N and N/n, such characters will vanish on C_n. Therefore, the probability that the joint kernel of a randomly and independently chosen s-tuple $(\chi_1, \chi_2, \ldots, \chi_s)$ contains the subgroup C_n equals $(1/n)^s$. Let $\Pr_s(C_n)$ be the probability that the joint kernel a randomly and independently chosen s-tuple of characters equals the subgroup C_n. Then using the fact that the lattice of subgroups of the cyclic group C_N is isomorphic to the partially ordered set of divisors of the integer N, ordered by divisibility, we have

$$n^{-s} = \sum_{d:\, n|\, d \text{ and } d|N} \Pr_s(C_d).$$

The lemma now follow by Möbius inversion. □

Using the change of variable $d = nj$, we obtain

$$\Pr_s(C_n) = n^{-s} \sum_j \mu(j) j^{-s},$$

where the variable j on the right ranges over some subset of divisors of the integer N, about which we will be deliberately vague. By the Möbius inversion formula,

$$\zeta(s) \left(\sum_{j=1}^{\infty} \mu(j) j^{-s} \right) = 1.$$

If we could change our combinatorial problem to get an unrestricted sum on the right-hand side, then we would have a probabilistic interpretation of $1/\zeta(s)$. We propose doing this by replacing the finite cyclic group C_n by a profinite cyclic group. The group C_∞ of rational numbers modulo 1 is such a profinite group. Every finitely generated subgroup of C_∞ is a finite cyclic group C_n. The character group C_∞^* of C_∞ is a compact group. It has a Haar measure which can be normalized to be a probability measure Pr. The set of all characters of the group C_∞ that vanish on a subgroup C_n has Haar measure equal to $1/n$. Thus, if we choose a sequence of s characters of C_∞ independently and at random, the probability that their joint kernel contains the group C_n equals $(1/n)^s$. Denoting again by $\mathrm{Pr}_s(C_n)$ the probability that the joint kernel of an s-tuple of characters equals the subgroup C_n, then we have the identity

$$\frac{1}{n^s} = \sum_{d:n|d} \mathrm{Pr}_s(C_d),$$

where the sum on the right is now infinite. By Möbius inversion, we obtain

$$\mathrm{Pr}_s(C_n) = \sum_{d:n|d} \mu(d/n)\frac{1}{d^s} = \frac{1}{n^s}\sum_{d=1}^{\infty}\frac{\mu(d)}{d^s} = \frac{1}{n^s\zeta(s)}.$$

This yields a probabilistic interpretation of the reciprocal of the Riemann zeta function for positive integers s.

Exercises

4.6.1. *The zeta function as a probability distribution.*

Let s be a real number exceeding 1. Define a probability Pr_s on the set $\{1, 2, 3, \ldots\}$ of positive integers as follows: if $A \subseteq \{1, 2, 3, \ldots\}$, then

$$\mathrm{Pr}_s(A) = \frac{1}{\zeta(s)} \sum_{n:n\in A} n^{-s}.$$

(a) Show that if $A_r = \{mr \colon 1 \le m < \infty\}$, and p and q are distinct primes, then

$$\mathrm{Pr}_s(A_p \cap A_q) = \mathrm{Pr}_s(A_p)\mathrm{Pr}_s(A_q) = \frac{1}{pq};$$

that is, the events that a random integer n is divisible by either of two primes p and q are independent relative to the probability Pr_s.

Subject to technical assumptions on the set A,

$$\lim_{s\to 1} \mathrm{Pr}_s(A) = \lim_{t\to\infty} \frac{|A \cap \{1, 2, \ldots, t\}|}{t}.$$

The quantity on the right is the *arithmetic density* of A. This shows that although arithmetic density is not a probability, it is, when restricted to suitable sets, the limit of probabilities.

(b) (Research problem) Find the technical assumptions.

4.6.2. (Research problem) Do the second half of this section rigorously. Find combinatorial interpretations of other results, for example, the functional equation, about the Riemann zeta function.

5

Symmetric Functions and Baxter Algebras

5.1 Symmetric Functions

The theory of symmetric functions is both a classical area of algebra and an active area of current research.[1] We will not follow well-trodden paths. Rather, we will focus on three topics: the interpretation of symmetric functions as generating functions of classes of functions, the theory of Baxter algebras, and the study of symmetric polynomials as polynomial functions over finite fields. Thus, our point of view is decidedly eccentric, at least at present. We remark that there are many conflicting notations for families of symmetric functions. We have chosen a notation that is closest to the earlier papers in the area.

Let $\mathbb{F}$ be a field and $x_1, x_2, \ldots, x_n$ be a set of n variables. A polynomial f in the variables $x_1, x_2, \ldots, x_n$ is a *symmetric polynomial* if it is an absolute invariant of the symmetric group $\mathfrak{S}_n$; that is, for all permutations γ of $\{1, 2, \ldots, n\}$,

$$f(x_{\gamma(1)}, x_{\gamma(2)}, \ldots, x_{\gamma(n)}) = f(x_1, x_2, \ldots, x_n).$$

To avoid iterated subscripts or superscripts, we use *letter-place* notation: x_i^{λ} is written $(i|\lambda)$, so that

$$(i_1|\lambda_1)(i_2|\lambda_2)\cdots(i_s|\lambda_s) = x_{i_1}^{\lambda_1} x_{i_2}^{\lambda_2} \cdots x_{i_s}^{\lambda_s},$$

where $\lambda_1, \lambda_2, \ldots, \lambda_s$ is a sequence of nonnegative integers and $i_1, i_2, \ldots, i_s$ are integers in $\{1, 2, \ldots, n\}$.

Let λ be the sequence $\lambda_1, \lambda_2, \ldots, \lambda_s$ with positive integer terms. Then λ is a *partition* of the positive integer n into s parts, briefly $\lambda \vdash n$, if $\lambda_1 \geq \lambda_2 \geq \cdots \geq \lambda_s > 0$ and $\lambda_1 + \lambda_2 + \cdots + \lambda_s = n$. Let a_i be the number of parts equal

[1] See Rota (1998a). Note that this paper is the transcript of an informal speech.

to i in λ, so that $a_1 1 + a_2 2 + \cdots = n$. We define $|\lambda|$ by

$$|\lambda| = a_1!a_2! \cdots a_n!.$$

Let $\lambda \vdash m$ and $\lambda = \lambda_1, \lambda_2, \ldots, \lambda_s$. The *monomial symmetric polynomial* k_λ indexed by λ is defined by

$$k_\lambda = \sum (i_1|\lambda_1)(i_2|\lambda_2) \cdots (i_s|\lambda_s),$$

where the sum ranges over all *distinct* monomials of the form $(i_1|\lambda_1)(i_2|\lambda_2) \cdots (i_s|\lambda_s)$. For example, $(1|3)(2|1)(3|1)$ and $(1|3)(3|1)(2|1)$ define the same monomial and $k_{3,1,1}(x_1.x_2, x_3) = x_1^3 x_2 x_3 + x_2^3 x_1 x_3 + x_3^3 x_1 x_2$.

5.1.1. Lemma. The monomial symmetric polynomials k_λ, where $\lambda \vdash n$, form a basis for all homogeneous symmetric polynomials of total degree n.

Proof. We use a Gröbner basis argument. Let f be a homogeneous symmetric polynomial of total degree n. Order the partitions of n lexicographically and let λ be the *leading partition*, that is, the maximum partition $\lambda \vdash n$ so that a monomial of the form $(i_1|\lambda_1)(i_2|\lambda_2) \cdots (i_s|\lambda_s)$ occurs with nonzero coefficient C in f. Since f is symmetric, the monomial $(\gamma(i_1)|\lambda_1)(\gamma(i_2)|\lambda_2) \cdots (\gamma(i_s)|\lambda_s)$ occurs with the same coefficient C for any permutation γ in $\mathfrak{S}_n$. Hence, the leading partition of the difference $f - Ck_\lambda$ is lexicographically smaller than λ. Repeating the process on $f - Ck_\lambda$, we can write f as a linear combination of monomial symmetric functions k_τ, where τ is a partition of n. □

If we have a fixed finite number $x_1, x_2, \ldots, x_n$ of variables, then k_λ would depend on the number n of variables. This can be avoided by taking an infinite set of variables and extending the definition of symmetric polynomials to symmetric formal power series. A *symmetric function* is a formal power series in countably many variables $x_1, x_2, \ldots,$ which is an absolute invariant of the "small" infinite symmetric group $\mathfrak{S}_\infty$ of all permutations on $\{1, 2, \ldots\}$ that fix all but finitely many elements. The *algebra of symmetric functions* $\mathcal{A}$ over a field $\mathbb{F}$ is the subalgebra of the algebra $\mathbb{F}[[x_1, x_2, \ldots]]$ of formal power series consisting of symmetric functions. If we set $x_i = 0$ for all $i > n$, then a symmetric function specializes to a symmetric polynomial in $x_1, x_2, \ldots, x_n$. The algebra $\mathcal{A}$ is graded by total degree, so that

$$\mathcal{A} = \mathcal{A}_0 \oplus \mathcal{A}_1 \oplus \cdots \oplus \mathcal{A}_d \oplus \cdots,$$

where $\mathcal{A}_d$ is the vector space of homogeneous symmetric functions of total degree d.

The monomial symmetric function k_λ is defined as before, except that the sequences $i_1, i_2, \ldots, i_s$ in the monomials in the sum have terms in the set

$\{1, 2, 3, \ldots\}$. The argument in the proof of Lemma 5.1.1 extends and the set $\{k_\lambda \colon \lambda \vdash d\}$ is a basis for $\mathcal{A}_d$. In particular, the subspace $\mathcal{A}_d$ has dimension $p(d)$, the number of partitions of d.

There are several families of symmetric functions. Many of them can be defined using monomial symmetric functions. We begin with the *elementary symmetric functions*. If m is a nonnegative integer and $11\ldots1$ is the partition of m into m parts (all equal to 1), then $a_m = k_{11\ldots1}$ or, equivalently,

$$a_m = \sum (i_1|1)(i_2|1)\cdots(i_m|1),$$

the sum ranging over all subsets $\{i_1, i_2, \ldots, i_m\}$ of size m of $\{1, 2, 3, \ldots\}$. Note that a_0 is the sum of one monomial, the empty product, and so $a_0 = 1$. The elementary symmetric functions are extended to partition indices multiplicatively: if $\lambda \vdash n$, then

$$a_\lambda = a_{\lambda_1} a_{\lambda_2} \cdots a_{\lambda_s}.$$

Two other families are defined in a similar way. First, we define the *complete homogeneous symmetric functions* h_λ. If m is a nonnegative integer, then

$$h_m = \sum_{\lambda\colon \lambda \vdash m} k_\lambda,$$

and if λ is a partition,

$$h_\lambda = h_{\lambda_1} h_{\lambda_2} \cdots h_{\lambda_s}.$$

Next, we define the *power-sum symmetric functions* s_λ. If m is a nonnegative integer, then

$$s_m = k_m = \sum_{i=1}^{\infty} (i|m),$$

and if λ is a partition, then

$$s_\lambda = s_{\lambda_1} s_{\lambda_2} \cdots s_{\lambda_s}.$$

Perhaps the most important family of symmetric functions are the Schur functions. We will discuss them peripherally as quotients of alternants in the exercises. This is not to ignore them. Although much has been done, we do not truly understand their combinatorics. In particular, why they are generating functions of character values of irreducible representations of the symmetric group remains as mysterious as ever.

Exercises

5.1.1. *Euler's partition formula.*
Show that

$$\prod_{i=1}^{\infty} \frac{1}{1-q^i} = \sum_{d=0}^{\infty} p(d)q^d.$$

Euler's formula gives a product formula for the *Hilbert function* $\sum_{d=0}^{\infty} \dim(\mathcal{A}_d) q^d$ of the algebra of symmetric functions.

5.2 Distribution, Occupancy, and the Partition Lattice

Partitions of integers occur in most definitions of symmetric functions. As integer partitions arise from set partitions, the partition lattice plays an important role in the theory of symmetric functions.

Let Π_n be the lattice of all partitions of the set $\{1, 2, \ldots, n\}$ ordered by reverse refinement. Let π be a partition of $\{1, 2, \ldots, n\}$ with blocks $B_1, B_2, \ldots, B_c$, listed so that the sizes $|B_i|$ are nonincreasing. The *type* $\lambda(\pi)$ of π is the integer partition $|B_1|, |B_2|, \ldots, |B_c|$ of n. If λ is the integer partition with parts $\lambda_1, \lambda_2, \ldots, \lambda_c$, then the number of set partitions of $\{1, 2, \ldots, n\}$ having type λ equals

$$\frac{n!}{\lambda_1!\lambda_2!\cdots\lambda_c!a_1!a_2!\cdots a_n!},$$

where a_i is the number of parts equal to i in λ; that is,

$$\frac{1}{|\lambda|}\binom{n}{\lambda}, \qquad \text{(N)}$$

where $\binom{n}{\lambda}$ is a multinomial coefficient and $|\lambda| = a_1!a_2!\cdots a_n!$.

The lattice Π_n is a geometric lattice of rank $n-1$. The minimum $\hat{0}$ is the partition $\{\{1\}, \{2\}, \ldots, \{n\}\}$ into n blocks (all having size 1). The maximum $\hat{1}$ is the partition $\{\{1, 2, \ldots, n\}\}$ into one block. The points or atoms are partitions into $n-1$ blocks, with one block of size 2 and $n-2$ blocks of size 1. In particular, there are $\binom{n}{2}$ atoms. The copoints are partitions into two blocks. There are $2^{n-1} - 1$ copoints. The rank function is given by $\operatorname{rank}(\sigma) = n - c(\sigma)$, where $c(\sigma)$ is the number of blocks in σ. An upper interval $[\sigma, \hat{1}]$ in Π_n is isomorphic to $\Pi_{n-c(\sigma)}$. To see this, think of each block B_i in σ as a single element. If $\sigma = \{B_1, B_2, \ldots, B_c\}$, then the lower interval $[\hat{0}, \sigma]$ is isomorphic to the Cartesian product

$$\Pi_{|B_1|} \times \Pi_{|B_2|} \times \cdots \times \Pi_{|B_c|}.$$

In particular, the Möbius function in Π_n can be computed if one knows $\mu(\hat{0}, \hat{1})$ in Π_n for all n.

5.2.1. Theorem.[2] In the partition lattice Π_n,

$$\mu(\hat{0}, \hat{1}) = (-1)^{n-1}(n-1)!.$$

Proof. We induct on n. The assertion is true for Π_1, the lattice with one element $\{\{1\}\}$, and Π_2, the lattice with two elements $\{\{1\}, \{2\}\}$ and $\{\{1, 2\}\}$. For the induction step, we use Weisner's theorem (3.1.6), choosing α to be the rank-$(n-2)$ partition consisting of the two blocks $\{1, 2, \ldots, n-1\}$ and $\{n\}$. Then

$$\mu(\hat{0}, \hat{1}) = - \sum_{\pi : \pi \wedge \alpha = \hat{0},\, \pi \neq \hat{0}} \mu(\pi, \hat{1}).$$

Next, observe that $\pi \wedge \alpha = \hat{0}$ and $\pi \neq \hat{0}$ if and only if π is a rank-1 partition of the form with a two-element block $\{j, n\}$ and $n-2$ one-element blocks, there are $n-1$ such partitions and the upper interval $[\pi, \hat{1}]$ is isomorphic to Π_{n-1}. Hence, by induction,

$$\mu(\hat{0}, \hat{1}) = -(n-1)[(-1)^{n-2}(n-2)!] = (-1)^{n-1}(n-1)!. \qquad \square$$

Another proof can be found in Exercise 3.5.2.

Most of the classical symmetric functions have interpretations as generating functions of classes of functions. Many of these classes are defined by conditions on their coimages.

Let D be a finite set and $X = \{x_1, x_2, x_3, \ldots\}$ be a countable set indexed by the positive integers, thought of as both a set of elements and a set of independent variables. Recall that the *coimage* of a function $f : D \to X$ is the partition of D whose blocks are those inverse images $f^{-1}(x_i)$ that are nonempty. For example, if $D = \{1, 2, 3, 4, 5, 6\}$ and $f(1) = f(3) = f(4) = x_2$, $f(5) = f(6) = x_3$, and $f(2) = x_8$, then coimage$(f) = \{\{1, 3, 4\}, \{2\}, \{5, 6\}\}$. The *image monomial* Gen(f) of a function $f : D \to X$ is defined by

$$\mathrm{Gen}(f) = \prod_{i : i \in D} f(i).$$

[2] Frucht and Rota (1963) and Rota, *Foundations I*.

For our example, $\mathrm{Gen}(f) = x_2^3 x_3^2 x_8$. If $\mathcal{F}$ is a family of functions, then its generating function $\mathrm{Gen}(\mathcal{F})$ is defined by

$$\mathrm{Gen}(\mathcal{F}) = \sum_{f:\, f \in \mathcal{F}} \mathrm{Gen}(f).$$

5.2.2. Theorem. Let π be a partition of D having type $\lambda = \lambda_1!\lambda_2! \cdots, \lambda_c!$, and

$$\begin{aligned} k_\pi &= \mathrm{Gen}\,\{f\colon \mathrm{coimage}(f) = \pi\}, \\ s_\pi &= \mathrm{Gen}\,\{f\colon \mathrm{coimage}(f) \geq \pi\}, \\ a_\pi &= \mathrm{Gen}\,\{f\colon \mathrm{coimage}(f) \wedge \pi = \hat{0}\}. \end{aligned}$$

Then

$$k_\pi = |\lambda| k_\lambda, \quad s_\pi = s_\lambda, \quad a_\pi = \lambda_1!\lambda_2! \cdots \lambda_c! a_\lambda.$$

Proof. Let $B_1, B_2, \ldots, B_c$ be the blocks of π. Then $\mathrm{coimage}(f) = \pi$ if and only if the restriction of f to B_i is constant and $f(x) \neq f(y)$ if x and y are in different blocks. Thus, $\mathrm{coimage}(f) = \pi$ if and only if

$$\mathrm{Gen}(f) = (i_1|\lambda_1)(i_2|\lambda_2) \cdots (i_c|\lambda_c), \tag{M}$$

where $i_1, i_2, \ldots, i_c$ are *distinct* indices. There are $|\lambda|$ distinct functions satisfying (M). Hence, $k_\pi = |\lambda| k_\lambda$.

Next, observe that $\mathrm{coimage}(f) \geq \pi$ if and only if

$$\mathrm{Gen}(f) = (i_1|\lambda_1)(i_2|\lambda_2) \cdots (i_c|\lambda_c),$$

where the indices $i_1, i_2, \ldots, i_c$ are not necessarily distinct. Hence,

$$\mathrm{Gen}\,\{f\colon \mathrm{coimage}(f) \geq \pi\} = \prod_{i=1}^{c} [(1|\lambda_i) + (2|\lambda_i) + (3|\lambda_i) + \cdots] = s_\lambda.$$

Finally, observe that $\mathrm{coimage}(f) \wedge \pi = \hat{0}$ if and only if the intersection of a block in $\mathrm{coimage}(f)$ and a block in π has size 0 or 1; that is to say, the restriction $f|B$ to a block B in π is injective (or one to one). Hence,

$$\mathrm{Gen}\,\{f\colon \mathrm{coimage}(f) \wedge \pi = \hat{0}\} = \prod_{i=1}^{c} \mathrm{Gen}\,\{f : B_i \to X,\ f \text{ is injective}\}.$$

A function $f : B \to X$ is injective if and only if $\mathrm{Gen}(f) = (i_1|1)(i_2|1) \cdots (i_{|B|}|1)$, where the indices $i_1, i_2, \ldots, i_{|B|}$ are distinct. Hence,

$$\mathrm{Gen}\,\{f : B_i \to X,\ f \text{ is injective}\} = |B|! a_{|B|}$$

and the last equation follows by multiplicativity. □

From Theorem 5.2.2, it follows immediately that

$$s_\pi = \sum_{\sigma:\sigma\geq\pi} k_\sigma,$$
$$a_\pi = \sum_{\sigma:\sigma\wedge\pi=\hat{0}} k_\sigma.$$

By Theorem 5.2.1 and the inversion formulas in Theorem 3.1.2 and Exercises 3.1.13, we obtain explicit formulas for k_π as linear combinations of s_π and a_π.

5.2.3. Theorem.

$$k_\pi = \sum_{\sigma:\sigma\geq\pi} \mu(\pi,\sigma)s_\sigma,$$
$$k_\pi = \sum_\sigma \left(\sum_{\tau:\tau\geq\pi\vee\sigma} \frac{\mu(\pi,\tau)\mu(\sigma,\tau)}{\mu(\hat{0},\tau)} \right) a_\sigma.$$

The next theorem gives the relationship between elementary and power-sum symmetric functions.

5.2.4. Theorem.

$$a_\pi = \sum_{\sigma:\sigma\leq\pi} \mu(\hat{0},\sigma)s_\sigma, \qquad s_\pi = \frac{1}{\mu(\hat{0},\pi)} \sum_{\sigma:\sigma\leq\pi} \mu(\sigma,\pi)a_\sigma.$$

Proof. The second formula is a Möbius inversion of the first. The first formula is proved by the following calculation:

$$\begin{aligned} a_\pi &= \sum_{\sigma:\sigma\wedge\pi=\hat{0}} k_\sigma \\ &= \sum_\sigma \left(\sum_{\tau:\tau\leq\sigma\wedge\pi} \mu(\hat{0},\tau) \right) k_\sigma \\ &= \sum_{\tau:\tau\leq\pi} \mu(\hat{0},\tau) \left(\sum_{\sigma:\sigma\geq\tau} k_\sigma \right) = \sum_{\tau:\tau\leq\pi} \mu(\hat{0},\tau)s_\tau. \end{aligned}$$ □

5.2.5. Waring's formula.

$$\sum_{m=0}^{\infty} a_m t^m = \exp\left(\sum_{k=1}^{\infty} \frac{(-1)^{k-1}s_k t^k}{k} \right).$$

Proof. By Theorems 5.2.1 and 5.2.4,

$$n!a_n = \sum_\sigma \prod_{B:B\in\sigma} (-1)^{|B|-1}(|B|-1)!s_{|B|},$$

the sum ranging over all partitions σ of $\{1, 2, \ldots, n\}$. Waring's formula now follows from the exponential formula (Theorem 4.1.1). □

Next, we discuss the two "fundamental" theorems of symmetric polynomials.

5.2.6. First fundamental theorem. Every symmetric polynomial can be written as a polynomial in the elementary symmetric polynomials.

First proof. By Lemma 5.1.1, it suffices to show that the monomial symmetric polynomials can be expressed as polynomials in the elementary symmetric functions. This is done explicitly in Theorem 5.2.3. □

Second proof. Apply the Gröbner basis argument directly to the elementary symmetric polynomials. Let $\lambda_1, \lambda_2, \ldots, \lambda_s$ be the leading partition of a symmetric polynomial f and C be the coefficient of a monomial with that partition. Then

$$f - Ca_1^{\lambda_1-\lambda_2}a_2^{\lambda_2-\lambda_3}\cdots a_{s-1}^{\lambda_{s-1}-\lambda_s}a_s^{\lambda_s}$$

has smaller leading partition. Iterating this, we construct an expression of f as a polynomial in the elementary symmetric polynomials. □

5.2.7. Second fundamental theorem. There is no nonzero polynomial $P(X_1, X_2, \ldots, X_n)$ in the elementary symmetric polynomials such that the polynomial

$$P(a_1(x_1, x_2, \ldots, x_n), a_2(x_1, x_2, \ldots, x_n), \ldots, a_n(x_1, x_2, \ldots, x_n)) \qquad \text{(I)}$$

is identically zero as a polynomial in the variables $x_1, x_2, \ldots, x_n$.

We will give three proofs:

First proof. The theorem is equivalent to the statement that the set of partition-indexed elementary symmetric functions a_π is a basis. Thus, it follows from Lemma 5.1.1 and Theorem 5.2.3.

Second proof. Suppose such a polynomial exists. Let N be the largest index so that X_N occurs in P and d be the highest exponent of X_N. Expand P as a polynomial

$$P_d(X_1, X_2, \ldots, X_{N-1})X_N^d + P_{d-1}(X_1, X_2, \ldots, X_{N-1})X_N^{d-1} + \cdots$$
$$+ P_0(X_1, X_2, \ldots, X_{N-1})$$

in X_N. Dividing by a power of X_N if necessary, we may assume that $P_0(X_1, X_2, \ldots, X_{N-1})$ is not identically zero.

Substitute in $a_N = x_1 x_2 \cdots x_N$ and set $x_N = 0$. This gives a relation

$$P_0(a_1, a_2, \ldots, a_{N-1}) = 0$$

among the smaller set $\{a_1, a_2, \ldots, a_{N-1}\}$. Since the single symmetric function a_1 in one variable equals x_1 and satisfies no nontrivial relations, the theorem follows by induction. □

Third proof. The last proof formalizes a philosophical argument and is valid only over an infinite field $\mathbb{F}$. There cannot be any nontrivial identities among the elementary symmetric polynomials; otherwise, only those polynomials with coefficients satisfying those identities can have zeros.

Suppose, for contradiction, that $P(X_1, X_2, \ldots, X_n)$ is a nonzero polynomial such that Equation (I) holds. Let $A_1, A_2, \ldots, A_n$ be elements in the infinite field $\mathbb{F}$ such that $P(A_1, A_2, \ldots, A_n) \neq 0$. In the algebraic closure of $\mathbb{F}$, the polynomial $x^n - A_1 x^{n-1} + A_2 x^{n-2} - \cdots \pm A_n$ has n zeros $z_1, z_2, \ldots, z_n$. Then, as Equation (I) holds,

$$\begin{aligned} &P(A_1, A_2, \ldots, A_N) \\ &\quad = P(a_1(z_1, z_2, \ldots, z_n), a_2(z_1, z_2, \ldots, z_n), \ldots, a_n(z_1, z_2, \ldots, z_n)) = 0, \end{aligned}$$

a contradiction. □

We end this section with one of the earliest theorems in the theory of symmetric polynomials.

5.2.8. Newton's identities. Let $m \geq 1$. Then

$$s_m - a_1 s_{m-1} + a_2 s_{m-2} - \cdots + (-1)^{m-1} a_{m-1} s_1 = (-1)^m m a_m,$$

where a_m equals 0 if the index m is greater than the number n of variables.

Proof. Let t be another variable. Then

$$\sum_{j=1}^{n} \frac{x_j}{1 - x_j t} = \sum_{i=0}^{\infty} [x_1^{i+1} + x_2^{i+1} + \cdots + x_n^{i+1}] t^i = \sum_{i=0}^{\infty} s_{i+1} t^i$$

and

$$f(t) = \prod_{k=1}^{n} (1 - x_k t) = \sum_{i=0}^{n} (-1)^i a_i t^i.$$

We multiply these equations. On the left side,

$$f(t) \sum_{j=1}^{n} \frac{x_j}{1 - x_j t} = \sum_{j=1}^{n} x_j \frac{f(t)}{1 - x_j t} = -f'(t),$$

where $f'(t)$ is the derivative of $f(t)$. Multiplying the power series and the polynomial on the right side and equating coefficients of powers of t, we obtain

$$a_1 = s_1, \quad -2a_2 = s_2 - a_1 s_1, \quad 3a_3 = s_3 - a_1 s_2 + a_2 s_1,$$

and so on. □

Newton's identities give an explicit algorithm for writing the power-sum symmetric polynomial s_j as a polynomial with integer coefficients in the elementary symmetric polynomial $a_1, a_2, \ldots, a_j$. For example, $s_1 = a_1$, $s_2 = a_1^2 - 2a_2$, and

$$s_3 = a_1(a_1^2 - 2a_2) - a_2 a_1 + 3a_3 = a_1^3 - 3a_2 a_1 + 3a_3.$$

Exercises

5.2.1. *Complete homogeneous symmetric functions.*

In this problem, we need the notion of a reluctant function (introduced in Section 4.3). A reluctant function $f : D \Rightarrow X$ is a *placing* if the connected component rooted at x_i of the graph of f is a union $P_{i,1} \cup P_{i,2} \cup \cdots \cup P_{i,s}$ of paths, such that each path has the root x_i at an endpoint and two paths have no vertices in common except for the root x_i. For example, if $D = \{1, 2, 3, 4, 5, 6\}$, then the functions $f, g : D \to D \cup X$ defined by

$$f(3) = 4, \; f(4) = x_2, \; f(1) = x_2, \; f(5) = 6, \; f(6) = x_3, \; f(2) = x_6,$$
$$g(1) = 4, \; g(4) = 3, \; g(3) = x_2, \; g(5) = x_3, \; g(6) = x_3, \; g(2) = x_6$$

give placings. Both placings have coimage $\{\{1, 3, 4\}, \{5, 6\}, \{2\}\}$.

Let π be a partition of a subset D' of D. Then $\mathcal{H}_\pi$ is the set of placings $f : D' \Rightarrow X$ such that for each block B of π and each nonempty inverse image $f^{-1}(x_i)$, either $B \cap f^{-1}(x_i) = \emptyset$ or $B \cap f^{-1}(x_i)$ equals the set of nonroot vertices in one of the paths $P_{i,j}$ in graph(f). For example, if $\pi = \{\{2, 3, 4\}, \{1, 5, 6\}\}$, then the first placing f is in $\mathcal{H}_\pi$ but the second placing g is not.

(a) Show that $\lambda(\pi)! h_{\lambda(\pi)} = \mathrm{Gen}(\mathcal{H}_\pi)$.

(b) Show that $h_\pi = \sum_\sigma \lambda(\sigma \wedge \pi)! k_\sigma$.

(c) Show that

$$1 + \sum_{n=1}^{\infty} h_n t^n = \exp\left(\sum_{n=1}^{\infty} \frac{s_n t^n}{n} \right).$$

5.2.2. *Alternating functions and alternants.*[3]

A polynomial $f(x_1, x_2, \ldots, x_n)$ is an *alternating function* if for all permutations $\sigma \in \mathfrak{S}_n$,

$$f(x_{\sigma(1)}, x_{\sigma(2)}, \ldots, x_{\sigma(n)}) = \operatorname{sign}(\sigma) f(x_1, x_2, \ldots, x_n).$$

The *van der Monde determinant* V, defined by

$$V = V(x_1, x_2, \ldots, x_n) = \det(x_i^{j-1})_{1 \le i,j \le n},$$

is an alternating function. The *alternant* $(x_1, x_2, \ldots, x_n | \alpha_1, \alpha_2, \ldots, \alpha_n)$ is defined by

$$(x_1, x_2, \ldots, x_n | \alpha_1, \alpha_2, \ldots, \alpha_n) = \det(x_i^{\alpha_j})_{1 \le i,j \le n},$$

where $\alpha_1, \alpha_2, \ldots, \alpha_n$ is a sequence of nonnegative integers.

(a) Show that an alternating polynomial f has the van der Monde determinant as a factor. Hence, f is alternating if and only if it equals Vg, where g is a symmetric polynomial.

(b) Show that every alternating polynomial is a linear combination of alternants.

(c) Let A be the $n \times n$ matrix whose ith row is

$$1 \quad x_i \quad x_i^2 \quad \ldots \quad x_i^{n-2} \quad \phi(x_1, x_2, \ldots, x_{i-1}, x_{i+1}, \ldots x_n),$$

where $\phi(x_1, x_2, \ldots, x_{i-1}, x_{i+1}, \ldots x_n)$ is a function in $n-1$ variables, with x_i omitted. Show that

$$\det A = cV(x_1, x_2, \ldots, x_n)$$

if and only if ϕ is a symmetric function in $n-1$ variables having total degree $n-1$.

(d) Let $\lambda = (\lambda_1, \lambda_2, \ldots, \lambda_n)$, where λ_i is a sequence of nondecreasing nonnegative integers. The Schur function S_λ is defined to be the following quotient of alternants:

$$\frac{(x_1, x_2, \ldots, x_n | \lambda_1 + n - 1, \lambda_2 + n - 2, \ldots, \lambda_{n-1} + 1, \lambda_n)}{V(x_1, x_2, \ldots, x_n)}.$$

Show that S_λ equals the sum of monomials, one for each way of filling in the Young diagram of $(\lambda_1, \lambda_2, \ldots, \lambda_n)$ so that rows are nondecreasing and columns are strictly increasing.

[3] Muir and Metzler (1933, chapter 11).

5.2.3. Define the *discriminant* $D(x_1, x_2, \ldots, x_n)$ by

$$D(x_1, x_2, \ldots, x_n) = \prod_{i:1 \le i < j \le n} (x_i - x_j)^2.$$

Express D as a polynomial in the elementary symmetric function.

5.2.4. *Crapo's formula for the permanent.*

This is a continuation of Exercise 3.1.9. Let $(x_{ij})_{1 \le i,j \le n}$ be an $n \times n$ matrix. If $R \subseteq \{1, 2, \ldots, n\}$ is a subset of row indices, let

$$\Phi(R) = \sum_{j=1}^{n} \left[\prod_{i:i \in R} x_{ij} \right].$$

If σ is a partition of $\{1, 2, \ldots, n\}$ with blocks $B_1, B_2, \ldots, B_c$, then

$$\Phi(\sigma) = \prod_{i=1}^{c} \Phi(B_i).$$

For example, if $n = 3$, then $\Phi(\{1, 2\}, \{3\})$ equals

$$(x_{11}x_{21} + x_{12}x_{22} + x_{13}x_{23})(x_{31} + x_{32} + x_{33}).$$

Show that

$$\operatorname{per} A = \sum_{\sigma:\sigma \in \Pi_n} \mu(\hat{0}, \sigma)\Phi(\sigma).$$

5.2.5. *Laws of symmetry.*[4]

(a) Let $a_\gamma = \sum_\delta c_{\gamma\delta} k_\delta$, where γ and the indices δ of summation are integer partitions. Show that $c_{\gamma\delta} = c_{\delta\gamma}$.

(b) Let $h_\gamma = \sum_\delta d_{\gamma\delta} k_\delta$, where γ and the indices δ of summation are integer partitions. Show that $d_{\gamma\delta} = d_{\delta\gamma}$.

5.2.6. *A theorem of Laguerre.*[5]

Show that the elementary symmetric polynomials $a_1, a_2, \ldots, a_n$ can be expressed as rational functions in the power-sum symmetric polynomials $s_1, s_3, s_5, \ldots, s_{2n-1}$ having odd degree.

5.2.7. *The co-invariant algebra.*[6]

Let $\mathbb{F}$ be an infinite field of characteristic 0 or positive characteristic greater than n and I be the ideal generated by the nonconstant elementary

[4] MacMahon (1887). The formulation given here follows P. Doubilet, *Foundations VII*.
[5] Pólya (1952).
[6] Chevalley (1955). In this elegant paper, Chevalley extends the fundamental theorems 5.2.6 and 5.2.7 (as well as the theorem in the exercise) to finite reflection groups.

symmetric polynomials $a_i(x_1, x_2, \ldots, x_n)$, $1 \leq i \leq n$, in the polynomial ring $\mathbb{F}[x_1, x_2, \ldots, x_n]$. The *co-invariant algebra* is the quotient $\mathbb{F}[x_1, x_2, \ldots, x_n]/I$.

Show *Chevalley's theorem for the symmetric group*. The co-invariant algebra $\mathcal{C}$ is a finite-dimensional algebra of dimension $n!$. In greater detail, the action of the symmetric group on $\mathcal{C}$ is isomorphic to the regular representation; $\mathcal{C}$ is graded by total degree (modulo I) and can be decomposed into $\bigoplus_d \mathcal{C}_d$, where $\mathcal{C}_d$ is the subspace of homogeneous polynomials of total degree d; its Hilbert function is given by the equation

$$\sum_{d=0}^{n(n-1)/2} \dim(\mathcal{C}_d)q^d = (1+q)(1+q+q^2)\cdots(1+q+q^2+\cdots+q^{n-1}).$$

5.2.8. *Modularly complemented geometric lattices.*[7]

A geometric lattice L is *modularly complemented* if for every point p in L, there exists a modular copoint c such that $p \not\leq c$. A geometric lattice L *splits* if there are proper flats X and Y such that for every point p, $p \leq X$ or $p \leq Y$.

(a) Show that a partition σ is modular in Π_n if and only if σ has at most one block of size greater than 1; that is, $\sigma = \{S\} \cup \{\{a\}: a \notin S\}$. Conclude that Π_n is modularly complemented.

(b) Show that if a rank-n geometric lattice L is modularly complemented and does not split, then there exists a subset of atoms such that the $\vee$-sublattice generated by the subset is isomorphic to the partition lattice Π_{n+1}. In this sense, the rank-n partition lattice is the minimal rank-n nonsplitting modularly complemented geometric lattice.

5.2.9. *Arrangements of hyperplanes.*[8]

Let $\mathcal{A}$ be a set of hyperplanes (that is, subspaces of dimension $n-1$) in the vector space $\mathbb{F}^n$, where $\mathbb{F}$ is a field. The *lattice* $L(\mathcal{A})$ *of intersections* consists of all subspaces in $\mathbb{F}^n$ formed by taking intersections of all subsets of $\mathcal{A}$, ordered by reverse set-containment. Denote by $[a_1x_1 + a_2x_2 + \cdots + a_nx_n]$ the hyperplane

$$\{(x_1, x_2, \ldots, x_n): a_1x_1 + a_2x_2 + \cdots + a_nx_n = 0\}.$$

(a) Show that $L(\mathcal{A})$ is a geometric lattice.

(b) Let $\mathcal{A}_{n-1}$ be the set of hyperplanes $\{[x_i - x_j] : 1 \leq i < j \leq n\}$. Show that $L(\mathcal{A}_{n-1})$ is isomorphic to Π_n.

[7] Kahn and Kung (1986). [8] Dowling (1973) and Orlik and Terao (1992).

(c) Let $\mathcal{H}_n(q)$ be the set of hyperplanes

$$\{[x_i]: 1 \leq i \leq n\} \cup \{[x_i - \alpha x_j]: 1 \leq i < j \leq n,\ \alpha \in \mathrm{GF}(q)\backslash\{0\}\}$$

in $\mathrm{GF}(q)^n$ and the *Dowling lattice* $Q_n(q)$ be the lattice of intersections of $\mathcal{H}_n(q)$. Show that $Q_n(q)$ is modularly complemented. If μ is the Möbius function of $Q_n(q)$, show that

$$\mu(\hat{0}, \hat{1}) = (-1)^n \prod_{i=0}^{n-1} ((q-1)i + 1).$$

5.3 Enumeration Under a Group Action

Let G be a group of permutations acting on a set S. Then G acts on the set $\mathrm{Fun}(S, X)$ of functions from S to another set X by the action: if $\gamma : S \to S$ is a permutation in G and $f : S \to X$ is a function, then γ sends f to the function $f\gamma$. The objective of *Pólya enumeration theory,* in its abstract form, is to derive information about the G-action on the function space $\mathrm{Fun}(S, X)$ from information about the G-action on the domain S.

We begin this section by setting up a framework for studying the combinatorics of group actions using concepts and results about Galois connections and incidence algebras from Sections 1.5 and 3.1. Using this framework, we develop an abstract Pólya enumeration theory from which classical Pólya theory is derived as a special case.[9]

We begin by defining a Galois coconnection (orb, per) between $L(G)$, the lattice of subgroups of a group G acting on a set S, and $\Pi(S)$, the lattice of partitions of S, under reverse refinement. If H is a subgroup of G, let $\mathrm{orb}(H)$ be the partition of S whose blocks are the orbits of H. More permutations make bigger orbits. Hence, $\mathrm{orb}: L(G) \to \Pi(S)$ is an order-preserving function. If π is a partition of S, let $\mathrm{per}(\pi)$ be the subgroup of (all) permutations in G leaving the blocks of π invariant. A coarser partition has bigger blocks and is left invariant by more permutations. Hence, $\mathrm{per}: \Pi(S) \to L(G)$ is order-preserving. Moreover, it is easy to check that

$$H \subseteq \mathrm{per}(\mathrm{orb}(H)) \quad \text{and} \quad \pi \geq \mathrm{orb}(\mathrm{per}(\pi)).$$

The functions (orb, per) form a Galois coconnection between $L(G)$ and $\Pi(S)$, and the composition $\mathrm{orb} \circ \mathrm{per}$ is a coclosure operator $\pi \mapsto \mathrm{ccl}(\pi)$ on $\Pi(S)$. A partition π is closed precisely when π equals $\mathrm{orb}(H)$ for some subgroup H of G. The closed partitions are called *periods.* Under the partial order of reverse refinement, the periods form a lattice $\mathcal{P}(G, S)$, the *lattice of periods* of G on S.

[9] Rota (1969b, pp. 330–334) and Rota and Smith (1977).

For example, let G be the symmetric group $\mathfrak{S}(S)$ of all permutations on the set S. If π is the partition $\{B_1, B_2, \ldots, B_c\}$ of S, then the *Young subgroup* $\mathfrak{S}(\pi)$ is the direct product defined by

$$\mathfrak{S}(\pi) = \mathfrak{S}(B_1) \times \mathfrak{S}(B_2) \times \cdots \times \mathfrak{S}(B_c).$$

By construction, $\mathfrak{S}(\pi)$ is the biggest subgroup H of $\mathfrak{S}(S)$ such that $\text{orb}(H) = \pi$. In particular, $\text{orb}(\mathfrak{S}(\pi)) = \pi$, every partition π is a period, and the lattice of periods of $\mathfrak{S}(S)$ is $\Pi(S)$.

Functions define partitions of S. If $f : S \to X$ is a function, then coimage(f) is the partition whose blocks are the nonempty sets among the inverse images $f^{-1}(x)$, $x \in X$. If γ is a permutation, then cycle(γ) is the partition whose blocks are the orbits of γ on S. A permutation γ *fixes* f if $f = f\gamma$ or, equivalently, $\text{cycle}(\gamma) \leq \text{coimage}(f)$. The *stabilizer* stab($f$) is the subgroup of permutations in G fixing f. The *period* per(f) of a function f is the partition orb(stab(f)), the partition whose blocks are the orbits of stab(f); in other words, per(f) is the coclosure of coimage(f).

Let x_{ij} be a set of variables, where i ranges over S and j ranges over X. The *coding monomial* Mon (f) is defined by

$$\text{Mon}\,(f) = \prod_{i:i \in S} x_{i,f(i)}.$$

The coding monomial retains all the information about the function f. If $\mathcal{F} \subseteq \text{Fun}(S, X)$, then

$$\text{Mon}\,(\mathcal{F}) = \sum_{f:f \in \mathcal{F}} \text{Mon}\,(f).$$

A subset $\mathcal{F}$ of Fun(S, X) is *G-closed* if $f \in \mathcal{F}$ and $\gamma \in G$ imply that $f\gamma$ is in $\mathcal{F}$.

Let $\mathcal{F}$ be a G-closed subset of functions in Fun(S, X). If π is a partition of S, let

$$\begin{aligned}
A(\mathcal{F}, \pi) &= \text{Mon}\,\{f\colon\ f \in \mathcal{F},\ \text{coimage}(f) = \pi\},\\
A(G, \mathcal{F}, \pi) &= \text{Mon}\,\{f\colon\ f \in \mathcal{F},\ \text{per}(f)\ =\ \pi\},\\
B(\mathcal{F}, \pi) &= \text{Mon}\,\{f\colon\ f \in \mathcal{F},\ \text{coimage}(f) \geq \pi\}.\\
B(G, \mathcal{F}, \pi) &= \text{Mon}\,\{f\colon\ f \in \mathcal{F},\ \text{per}(f) \geq \pi\}.
\end{aligned}$$

In general, these generating functions are formal power series in the variables x_{ij}. If X is finite, then they are polynomials.

5.3.1. Lemma. Let $\mathcal{F}$ be a G-closed class of functions and π be a period in $\Pi(S)$. Then

$$A(G,\mathcal{F},\pi) = \sum_{\sigma:\sigma\geq\pi \text{ in } \mathcal{P}(G,S)} \mu_{\mathcal{P}(G,S)}(\pi,\sigma)B(\mathcal{F},\sigma) = \sum_{\tau:\mathrm{ccl}(\tau)=\pi \text{ in } \Pi(S)} A(\mathcal{F},\tau).$$

Proof. In the lattice $\mathcal{P}(G, S)$ of periods,

$$B(G,\mathcal{F},\pi) = \sum_{\sigma:\sigma\geq\pi} A(G,\mathcal{F},\sigma).$$

By Möbius inversion,

$$A(G,\mathcal{F},\pi) = \sum_{\sigma:\sigma\geq\pi} \mu_{\mathcal{P}(G,S)}(\pi,\sigma)B(G,\mathcal{F},\sigma).$$

Now note that if σ is a period, then coimage$(f) \geq \sigma$ implies that per$(f) \geq \sigma$. Hence, for periods σ, $B(G,\mathcal{F},\sigma) = B(\mathcal{F},\sigma)$. The first part of the equation now follows.

Since per(f) is the coclosure of coimage(f),

$$\begin{aligned} A(G,\mathcal{F},\pi) &= \sum_{f:\mathrm{per}(f)=\pi} \mathrm{Mon}(f) \\ &= \sum_{\tau:\mathrm{ccl}(\tau)=\pi} \left(\sum_{f:\mathrm{coimage}(f)=\tau} \mathrm{Mon}(f) \right) \\ &= \sum_{\tau:\mathrm{ccl}(\tau)=\pi} A(\mathcal{F},\tau). \end{aligned}$$

This proves the second part. □

Next, we derive Pólya's enumeration theorem.[10] Let x_j be new variables, one for each element j in X and $\tilde{A}(G,\mathcal{F},\pi)$ be the formal power series obtained from $A(G,\mathcal{F},\pi)$ by the substitutions $x_{ij} = x_j$. Under these substitutions, Mon(f) becomes Gen(f), where Gen$(f) = \prod_{i:i\in S} x_{f(i)}$, the image monomial defined in Section 5.2. In particular,

$$\tilde{A}(G,\mathcal{F},\pi) = \sum_{f:\, f\in\mathcal{F},\, \mathrm{per}(f)=\pi} \mathrm{Gen}(f).$$

The formal power series $\tilde{A}(\mathcal{F},\pi)$, $\tilde{B}(\mathcal{F},\pi)$, and $\tilde{B}(G,\mathcal{F},\pi)$ are defined analogously.

The G-action on a G-closed subset $\mathcal{F}$ of functions partitions $\mathcal{F}$ into orbits. If two functions f_1 and f_2 are in the same orbit, then Gen$(f_1) =$ Gen(f_2). If

[10] For the standard treatment of Pólya enumeration theory, see, for example, Berge (1971, chapter 5), Harary and Palmer (1973), and Pólya and Read (1987).

$\mathcal{O}$ is an orbit, we define $\text{Gen}(\mathcal{O})$ to equal $\text{Gen}(f)$, where f is any function in $\mathcal{O}$. The *inventory* $\text{Inv}(G, \mathcal{F})$ of the G-action on $\mathcal{F}$ is the generating function defined by

$$\text{Inv}(G, \mathcal{F}) = \sum_{\mathcal{O}} \text{Gen}(\mathcal{O}),$$

where the sum ranges over all orbits in $\mathcal{F}$.

The inventory of $\mathcal{F}$ can be expressed as a linear combination in $\tilde{A}(G, \pi)$. To do this, we need the orbit-stabilizer lemma[11] from elementary group theory: if f is a function in the orbit $\mathcal{O}$, then $|\mathcal{O}| = |G|/|\text{stab}(f)|$. In particular, if $\pi = \text{per}(f)$, then $\text{per}(\pi) = \text{stab}(f)$. Hence, if $\mathcal{O}$ is the orbit containing f, then

$$|\mathcal{O}| = |G|/|\text{per}(\pi)|$$

and the size of the orbit of f depends only on the partition $\text{per}(f)$. Hence,

$$\begin{aligned}\text{Gen}(\mathcal{O}) &= \frac{1}{|\mathcal{O}|} \sum_{f:\, f \in \mathcal{O}} \text{Gen}(f) \\ &= \frac{1}{|G|} \sum_{f:\, f \in \mathcal{O}} |\text{stab}(f)| \text{Gen}(f).\end{aligned}$$

From this, we conclude that

$$\begin{aligned}\text{Inv}(G, \mathcal{F}) &= \sum_{\mathcal{O}} \left(\frac{1}{|G|} \sum_{f:\, f \in \mathcal{O}} |\text{stab}(f)| \text{Gen}(f) \right) \\ &= \frac{1}{|G|} \sum_{f:\, f \in \mathcal{F}} |\text{stab}(f)| \text{Gen}(f) \\ &= \frac{1}{|G|} \sum_{\pi:\, \pi \in \mathcal{P}(G,S)} |\text{per}(\pi)| \left(\sum_{f:\, \text{per}(f) = \pi} \text{Gen}(f) \right) \\ &= \frac{1}{|G|} \sum_{\pi:\, \pi \in \mathcal{P}(G,S)} |\text{per}(\pi)| \tilde{A}(G, \mathcal{F}, \pi).\end{aligned}$$

To state the next theorem, we need one more definition. If σ is a partition, let

$$\phi(\sigma) = |\{\gamma \colon \gamma \in G,\ \text{cycle}(\gamma) = \sigma\}|.$$

Recalling that $\text{per}(\sigma) = \{\gamma \colon \gamma \in G,\ \text{cycle}(\gamma) \leq \sigma\}$, we have

$$|\text{per}(\sigma)| = \sum_{\pi:\, \pi \leq \sigma} \phi(\pi).$$

[11] This lemma is proved by a bijection between elements of the orbit and cosets of the stabilizer.

By Möbius inversion, if $\sigma \in \mathcal{P}(G, S)$, then

$$\phi(\sigma) = \sum_{\pi : \pi \leq \sigma} |\mathrm{per}(\pi)| \mu_{\mathcal{P}(G,S)}(\pi, \sigma).$$

5.3.2. Theorem.

$$\sum_{\pi : \pi \in \mathcal{P}(G,S)} |\mathrm{per}(\pi)| \tilde{A}(G, \mathcal{F}, \pi) = \sum_{\sigma : \sigma \in \mathcal{P}(G,S)} |\phi(\sigma)| \tilde{B}(\mathcal{F}, \sigma).$$

Proof.

$$\begin{aligned}
\sum_{\pi : \pi \in \mathcal{P}(G,S)} |\mathrm{per}(\pi)| \tilde{A}(G, \mathcal{F}, \pi) &= \sum_{\pi : \pi \in \mathcal{P}(G,S)} \sum_{\sigma : \sigma \geq \pi} |\mathrm{per}(\pi)| \mu_{\mathcal{P}(G,S)}(\pi, \sigma) \tilde{B}(\mathcal{F}, \sigma) \\
&= \sum_{\sigma : \sigma \in \mathcal{P}(G,S)} \sum_{\pi : \pi \leq \sigma} |\mathrm{per}(\pi)| \mu_{\mathcal{P}(G,S)}(\pi, \sigma) \tilde{B}(\mathcal{F}, \sigma) \\
&= \sum_{\sigma : \sigma \in \mathcal{P}(G,S)} \phi(\sigma) \tilde{B}(\mathcal{F}, \sigma).
\end{aligned}$$

□

The right side of the equation in Theorem 5.3.2 is relatively easy to calculate from the group G. In particular, since $\phi(\sigma) = 0$ unless σ equals $\mathrm{cycle}(\gamma)$ for a permutation γ in G, it is not necessary to know what other partitions are periods. Also, although the proof involves a Möbius inversion, the Möbius function involved is transferred from one function to another and does not appear in the final result. Thus, the chore of calculating a Möbius function explicitly is avoided.

5.3.3. Corollary. Let $\mathcal{F}$ be a G-closed subset of functions. Then

$$\mathrm{Inv}(G, \mathcal{F}) = \frac{1}{|G|} \sum_{\gamma : \gamma \in G} \tilde{B}(\mathcal{F}, \mathrm{cycle}(\gamma)).$$

Proof. As observed earlier, $\phi(\sigma) = 0$ unless there is a permutation γ in G such that $\mathrm{cycle}(\gamma) = \sigma$. Hence,

$$\sum_{\sigma : \sigma \in \mathcal{P}(G,S)} |\phi(\sigma)| \tilde{B}(\mathcal{F}, \mathrm{cycle}(\gamma)) = \sum_{\gamma : \gamma \in G} \tilde{B}(\mathcal{F}, \mathrm{cycle}(\gamma)).$$

□

If $X = \{1, 2, \ldots, m\}$, $\mathcal{F} = \mathrm{Fun}(S, X)$, and $a_i(\gamma)$ is the number of cycles of length i in the cycle decomposition of γ, then

$$\tilde{B}(\mathcal{F}, \mathrm{cycle}(\gamma)) = \prod_{i=1}^{n} [x_1^i + x_2^i + \cdots + x_m^i]^{a_i(\gamma)}.$$

Hence, we obtain Pólya's enumeration theorem as a consequence of Corollary 5.3.3.

5.3.4. Pólya's enumeration theorem.

$$\mathrm{Inv}(G, \mathrm{Fun}(S, X)) = \frac{1}{|G|} \sum_{\gamma:\gamma \in G} \prod_{i=1}^{n} [x_1^i + x_2^i + \cdots + x_m^i]^{a_i(\gamma)}.$$

Pólya's theorem is usually stated in terms of cycle indices. If G is a permutation group acting on a finite set S of size n, then its *cycle index* $P(G; u_1, u_2, \ldots, u_n)$ is the polynomial, in yet another set of variables $u_1, u_2, \ldots, u_n$, defined by

$$P(G; u_1, u_2, \ldots, u_n) = \frac{1}{|G|} \sum_{\gamma:\gamma \in G} \prod_{i=1}^{n} u_i^{a_i(\gamma)}.$$

In this notation,

$$\mathrm{Inv}(G, \mathrm{Fun}(S, X)) = P\left(G; \sum_{i=1}^{m} x_i, \sum_{i=1}^{m} x_i^2, \sum_{i=1}^{m} x_i^3, \ldots, \sum_{i=1}^{m} x_i^n\right).$$

Exercises

5.3.1. Describe the lattice of periods of the alternating group $\mathfrak{A}(S)$ of even permutations on a set S.

5.3.2. Let $S = \{1, 2, \ldots, n\}$ and C_n be the group generated by the cyclic permutation $1 \mapsto 2, 2 \mapsto 3, \ldots, n-1 \mapsto n, n \mapsto 1$.

(a) Show that for each divisor d of n, the partition with n/d blocks

$$\{i, i + n/d, i + 2n/d, \ldots, i + (d-1)(n/d)\}$$

is a period and every period is of this form. Conclude that the lattice of periods is isomorphic to the lattice of divisors of n (ordered by divisibility). Note that the lattice of subgroups of C_n is also isomorphic to the lattice of divisors of n.

Since lattices of divisors are Cartesian product of chains, chains and product of chains are lattices of periods.

(b) Show that the Cartesian product of two lattices of periods is a lattice of periods.

(c) (Research problem) Is every finite lattice isomorphic to a lattice of periods?

5.3.3. (Research problem) Let $\mathrm{GF}(q)$ be a finite field. Consider the general linear group $\mathrm{GL}(n, q)$ as a permutation group on the vector space $\mathrm{GF}(q)^n$. Is there a good description of the lattice of periods?

5.3.4. (Research problem) Are the Dowling lattices in Exercise 5.2.9 "naturally" lattices of periods of a permutation group? Construct the lattice of periods of a wreath product of a permutation groups with another group.

5.3.5. *The cycle index of the symmetric group.*

Show that

$$\sum_{n=0}^{\infty} P(\mathfrak{S}_n; x_1, x_2, \ldots, x_n)t^n = \exp\left(\sum_{k=1}^{\infty} \frac{x_k t^k}{k}\right).$$

5.3.6. *Congruences from group actions.*[12]

Let S and X be finite sets and G be a permutation group acting on S.

(a) A function $f : S \to X$ is *aperiodic* if per(f) is the minimum partition $\hat{0}$. Show that the number of aperiodic functions is divisible by the order $|G|$ of the group G.

(b) Let $\mathcal{F}$ be a G-closed subset in Fun(S, X). Show that $\tilde{A}(G, \mathcal{F}, \hat{0}) \equiv 0 \bmod |G|$, and hence, by Theorem 5.3.1,

$$\sum_{\sigma:\sigma \geq \pi \text{ in } \mathcal{P}(G,S)} \mu_{\mathcal{P}(G,S)}(\pi, \sigma)\tilde{B}(\mathcal{F}, \sigma) \equiv 0 \mod |G|.$$

(c) Consider the case when G is the cyclic group C_n and $X = \{1, 2, \ldots, m\}$. Using Exercise 5.3.2 and (b), show that

$$\sum_{d:d|n} \mu(d)(x_1^d + x_2^d + \cdots + x_m^d)^{n/d} \equiv 0 \mod n.$$

Conclude that for any m-tuple of nonnegative integers $i_1, i_2, \ldots, i_m$ such that $i_1 + i_2 + \cdots + i_m = n$,

$$\sum_{d:d|n} \mu(d)\binom{n/d}{i_1/d, i_2/d, \ldots, i_m/d} \equiv 0 \mod n,$$

where a multinomial coefficient is defined to be zero if any of its parameters is not an integer. In particular, conclude that

$$\binom{p^r}{i_1, i_2, \ldots, i_m} \equiv \binom{p^{r-1}}{i_1/p, i_2/p, \ldots, i_m/p} \mod p^r$$

if p divides each of the indices $i_1, i_2, \ldots, i_m$, and

$$\binom{p^r}{i_1, i_2, \ldots, i_m} \equiv 0 \mod p^r,$$

otherwise.

5.3.7. (Research problem) Find a vector space analog of abstract Pólya enumeration theory.

[12] Rota and Sagan (1980). For further work, see Deutsch and Sagan (2006), Postnikov and Sagan (2007), and the references in these papers.

5.4 Baxter Operators

In the next three sections, we shall give a selective account of the theory of Baxter algebras.[13] The relevance of Baxter algebras to symmetric functions will become clear at the end of this section.

Let $\mathcal{B}$ be an algebra[14] over a field $\mathbb{F}$ and ϑ be an element in $\mathbb{F}$. An $\mathbb{F}$-linear operator $P : \mathcal{B} \to \mathcal{B}$ is a *Baxter operator* (*with parameter* ϑ) if it satisfies the *Baxter identity*: for all x and y in $\mathcal{B}$,

$$(Px)(Py) + \vartheta P(xy) = P(x(Py)) + P((Px)y).$$

A *Baxter algebra* $(\mathcal{B}, P)$ is a pair, where $\mathcal{B}$ is an $\mathbb{F}$-algebra and P is a Baxter operator on $\mathcal{B}$. Note that if P is a Baxter operator with nonzero parameter ϑ, then $-\vartheta^{-1}P$ is a Baxter operator with parameter -1. Thus, if $\vartheta \neq 0$, we may choose the parameter ϑ to be -1 (and indeed, any convenient nonzero value). The theory when $\vartheta = 0$ is slightly different.

A subset $\mathcal{D}$ of a Baxter algebra $(\mathcal{B}, P)$ is a *Baxter subalgebra* if $\mathcal{D}$ is an $\mathbb{F}$-subalgebra closed under P. (That is, if $x \in \mathcal{D}$, then $Px \in \mathcal{D}$.) A function f from a Baxter algebra $(\mathcal{A}, Q)$ to a Baxter algebra $(\mathcal{B}, P)$ is a *Baxter algebra homomorphism* if f is an algebra homomorphism and f commutes with the Baxter operators; that is, $fQ = Pf$.

5.4.1. Lemma. Let $E : \mathcal{A} \to \mathcal{A}$ be an endomorphism so that the endomorphism $I - E$ has an inverse and

$$P = E(I - E)^{-1}.$$

Then P is a Baxter operator with parameter -1.

Proof. Write $(I - E)^{-1}$ as the infinite sum $I + E + E^2 + \cdots$, so that

$$P = \sum_{i=1}^{\infty} E^i.$$

Then

$$\begin{aligned} P(x(Py)) &= P\left(\sum_{i=1}^{\infty} xE^i(y)\right) \\ &= \sum_{j=1}^{\infty} E^j(x)\left(\sum_{i=1}^{\infty} E^{j+i}(y)\right) \\ &= \sum_{i,j:\, j<i} E^j(x)E^i(y). \end{aligned}$$

[13] Rota (1969b) and Rota and Smith (1972). For a survey, see Rota (1995).

[14] We do not assume that $\mathcal{B}$ has an multiplicative identity 1. Most concrete examples coming from probability theory or functional analysis do not have identities.

Similarly,

$$(Px)(Py) = \sum_{i,j=1}^{\infty} E^i(x)E^j(y),$$

$$P(xy) = \sum_{i=1}^{\infty} E^i(x)E^i(y),$$

$$P((Px)y) = \sum_{i,j:\, j>i} E^j(x)E^i(y),$$

and Baxter's identity follows. □

A particularly important example of a Baxter algebra is the standard algebra. Let $\mathbb{F}$ be a field, $\mathcal{A}$ an $\mathbb{F}$-algebra, and $\mathcal{A}_\infty$ the $\mathbb{F}$-algebra of all sequences $(a_i)_{1\le i<\infty}$ with terms a_i in $\mathcal{A}$, under termwise addition, multiplication, and scalar multiplication. Then the operator

$$P : (a_1, a_2, a_3, \dots) \mapsto (0, a_1, a_1 + a_2, a_1 + a_2 + a_3, \dots)$$

is a Baxter operator with parameter -1. This follows from Lemma 5.4.1, taking E to be the *shift*

$$(a_1, a_2, a_3, \dots) \mapsto (0, a_1, a_2, a_3, \dots).$$

Let x_{ij}, $1 \le i \le n, 1 \le j < \infty$, be infinitely many indeterminates and $\mathbb{F}(x_{ij})$ be the field of rational functions in the indeterminates x_{ij}. Let $\underline{x}_i$ be the sequence $(x_{i1}, x_{i2}, x_{i3}, \dots)$. The *standard Baxter algebra* $(\mathcal{S}_n, P)$ on n generators is the intersection of all Baxter subalgebras in $\mathbb{F}(x_{ij})_\infty$ containing the sequences $\underline{x}_1, \underline{x}_2, \dots, \underline{x}_n$.

To see the connection with symmetric functions, do the following computation: let $n = 1$ and $\underline{x}$ be the sequence $(x_1, x_2, x_3, \dots)$. Then

$$\begin{aligned}
P\underline{x} &= (0, x_1, x_1 + x_2, x_1 + x_2 + x_3, x_1 + x_2 + x_3 + x_4, \dots),\\
\underline{x}(P\underline{x}) &= (0, x_1x_2, x_1x_3 + x_2x_3, x_1x_4 + x_2x_4 + x_3x_4,\\
&\qquad x_1x_5 + x_2x_5 + x_3x_5 + x_4x_5, \dots),\\
P(\underline{x}P(\underline{x})) &= (0, 0, x_1x_2, x_1x_2 + x_1x_3 + x_2x_3,\\
&\qquad x_1x_2 + x_1x_3 + x_2x_3 + x_1x_4 + x_2x_4 + x_3x_4, \dots),\\
\underline{x}P(\underline{x}P(\underline{x})) &= (0, 0, x_1x_2x_3, x_1x_2x_4 + x_1x_3x_4 + x_2x_3x_4,\\
&\qquad x_1x_2x_5 + x_1x_3x_5 + x_2x_3x_5 + x_1x_4x_5 + x_2x_4x_5 + x_3x_4x_5, \dots),\\
P(\underline{x}P(\underline{x}P(\underline{x}))) &= (0, 0, 0, x_1x_2x_3, x_1x_2x_3 + x_1x_2x_4 + x_1x_3x_4 + x_2x_3x_4, \dots),
\end{aligned}$$

and so on. Thus, the $(j+1)$st term in

$$\underbrace{P(\underline{x}P(\underline{x} \cdots P(\underline{x}P(\underline{x}))))}_{m \text{ times}}$$

is the elementary symmetric polynomial $a_m(x_1, x_2, \dots, x_j)$.

Similar computations show that if $\underline{y}_i$ are monomials in $\underline{x}_1, \underline{x}_2, \ldots, \underline{x}_n$, then in the sequence

$$P(\underline{y}_1(P(\underline{y}_2 \cdots P(\underline{y}_{m-1}P(\underline{y}_m))))),$$

the first m terms are zero, the $(m+1)$st term is a nonzero polynomial in the indeterminates x_{ij}, $1 \leq i \leq n, 1 \leq j < m+1$, and in general, if $l \geq m+1$, the lth term is a nonzero polynomial in x_{ij}, $1 \leq i \leq n$, $1 \leq j < l$.

Exercises

5.4.1. Give a proof of Lemma 5.4.1 not using an infinite expansion.

5.4.2. *Integration by parts.*[15]

Let $C(0, \infty)$ be the algebra of all real-valued continuous functions on the real half-line $[0, \infty)$ and let

$$Pf(x) = \int_0^x f(\xi)d\xi.$$

Then the integration-by-parts formula

$$\int udv = uv - \int vdu,$$

applied to $u = Pf$ and $v = Pg$, implies that P satisfies the Baxter identity

$$(Pf)(Pg) = P((Pf)g) + P(f(Pg))$$

with parameter 0.

Consider the initial value problem

$$\frac{dy}{dx} = \lambda\varphi(x)y, \quad y(0) = 1,$$

where $\varphi(x)$ is a continuous function. Then by a simple integration, the solution is given by

$$y = \exp(\lambda P(\varphi)) = 1 + \lambda P(\varphi) + \frac{1}{2!}\lambda^2 P(\varphi)^2 + \frac{1}{3!}\lambda^3 P(\varphi)^3 + \cdots.$$

Another way to obtain a solution is to integrate both sides of the differential equation, obtaining the operator equation

$$y = 1 + \lambda \int_0^x \varphi(\xi)y(\xi)d\xi = 1 + \lambda P(\varphi y).$$

[15] Baxter (1960).

This equation can be iterated, obtaining

$$\begin{aligned} y &= 1 + \lambda P(\varphi[1 + \lambda P(\varphi y)] \\ &= 1 + \lambda P(\varphi) + \lambda^2 P(\varphi P(\varphi[1 + \lambda P(\varphi y)]) \\ &= 1 + \lambda P(\varphi) + \lambda^2 P(\varphi P(\varphi)) + \lambda^3 P(\varphi(P(\varphi[1 + \lambda P(\varphi y)]) \\ &\vdots \\ &= 1 + \lambda P(\varphi) + \lambda^2 P(\varphi P(\varphi)) + \lambda^3 P(\varphi(P(\varphi P(\varphi)) + \cdots . \end{aligned}$$

Since the initial value problem has a unique solution, the two solutions are equal. Equating coefficients of powers of λ, we obtain a sequence of identities:

$$P(\varphi)^n = n! P(\varphi P(\varphi P(\cdots \varphi P(\varphi)))). \tag{B}$$

The case $n = 2$ is

$$2P(\varphi P(\varphi)) = P(\varphi)^2,$$

the special case of the Baxter identity with $\vartheta = 0$ and $x = y = \varphi$.

Show that all the identities (B) are implied by special case $n = 2$.

5.4.3. *Shuffle identities and convolutional algebras.*

Let $(\mathcal{A}, P)$ be a Baxter algebra with parameter 0.

(a) Show that P satisfies the *shuffle identities:* for elements $f_1, f_2, \ldots, f_n, g_1, g_2, \ldots, g_m$ in $\mathcal{A}$,

$$\begin{aligned} &P(f_1(P(f_2 P(f_3(P(\cdots Pf_n)))))) \, P(g_1(P(g_2 P(g_3(P(\cdots Pg_m)))))) \\ &\quad = \sum P(h_1(P(h_2 P(h_3(P(\cdots Ph_{n+m})))))), \end{aligned}$$

where the sum ranges over all *shuffles,* that is, a sequence h_k of length $m + n$ so that each f_i and g_j occurs exactly once and the subsequence formed by restricting h_k to the terms f_i (respectively, g_j) is the sequence $f_1, f_2, \ldots, f_n$ (respectively, $g_1, g_2, \ldots, g_m$) in the original order.

The shuffle identity is satisfied by the integral operator P defined in Exercise 5.4.2. Although the study of derivations, that is, *differential algebra*, has been much studied, the countertheory, *integration algebra* or the *theory of indefinite integration*, has been neglected, except by pioneers like K.-T. Chen.[16]

(b) (Research problem) Find an elementary proof of the *Titchmarsh convolution theorem*: let $f(x)$ and $k(x)$ be real-valued integrable functions on the interval $(0, \gamma)$ and

$$\int_0^\gamma f(y)k(x-y)dy = 0$$

[16] See Chen (2001).

for almost all x in $(0, \gamma)$. Then there exist α and β such that $f(x) = 0$ for almost all x in $(0, \alpha)$ and $k(x) = 0$ for almost all x in $(0, \beta)$, where $\alpha + \beta = \gamma$.

Titchmarsh's theorem implies that the ring of continuous real-valued function on $(0, \infty)$ is an integral domain. An elementary proof would be a good test case for the theory of indefinite integration.[17]

5.5 Free Baxter Algebras

The main result in this section is a description of free Baxter algebras. We begin with an informal discussion of free algebras. If $\mathbb{F}$ is a field, then the free (commutative) $\mathbb{F}$-algebra on n generators is the polynomial algebra $\mathbb{F}[x_1, x_2, \ldots, x_n]$. The polynomial algebra is free in two senses. It is the algebra of all possible expressions constructed from the generators x_i using the three algebra operations: addition, multiplication, and scalar multiplication. The polynomial algebra also satisfies a *universal property*: if $\mathcal{R}$ is an $\mathbb{F}$-algebra which can be generated by n generators $a_1, a_2, \ldots, a_n$, then the function $x_i \mapsto a_i$ extends to an $\mathbb{F}$-algebra homomorphism $\mathbb{F}[x_1, x_2, \ldots, x_n] \to \mathcal{R}$. The universal property is easily shown in this case. The homomorphism property forces us to send a polynomial $f(x_1, x_2, \ldots, x_n)$ to $f(a_1, a_2, \ldots, a_n)$. Since polynomials satisfy no relations (other than those implied by the $\mathbb{F}$-algebra axioms), this extension is a well-defined homomorphism.

From this example, we see that there are two ways to explicitly describe free algebras: by construction, that is, finding a way to write down "all possible elements," or by showing that a specific concretely defined algebra satisfies the universal property. Roughly speaking, the first approach focuses on the syntax and the second focuses on a particular semantics or a specific model.

The construction of the free Baxter algebra over a field $\mathbb{F}$ begins with the construction of the free algebra $\mathcal{F}_n$ with a linear operator T (not satisfying any identity) on n generators $\xi_1, \xi_2, \ldots, \xi_n$. The elements in $\mathcal{F}_n$ can be defined recursively by

(1) The elements in $\mathbb{F}$ and the generators $\xi_1, \xi_2, \ldots, \xi_n$ are in $\mathcal{F}_n$.

(2) If u and v are in $\mathcal{F}_n$, then $u + v$, uv, and Tu are in $\mathcal{F}_n$.

The recursive definition certainly produces all possible expressions, but an element is always produced several times. For example, the element $T((T\xi_1)\xi_2 + T(\xi_1\xi_4)T(\xi_1)\xi_5)$ is produced. Since T is linear, this element equals $T((T\xi_1)\xi_2) + T(T(\xi_1\xi_4)T(\xi_1)\xi_5)$. The same "crisis" occurs for polynomials and is resolved by a tacit agreement as to when two expressions specify the

[17] See problem 3 in Rota (1998b). Titchmarsh's theorem can be found in Titchmarsh (1986).

same polynomial (see Exercise 1.3.8 for a formal treatment of "equality" in the case of lattice polynomials).

A *monomial* in $\mathcal{F}_n$ is an expression obtained using only multiplication and the operator T. The *occurrence* $\operatorname{occ}(m)$ of the monomial m is the number of times T occurs in it. For example, $\operatorname{occ}(T(\xi_1^2 T(\xi_1\xi_2^5\xi_3 T(\xi_3\xi_2^2)))) = 3$. If $t \in \mathcal{F}_n$, then t can be written (irredundantly) as a linear combination $\sum \lambda_i a_i$ of monomials a_i with coefficients λ_i in $\mathbb{F}$. We define $\operatorname{occ}(t)$ to be the maximum of the occurrences $\operatorname{occ}(a_i)$.

A subset I in $\mathcal{F}_n$ is a T-ideal if it is an ideal and closed under the operator T. (That is, if $u \in I$, then $Tu \in I$.) The *free Baxter algebra* $(\mathcal{B}_n, T)$ on n generators with parameter ϑ can be defined as the quotient of $\mathcal{F}_n$ by the T-ideal H generated by all elements of the form

$$(Tx)(Ty) + \vartheta T(xy) - T(x(Ty)) - T((Tx)y),$$

where $x, y \in \mathcal{F}_n$. By slightly more general versions of the homomorphism theorems for $\mathbb{F}$-algebras, $(\mathcal{B}_n, T)$ satisfies the universal property. Hence, it is indeed free.

Having defined the free Baxter algebra, we will provide two descriptions. We begin with a concrete description due to Rota.

5.5.1. Theorem. The standard Baxter algebra $(\mathcal{S}_n, P)$ is isomorphic to the free Baxter algebra $(\mathcal{B}_n, T)$ with parameter -1.

Proof. The assignment $\xi_i \mapsto \underline{x}_i$ extends to a homomorphism φ from the free algebra $\mathcal{F}_n$ with operator T onto the standard algebra $(\mathcal{S}_n, P)$ satisfying $\varphi T = P\varphi$. Since $(\mathcal{S}_n, P)$ is a Baxter algebra, the kernel of φ contains the T-ideal H (with $\vartheta = -1$). We will prove the theorem by showing that the kernel of φ equals H.

5.5.2. Proposition. For all t in $\mathcal{F}_n$, $\varphi(t) = 0$ implies that $t \in H$.

To do this, we proceed by induction on $\operatorname{occ}(t)$. We begin with a lemma.

5.5.3. Lemma. Every element t of $\mathcal{F}_n$ can be written as a sum $r + s$, where r is in the T-ideal H and s is a linear combination of monomials of the form

$$aTb, \quad a, \quad Tb,$$

with a and b monomials, $\operatorname{occ}(a) = 0$, and $\operatorname{occ}(b) < \operatorname{occ}(t)$.

Proof. It suffices to prove the lemma for a monomial t. We argue by induction. The lemma holds trivially if $\operatorname{occ}(t) = 0$. If t has two factors of the form Tc

and Td, then, choosing c and d suitably, we can write $t = e(Tc)(Td)$. Then $t = r + s$, where

$$r = e[(Tc)(Td) - T(cd) - T(c(Td)) - T((Tc)d)]$$

is in the T-ideal H and

$$s = eT(cd) + ec(Td) + ed(Tc).$$

In addition, occ(e), occ(cd), occ(ec), occ(d) occ(ed), and occ(c) are all strictly smaller than occ(t) − 1. The lemma now follows by induction. □

By the lemma, each element t of $\mathcal{F}_n$ can be written as a sum $r + s$, with $r \in H$ and

$$s = \sum_i \alpha_i a_i T b_i + \sum_j \beta_j T c_j + \sum_k \gamma_k d_k = u + v + w, \qquad \text{(S)}$$

where u, v, and w denote the three linear combinations, and a_i and d_k are monomials in $\xi_1, \xi_2, \ldots, \xi_n$ formed using only multiplication. To show that $\ker \varphi = H$, it suffices to show that whenever s has the special form (S), $\varphi(s) = 0$ implies that $s \in H$.

Since $\varphi T = P\varphi$, we have

$$\varphi(s) = \sum_i \alpha_i \varphi(a_i) P\varphi(b_i) + \sum_j \beta_j P\varphi(c_j) + \sum_k \gamma_k \varphi(d_k) = \varphi(u) + \varphi(v) + \varphi(w).$$

We will now show that $\varphi(s) = 0$ implies that $w = 0$. To do this, we need two observations: First, each monomial d_k in w is formed using only multiplication. Thus, the jth term in $\varphi(w)$ equals $p(x_{1j}, x_{2j}, \ldots, x_{nj})$ for some polynomial p (not depending on j). Second, the image of P is contained in the set of sequences with first term equal to 0. Since the monomials in both u and v use P at least once and $\varphi(s) = 0$, the first term of the sequence $\varphi(w)$ is 0. These two observations imply that $p(x_{11}, x_{21}, \ldots, x_{n1}) = 0$, where $x_{11}, x_{21}, \ldots, x_{n1}$ are indeterminates; that is, p is the zero polynomial. We conclude that $w = 0$.

We note that if occ(t) = 0, then $u = v = 0$. Thus, we have also established the case occ(t) = 0 of Proposition 5.5.2.

We can now write $t = r + s$, where $s = u + v$. We can also assume, by induction, that if occ(t') < occ(t), then $\varphi(t') = 0$ implies that $t' \in H$.

Let term$_m(\underline{y})$ be the mth term in a sequence $\underline{y}$ in $(\mathcal{S}_n, P)$. Observe that if term$_m(\underline{y})$ is the first nonzero term in $\underline{y}$, then the $(m+1)$st term in $P\underline{y}$ equals y_m and is nonzero. Hence, in $\mathcal{S}_n$, $P\underline{y} = 0$ implies that $\underline{y} = 0$. In particular, since $\varphi(Tb) = P(\varphi b)$, $\varphi(Tb) = 0$ implies that $\varphi(b) = 0$.

We apply this argument to a monomial c_j in v. If $\varphi(Tc_j) = 0$, then $\varphi(c_j) = 0$. Since $\mathrm{occ}(c_j) < \mathrm{occ}(t)$, this implies that $c_j \in H$. Similarly, consider a monomial $a_i T b_i$ in u. Since $\mathrm{occ}(a_i) = 0$, a_i is a nonzero ordinary monomial in ξ_i and the jth term in $\varphi(a_i)$ is the same nonzero monomial, with x_{ij} substituted for ξ_i. As φ is a homomorphism, $\varphi(a_i T b_i) = \varphi(a_i) P \varphi(b_i)$. Suppose that $\varphi(a_i T b_i) = 0$. Then since $\varphi(a_i)$ has no nonzero terms, $P\varphi(b_i) = 0$, and we can conclude, as earlier, that $b_i \in H$.

The preceding argument allows us to tidy up and assume that

$$s = \sum_i \alpha_i a_i T b_i + \sum_j \beta_j T c_j = u + v,$$

where the monomials a_i are nonconstant, the terms $a_i T b_i$ and $T c_j$ are distinct, and none of the terms is in the T-ideal H.

The next step is to show that $\varphi(s) = 0$ implies that $\varphi(u) = 0$ and $\varphi(v) = 0$ separately, or, put another way, there is no cancellation of terms in $\varphi(u)$ and $\varphi(v)$.

We first show that $\varphi(u)$ cannot be nonzero. Suppose the contrary, that

$$\sum_i \alpha_i \varphi(a_i) P \varphi(b_i) \neq 0.$$

Let m be the index of the first nonzero term in the sequence $\varphi(u)$ and $\mathrm{term}_m(\varphi(a_i)P\varphi(b_i))$ be a summand from the linear combination contributing nontrivially to $\mathrm{term}_m(\varphi(u))$. Since a_i is a nonconstant monomial, $\mathrm{term}_m(\varphi(a_i))$ uses at least one of the indeterminates x_{lm}. Hence, $\mathrm{term}_m(\varphi(u))$ is a nonzero polynomial using at least one of the indeterminate x_{lm}. As $\varphi(s)$ is assumed to be zero, $\mathrm{term}_m(\varphi(v)) = -\mathrm{term}_m(\varphi(u))$. However, because P is applied to all the terms in $\varphi(v)$, $\mathrm{term}_m(\varphi(v))$ is a polynomial using only the indeterminates x_{lj}, $j \leq m-1$, a contradiction. We conclude that $\varphi(u) = 0$, and hence, $\varphi(v) = 0$ as well.

Next, we show that $\varphi(v) = 0$ implies $v \in H$. But since

$$\varphi(v) = \sum_j \beta_j P \varphi(c_j) = P\varphi\left(\sum_j \beta_j c_j\right) = 0,$$

$\varphi(\sum_j \beta_j c_j) = 0$. As $\mathrm{occ}(c_j) < \mathrm{occ}(t)$, we conclude, by induction, that $\sum_j \beta_j c_j \in H$, and hence, $v \in H$.

The final task is to show that $\varphi(u) = 0$ implies $u \in H$. To do this, we regroup the linear combination u according to the monomials a_i and redefine a_i and b_i so that

$$u = \sum a_i T b_i,$$

where the monomials a_i are distinct (and nonconstant) and b_i are linear combinations of a monomials. For each index m, $\mathrm{term}_m(\varphi(a_i))$ is a monomial in x_{im}, whereas $\mathrm{term}_m(P\varphi(b_i))$ is polynomial in x_{lj}, $j < m$. Since the monomials $\mathrm{term}_m(\varphi(a_i))$ are distinct, this implies that $\mathrm{term}_m(P\varphi(b_i)) = 0$; otherwise, we have a nontrivial algebraic relation among the indeterminates x_{lj}. Thus, $P\varphi(b_i)$ is the zero sequence. As in the earlier cases, this implies that each b_i is in H, and hence, $u \in H$.

We can now conclude that s and, thus, t are in H. □

We remark that we have restricted our discussion to standard algebras $(\mathcal{S}_n, P)$ with a finite number of generators. This was done to keep the notation simple. Standard algebras can be defined verbatim for any set of generators. Theorem 5.5.1 and its proof are valid with minor changes.

The second description of the free Baxter algebra is due to Cartier.[18] Cartier introduced a *square bracket* notation for calculations in Baxter algebras. Let $(\mathcal{B}, P)$ be a Baxter algebra. Recursively, we define $[a] = P(a)$, and if $n \geq 2$,

$$[a_1, a_2, a_3, \ldots, a_n] = P(a_1[a_2, a_3, \ldots, a_n]),$$

where a and a_n are elements in $\mathcal{B}$ and $a_1, a_2, \ldots, a_{n-1}$ may be an element in $\mathcal{B}$ or the *empty* element $|$. The empty element acts like an identity, but is not allowed at the extreme right of a bracket. The recursive definition specifies that for every square bracket, an application of P is made. For example,

$$[a_1, a_2, a_3] = P(a_1[a_2, a_3]) = P(a_1 P(a_2[a_3]) = P(a_1 P(a_2 P(a_3)),$$

$[a_1, |, a_3] = P(a_1 P^2(a_3))$, and $[|, |, a] = P^3(a)$. In this notation, the Baxter identity becomes

$$[a][b] = -\vartheta[ab] + [a, b] + [b, a].$$

The next lemma allows us to express a product of brackets as a linear combination of brackets.

5.5.4. Cartier's identity.

$$\prod_{j=1}^{m}[a_{j1}, a_{j2}, \ldots, a_{jr_j}] = \sum_{n, I_1, I_2, \ldots, I_m} (-\vartheta)^{r_1+r_2+\cdots+r_m-n}[c_1, c_2, \ldots, c_n],$$

[18] Cartier (1972).

where the sum ranges over all n, $I_1, I_2, \ldots, I_m$, where n is an integer such that $1 \leq n \leq r_1 + r_2 + \cdots + r_m$, $|I_j| = r_j$, $I_1 \cup I_2 \cup \cdots \cup I_m = \{1, 2, \ldots, n\}$, and

$$c_k = \prod_{l : k \in I_l} a_{l\alpha_l},$$

where the product is taken over all l such that $k \in I_l$ and k is the α_lth number in I_l when I_l is put in increasing order.

The following example, involving the triple product $[a_1, a_2, a_3, a_4][b_1, b_2, b_3][c_1, c_2]$, where we write, say, b_3 instead of a_{23}, shows how a term in the sum on the right is formed. In this case, $m = 3$, $r_1 = 4$, $r_2 = 3$, and $r_3 = 2$. Taking $n = 6$, $I_1 = \{1, 3, 4, 5\}$, $I_2 = \{2, 4, 6\}$, and $I_3 = \{4, 6\}$, we obtain the term

$$(-\vartheta)^3 [a_1, b_1, a_2, a_3 b_2 c_1, a_4, b_3 c_2]$$

on the right-hand side.

We sketch the proof of Cartier's identity. By induction on number of brackets, it suffices to prove the lemma for two brackets. Let

$$Y = [a_1, a_2, \ldots, a_r][b_1, b_2, \ldots, b_s]$$

be a product of two brackets. We induct on the *combined length* $r + s$ of the two brackets. Write the product Y as

$$P(a_1 [a_2, a_3, \ldots, a_r] P(b_1 [b_2, b_3, \ldots, b_s]).$$

By Baxter's identity,

$$\begin{aligned} Y = &-\vartheta P(a_1 b_1 [a_2, a_3, \ldots, a_r][b_2, b_3, \ldots, b_s]) \\ &+ P(a_1 [a_2, a_3, \ldots, a_r][b_1, b_2, b_3, \ldots, b_s]) \\ &+ P(b_1 [a_1, a_2, a_3, \ldots, a_r][b_2, b_3, \ldots, b_s]). \end{aligned}$$

The three products of two brackets on the right side all have combined length less than $r + s$. Hence by induction, they are linear combinations of brackets. Thus,

$$\begin{aligned} P(a_1 b_1 [a_2, a_3, \ldots, a_r][b_2, b_3, \ldots, b_s]) &= P\left(\sum (-\vartheta)^j a_1 b_1 [c_{i1}, c_{i2}, \ldots, c_{ir_i}]\right) \\ &= \sum (-\vartheta)^j [a_1 b_1, c_{i1}, c_{i2}, \ldots, c_{ir_i}], \end{aligned}$$

where the exponent j in the coefficient $(-\vartheta)^j$ depends on the summand. Similar calculations show that the other two terms are linear combinations of brackets. Cartier's identity follows by induction and careful bookkeeping.

Cartier's identity says that in any Baxter algebra, the set of elements of the form

$$a_0 [a_1, a_2, \ldots, a_n]$$

spans. This fact underlies Cartier's construction of free Baxter algebras. The *free Baxter algebra* $(\mathcal{C}(X), P)$ over the field $\mathbb{F}$ on the set X of generators is defined in the following way: consider all expressions of the form $a_0[a_1, a_2, \ldots, a_n]$, where $a_0, a_1, \ldots, a_{n-1}$ are monomials (of positive degree) in the generators or the empty element, and a_n is a positive-degree monomial. Form the vector space $\mathcal{C}(X)$ of all formal linear combinations of these expressions. Define a multiplication by

$$a_0[a_1, a_2, \ldots, a_r]b_0[b_1, b_2, \ldots, b_s] = \sum_{n, I_1, I_2, \ldots, I_m} (-\vartheta)^{r_1+r_2+\cdots+r_m-n} a_0 b_0[c_1, c_2, \ldots, c_n],$$

where the sum on the left is the same as in Cartier's identity. The Baxter operator P on $\mathcal{C}(X)$ is defined by

$$P(a_0[a_1, a_2, \ldots, a_r]) = [a_0, a_1, a_2, \ldots, a_r]$$

on square brackets and extended by linearity. It is complicated but straightforward to show that $(\mathcal{C}(X), P)$ is free.

Exercises

5.5.1. Show that

$$[a][b_1, b_2, \ldots, b_r] = -\sum_{i=1}^{r} \vartheta[b_1, \ldots, b_{i-1}, ab_i, b_{i+1} \ldots, b_r]$$
$$+ [a, b_1, b_2, \ldots, b_r] + \sum_{i=2}^{r} [b_1, \ldots, b_{i-1}, a, b_i, \ldots, b_r]$$
$$+ [b_1, b_2, \ldots, b_r, a].$$

5.5.2. Do the bookkeeping required to prove the exact form of Cartier's identity.

5.5.3. *Free noncommutative Baxter algebras.*[19]
Find explicit descriptions of free noncommutative Baxter algebras.

5.5.4. (Research problem) Are there simple Baxter algebra expressions, perhaps in terms of Cartier brackets, for other classical symmetric functions?

[19] Aguiar and Moreira (2006) and Ebrahimi-Fard and Guo (2008).

5.6 Identities in Baxter Algebras

Since every Baxter algebra generated by n elements is a homomorphic image of the standard algebra $(\mathcal{S}_n, P)$, Theorem 5.5.1 and the universal property for free algebras imply that if an identity holds in $(\mathcal{S}_n, P)$, then it holds (after renormalizing the parameter) in a Baxter algebra with nonzero parameter.

5.6.1. Theorem. Let P be a Baxter operator with nonzero parameter ϑ. Then

$$\sum_{m=0}^{\infty} \underbrace{P(\underline{x}P(\underline{x}\cdots P(\underline{x}P(\underline{x}))))}_{m \text{ times}} t^m = \exp\left(\sum_{k=1}^{\infty} \frac{\vartheta^{k-1} P(x^k)t^k}{k}\right)$$

as a formal power series identity in the variable t.

Proof. We begin by proving this identity in the standard algebra $(\mathcal{S}_1, P)$. As observed at the end of Section 5.4, the $(j+1)$st term in the sequence $P(\underline{x}\cdots P(\underline{x}P(\underline{x})))$, with m occurrences of P, is the elementary symmetric polynomial $a_m(x_1, x_2, \ldots, x_j)$. Next, observe that

$$P(\underline{x}^k) = (0, x_1^k, x_1^k + x_2^k, x_1^k + x_2^k + x_3^k, \ldots);$$

that is, the $(j+1)$st term in $P(\underline{x}^k)$ is the power-sum symmetric polynomial $S_k(x_1, x_2, \ldots, x_j)$.

By Waring's identity (5.2.5),

$$\sum_{m=0}^{\infty} \text{term}_{j+1}(P(\underline{x}\cdots P(\underline{x}P(\underline{x}))))t^m = \exp\left(\sum_{k=1}^{\infty} \frac{(-1)^{k-1}\text{term}_{j+1}(P(\underline{x}^k))t^k}{k}\right).$$

Thus, as a formal power series with coefficients which are sequences of polynomials,

$$\sum_{m=0}^{\infty} P(\underline{x}\cdots P(\underline{x}P(\underline{x})))t^m = \exp\left(\sum_{k=0}^{\infty} \frac{(-1)^{k-1}P(\underline{x}^k)t^k}{k}\right).$$

The theorem when the parameter is -1 now follows from the universal property. To obtain the general case, replace P by $-\vartheta^{-1}P$ and t by $-\vartheta t$. □

Two concrete instances of Theorem 5.6.1 are given in Exercises 5.6.1 and 5.6.2.

We turn now to identities obtained using the theory of group action developed in Section 5.3. Let $S = \{1, 2, \ldots, n\}$, G be a permutation group acting on S, and $X = \{1, 2, 3, \ldots\}$. Consider the generating functions $A(\mathcal{F}, \pi)$, $A(G, \mathcal{F}, \pi)$, $B(\mathcal{F}, \pi)$, and $B(G, \mathcal{F}, \pi)$. As X is infinite, they are formal power series. Define $\underline{A}(G, \mathcal{F}, \pi)$ to be the sequence whose jth term is the polynomial $A_j(G, \mathcal{F}, \pi)$

obtained from $A(G, \mathcal{F}, \pi)$ by setting $x_{ik} = 0$, $k \geq j$. The sequences $\underline{A}(\mathcal{F}, \pi)$, $\underline{B}(\mathcal{F}, \pi)$, and $\underline{B}(G, \mathcal{F}, \pi)$ are defined analogously.

5.6.2. Lemma. If the partition π is a period, then the sequences $\underline{A}(\mathrm{Fun}(S, X), \pi)$, $\underline{A}(G, \mathrm{Fun}(S, X), \pi)$, $\underline{B}(\mathrm{Fun}(S, X), \pi)$, and $\underline{B}(G, \mathrm{Fun}(S, X), \pi)$ are in the standard Baxter algebra $(\mathcal{S}_n, P)$.

Proof. If $D \subseteq S$, let

$$\underline{x}(D) = \prod_{d:d \in D} \underline{x}_d.$$

If π is the partition $\{D_1, D_2, \ldots, D_c\}$, then a straightforward calculation (similar to that at the end of Section 5.4) yields

$$\begin{aligned}&\underline{A}(\mathrm{Fun}(S, X), \pi)\\ &\quad = \sum_{\gamma:\gamma \in \mathfrak{S}_c} P(\underline{x}(D_{\gamma(1)}) P\underline{x}(D_{\gamma(2)}) \cdots P(\underline{x}(D_{\gamma(c-1)}) P(\underline{x}(D_{\gamma(c)})))),\end{aligned}$$

the sum ranging over all permutations γ of $\{1, 2, \ldots, c\}$. Another calculation yields

$$\underline{B}(\mathrm{Fun}(S, X), \pi) = P(\underline{x}(D_1)) P(\underline{x}(D_2)) \cdots P(\underline{x}(D_c)).$$

We conclude that $\underline{A}(\mathrm{Fun}(S, X), \pi)$, $\underline{B}(\mathrm{Fun}(S, X), \pi)$, and hence, as π is a period, $\underline{A}(G, \mathrm{Fun}(S, X), \pi)$ and $\underline{B}(G, \mathrm{Fun}(S, X), \pi)$ are in $(\mathcal{S}_n, P)$. □

If π is the partition $\{D_1, D_2, \ldots, D_c\}$ of $\{1, 2, \ldots, n\}$ and $x_1, x_2, \ldots, x_n$ are elements in a Baxter algebra $(\mathcal{B}, P)$ with nonzero parameter ϑ, let

$$\begin{aligned}&A(\pi, P; x_1, x_2, \ldots, x_n)\\ &\quad = \sum_{\gamma:\gamma \in \mathfrak{S}_c} (-\vartheta)^{n-c} P(x(D_{\gamma(1)}) P(x(D_{\gamma(2)}) \cdots P(x(D_{\gamma(c)}))))\end{aligned}$$

and

$$B(\pi, P; x_1, x_2, \ldots, x_n) = (-\vartheta)^{n-c} P(x(D_1)) P(x(D_2))) \cdots P(x(D_c)),$$

where $x_D = \prod_{d:d \in D} x_d$.

5.6.3. Theorem. Let G be a permutation group acting on a finite set S and π be a partition in the lattice $\mathcal{P}(G, S)$ of periods. Then in a Baxter algebra with nonzero parameter ϑ,

$$\sum_{\tau:\,\mathrm{ccl}(\tau)=\pi \text{ in } \Pi(S)} A(\tau, P; x_1, x_2, \ldots, x_n)$$
$$= \sum_{\sigma:\,\sigma\geq\pi} \mu_{\mathcal{P}(G,S)}(\pi, \sigma) B(\pi, P; x_1, x_2, \ldots, x_n).$$

Proof. By Lemma 5.3.1, the identity holds in the standard algebra. By the universal property, these identities hold in all Baxter algebras with parameter -1 and after renormalization, in all Baxter algebras with nonzero parameter. □

We note the following special case when G is the symmetric group $\mathfrak{S}_n$ acting on $\{1, 2, \ldots, n\}$ and π is the minimum partition $\hat{0}$.

5.6.4. Corollary. In a Baxter algebra with nonzero parameter ϑ,

$$\sum_{\gamma:\,\gamma\in\mathfrak{S}_n} P(x_{\gamma(1)}(Px_{\gamma(2)} \cdots P(x_{\gamma(n)})))$$
$$= \sum_{\{D_1, D_2, \ldots, D_c\}} \vartheta^{n-c} \prod_{j=1}^{c} (|D_i| - 1)! P(x(D_i)),$$
$$= \sum_{\gamma:\,\gamma\in\mathfrak{S}_n} \vartheta^{n-c} \prod_{D:\,D\in\mathrm{cycle}(\gamma)} P(x(D)),$$

where the second sum ranges over all partitions $\{D_1, D_2, \ldots, D_c\}$ of $\{1, 2, \ldots, n\}$ and for a permutation γ, cycle(γ) is the partition given by the cycle decomposition of γ.

Proof. The lattice of periods of $\mathfrak{S}_n$ acting on $\{1, 2, \ldots, n\}$ is the lattice of partitions Π_n. By Theorem 5.2.1,

$$\mu(\hat{0}, \{D_1, D_2, \ldots, D_c\}) = (-1)^{n-c} \prod_{i=1}^{c} (|D_i| - 1)!.$$

This proves the first equality. The second equality follows by observing that the number of permutations γ such that cycle$(\gamma) = \{D_1, D_2, \ldots, D_c\}$ equals $\prod_{i=1}^{c} (|D_i| - 1)!$. □

Theorem 5.6.3 was motivated by an identity of Spitzer and Bohnenblust.[20] To describe this identity, we need another Baxter algebra. Let $\mathcal{M}$ be the $\mathbb{R}$-algebra of functions $f : \mathbb{R} \to \mathbb{R}$ with finite support (that is, $f(x) = 0$ except

[20] Spitzer (1956).

for a finite set of real numbers), with *convolution*

$$fg(x) = \sum_{y:\, y \in \mathbb{R}} f(y)g(x-y) = \sum_{y,z:\, y+z=x} f(y)g(z)$$

as the product. If x is a real number, let

$$x^+ = \frac{|x|+x}{2} = \max\{0, x\}.$$

Let $P: \mathcal{M} \to \mathcal{M}$ be the linear operator

$$Pf(x) = \sum_{y:y^+=x} f(y).$$

Put another way, $Pf(x) = f(x)$ if x is positive and $Pf(0) = \sum_{y:\, y \le 0} f(y)$.

5.6.5. Lemma. The operator $P : \mathcal{M} \to \mathcal{M}$ is a Baxter operator with parameter 1.

Proof. Observe that

$$P(fPg + gPf)(x) = \sum_{y,z:\, (y^+ + z)^+ = x \text{ or } (y+z^+)^+ = x} f(y)g(z)$$

$$(P(fg) + (Pg)(Pf))(x) = \sum_{y,z:\, (y+z)^+ = x \text{ or } y^+ + z^+ = x} f(y)g(z).$$

Thus, Baxter's identity follows if for every pair y, z of real numbers, the two multisets

$$\{(y^+ + z)^+,\ (y + z^+)^+\}$$

and

$$\{(y+z)^+,\ y^+ + z^+\}$$

are equal. This can be easily seen by considering the possible cases. □

Let $\underline{x}$ be the sequence $x_1, x_2, \ldots, x_n$ of real numbers. Let $S(\underline{x})$ and $T(\underline{x})$ be the multisets defined by

$$S(\underline{x}) = \{((((x^+_{\gamma(n)} + x_{\gamma(n-1)})^+ + \cdots)^+ + x_{\gamma(2)})^+ + x_{\gamma(1)})^+ \colon \gamma \in \mathfrak{S}_n\},$$

$$T(\underline{x}) = \left\{ \sum_{D:\, D \in \mathrm{cycle}(\gamma)} \left(\sum_{i:\, i \in D} x_i \right)^+ : \gamma \in \mathfrak{S}_n \right\}.$$

Let $h_i : \mathbb{R} \to \mathbb{R}$ be the functions defined by $h_i(y) = 1$ if $y = x_i$ and 0 otherwise. Then the value of the function

$$\sum_{\gamma:\, \gamma \in \mathfrak{S}_n} P(h_{\gamma(1)}(Ph_{\gamma(2)} \cdots P(h_{\gamma(n-1)} P(h_{\gamma(n)}))))$$

at the real number x is the multiplicity of x in the multiset $S(\underline{x})$. Similarly, the value of

$$\sum_{\gamma:\gamma\in\mathfrak{S}_n} \prod_{D:\,D\in\mathrm{cycle}(\gamma)} P(h(D))$$

at x is the multiplicity of x in $T(\underline{x})$. Theorem 5.6.6 implies the following result.

5.6.7. The Bohnenblust–Spitzer identity. If $\underline{x}$ is a length-n sequence of real numbers, then the two multisets $S(\underline{x})$ and $T(\underline{x})$ are equal.

Exercises

5.6.1. *q-Integration.*

Let q be another parameter, $\mathcal{A}$ be an algebra of functions in the variable x, and $E:\mathcal{A}\to\mathcal{A}$ be the endomorphism defined by

$$Ef(x) = f(qx).$$

Then the operator $P = \sum_{i=1}^{\infty} E^i$, acting on functions by

$$(Pf)(x) = f(qx) + f(q^2x) + f(q^3x) + \cdots,$$

is a Baxter operator with parameter -1 called the *q-integral.*

(a) Use Theorem 5.6.1 to obtain a proof of Euler's identity

$$\sum_{n=1}^{\infty} \frac{(-1)^n t^n q^{n(n+1)/2}}{(1-q)(1-q^2)\cdots(1-q^n)} = \prod_{k=1}^{\infty}(1+q^k t).$$

(b) (Research problem) Is there a unified theory of hypergeometric and q-hypergeometric functions based on Baxter operators?

5.6.2. *Spitzer's formula*[21]

Let $\mathcal{A}$ be the Banach algebra of functions $\varphi(t)$ of the form

$$\varphi(t) = \int_{-\infty}^{\infty} e^{itx}\,dF(x),$$

where F is a function of bounded variation such that $\lim_{x\to\infty} F(x)$ exists. Let

$$P\varphi(t) = \int_0^{\infty} e^{itx}\,dF(x) \; + F(0) - \lim_{x\to\infty} F(x).$$

(a) Show that P is a Baxter operator with parameter 1.

[21] Spitzer (1956).

Note that if X is a random variable and $\varphi(t)$ the characteristic function of X, then $P\varphi(t)$ is the characteristic function of $\max\{0, X\}$. Let $X_1, X_2, X_3, \ldots$ be a sequence of independent identically distributed random variables, $S_n = X_1 + X_2 + \cdots + X_n$,

$$M_n = \max\{0, S_1, S_2, \ldots, S_n\},$$

$F_n(x)$ is the probability distribution function $\Pr(M_n \le x)$, and $H_k(x)$ is $\Pr(\max(0, S_k))$

(b) Prove Spitzer's formula:

$$\sum_{n=0}^{\infty} \left(\int_0^{\infty} e^{itx} dF_n(x) \right) \lambda^n = \exp\left(\sum_{k=1}^{\infty} \left(H_k(0) + \int_0^{\infty} e^{itx} dH_k(x) \right) \frac{\lambda^k}{k} \right).$$

5.6.3. Prove the combinatorial version of Spitzer's identity in Exercise 5.6.2: the multisets $T(\underline{x})$ and

$$\left\{ \max_{k:1\le k\le n} \left(\sum_{i=1}^{k} x_{\gamma(i)} \right)^{+} : \gamma \in \mathfrak{S}_n \right\}$$

are equal.

5.6.4. If $\underline{x}$ is the sequence $x_1, x_2, \ldots, x_n$ of real numbers or points in $\mathbb{R}^n$, let $s_0(\underline{x}) = 0$ and $s_k(\underline{x}) = x_1 + x_2 + \cdots + x_k$. If $\gamma \in \mathfrak{S}_n$, then $\gamma(\underline{x})$ is the sequence $x_{\gamma(1)}, x_{\gamma(2)}, \ldots, x_{\gamma(n)}$.[22]

(a) Prove *Kac's identity:*

$$\sum_{\gamma:\gamma\in\mathfrak{S}_n} \max_{k:0\le k\le n} \{s_k(\gamma(\underline{x}))\} = \sum_{\gamma:\gamma\in\mathfrak{S}_n} x_{\gamma(1)} N(\gamma(\underline{x})),$$

where $N(\gamma(\underline{x}))$ is the number of positive terms in the sequence $s_1(\gamma(\underline{x}))$, $s_2(\gamma(\underline{x})), \ldots, s_n(\gamma(\underline{x}))$.

(b) Prove that

$$\sum_{\gamma:\gamma\in\mathfrak{S}_n} \left(\max_{k:0\le k\le n} \{s_k(\gamma(\underline{x}))\} - \min_{k:0\le k\le n} \{s_k(\gamma(\underline{x}))\} \right) = \sum_{\gamma:\gamma\in\mathfrak{S}_n} \sum_{k=1}^{n} \frac{|s_k(\gamma(\underline{x}))|}{k}.$$

(c) Let $\underline{u}$ be the sequence $u_1, u_2, \ldots, u_n$ of points in $\mathbb{R}^2$ and $L(\sigma(\underline{u}))$ be the length of the (piecewise linear) boundary of the convex hull of the points

[22] Barndorff-Nielsen and Baxter (1963), Kac (1954), and Spitzer and Widom (1961).

$0, s_1(\sigma(\underline{u})), s_2(\sigma(\underline{u})), \ldots, s_n(\sigma(\underline{u}))$. Show that

$$\sum_{\sigma:\sigma\in\mathfrak{S}_n} L(\sigma(\underline{u})) = 2 \sum_{\sigma:\sigma\in\mathfrak{S}_n} \left(\sum_{k=1}^{n} \frac{1}{k} \|s_k(\sigma)\| \right).$$

(If u is the vector (x, y), then $\|u\| = \sqrt{x^2 + y^2}$.)

5.7 Symmetric Functions Over Finite Fields

In this section, we find the algebraic dependencies among elementary symmetric polynomials, regarded as functions over a finite field of prime order.[23]

A polynomial $f(x_1, x_2, \ldots, x_n)$ with coefficients in a field $\mathbb{F}$ defines a function from $\mathbb{F}^n$ to $\mathbb{F}$ by *evaluating* a point $(b_1, b_2, \ldots, b_n)$ in $\mathbb{F}^n$ to obtain the function value $f(b_1, b_2, \ldots, b_n)$. If the field $\mathbb{F}$ is infinite (of any characteristic), then distinct polynomials define distinct functions. This is not hard to show from first principles. It also follows from the fact that the set of zeros of a nonzero polynomial has dimension strictly smaller than n. Hermann Weyl named this fact "the principle of the irrelevance of algebraic dependencies." In particular, in many cases, one may ignore polynomial conditions when giving a proof. However, when the field is finite, two distinct polynomials may define the same function.

Let GF(q) be the finite field of order q, where q is a power of a prime p. We first show that all functions are polynomial functions.

5.7.1. Lemma. Let $f : \mathrm{GF}(q)^n \to \mathrm{GF}(q)$ be a function. Then there exists a polynomial $\hat{f}(x_1, x_2, \ldots, x_n)$ with coefficients in GF(q) so that f equals the function defined by evaluating the polynomial $\hat{f}$.

There are many ways to prove Lemma 5.7.1. Since the domain $\mathrm{GF}(q)^n$ is finite, one way is to use the standard proof of the Lagrange interpolation formula. If $\underline{b}$ is the n-tuple $(b_1, b_2, \ldots, b_n)$ in $\mathrm{GF}(q)^n$, let

$$\ell_{\underline{b}}(x_1, x_2, \ldots, x_n) = \left(\prod_{a_1:\, a_1 \neq b_1} \frac{x_1 - a_1}{b_1 - a_1} \right) \left(\prod_{a_2:\, a_2 \neq b_2} \frac{x_2 - a_2}{b_2 - a_2} \right) \cdots \left(\prod_{a_n:\, a_n \neq b_n} \frac{x_n - a_n}{b_n - a_n} \right).$$

[23] This section is an exposition of Aberth (1964), which followed up work done in Fine (1950).

Then $\ell_{\underline{b}}$ behaves as a "delta" function: it equals 0 if $(x_1, x_2, \ldots, x_n) \neq (b_1, b_2, \ldots, b_n)$ and 1 otherwise. In particular, for a function $f : \mathrm{GF}(q)^n \to \mathrm{GF}(q)$,

$$f(x_1, x_2, \ldots, x_n) = \sum_{\underline{b}:\, \underline{b} \in \mathrm{GF}(q)^n} f(\underline{b}) \ell_{\underline{b}}(x_1, x_2, \ldots, x_n).$$

Another polynomial expression for the function $\ell_{\underline{b}}$ can be obtained using Fermat's little theorem: that $x^{q-1} = 1$ if and only if $x \neq 0$. Thus,

$$\ell_{\underline{b}}(x_1, x_2, \ldots, x_n) = \prod_{i=1}^{n} (1 - (x_i - b_i)^{q-1}).$$

The two polynomial expressions for $\ell_{\underline{b}}$ look quite different. The next lemma shows that they can be transformed into one another by algebraic relations implied by $x_i^q = x_i$. Let $\mathcal{F}$ be the set of functions $\mathrm{GF}(q)^n \to \mathrm{GF}(q)$. This forms a GF$(q)$-algebra with multiplication given by $f \cdot g(x) = f(x)g(x)$.

5.7.2. Lemma. The algebra $\mathcal{F}$ is isomorphic to the quotient

$$\mathrm{GF}(q)[x_1, x_2, \ldots, x_n]/(x_1^p - x_1, x_2^q - x_2, \ldots, x_n^q - x_n).$$

Proof. By Lemma 5.7.1, the evaluation ϵ defines a surjective GF(q)-algebra homomorphism from $\mathrm{GF}(q)[x_1, x_2, \ldots, x_n]$ to $\mathcal{F}$. By Fermat's little theorem, $b^q = b$ for every element b in GF(q). Thus, the polynomials $x_i^q - x_i$ define the zero function and are in the kernel of ϵ. Hence, ϵ defines a surjective homomorphism from the quotient $\mathrm{GF}(q)[x_1, x_2, \ldots, x_n]/I$ to $\mathcal{F}$, where I is the ideal generated by $x_1^q - x_1,\ x_2^q - x_2,\ \ldots,\ x_n^q - x_n$. The relations $x_i^q = x_i$ allow us to reduce any monomial modulo I to a monomial of the form $x_1^{a_1} x_2^{a_2} \cdots x_n^{a_n}$, where $0 \leq a_i \leq q - 1$. In particular, these monomials span $\mathrm{GF}(q)[x_1, x_2, \ldots, x_n]/I$, implying that the quotient is finite and has size at most $q^{(q^n)}$. On the other hand, $|\mathcal{F}| = q^{(q^n)}$. Since ϵ is a surjective GF(q)-algebra homomorphism, we conclude that $\mathrm{GF}(q)[x_1, x_2, \ldots, x_n]/I$ and $\mathcal{F}$ are isomorphic. □

In the remainder of this section, we work over the finite field GF(p) of integers modulo a prime p. Let e be an element in the field GF(p) and $\alpha_e : \mathrm{GF}(p)^n \to \{0, 1, 2, \ldots\}$ be the function defined by

$$\alpha_e(b_1, b_2, \ldots, b_n) = |\{j\colon b_j = e\}|;$$

that is, $\alpha_i(\underline{b})$ is the number of occurrences of e in the n-tuple $\underline{b}$. Note that $\alpha_0(\underline{b}) + \alpha_1(\underline{b}) + \cdots + \alpha_{p-1}(\underline{b}) = n$, and so $\alpha_0(\underline{b})$ is deducible from the other values $\alpha_e(\underline{b})$.

The functions α_e are integer-valued symmetric functions. If $f : \mathrm{GF}(p)^n \to \mathrm{GF}(p)$ is a symmetric function, then the value of f on an n-tuple $\underline{b}$ depends only on the $(p-1)$-tuple $(\alpha_1(\underline{b}), \alpha_2(\underline{b}), \ldots, \alpha_{p-1}(\underline{b}))$ of nonnegative integers. For example,

$$a_2(\underline{b}) = \sum_{e:\, e \in \mathrm{GF}(p)^\times} \binom{\alpha_e(\underline{b})}{2} e^2 + \sum_{d,e:\, d \neq e, d,e \in \mathrm{GF}(p)^\times} \alpha_d(\underline{b})\alpha_e(\underline{b}) de,$$

where $\mathrm{GF}(p)^\times = \mathrm{GF}(p) \backslash \{0\}$.

5.7.3. Lemma. There is a bijection between the set of $(p-1)$-tuples $(\alpha_1, \alpha_2, \ldots, \alpha_{p-1})$ of integers satisfying $0 \leq \alpha_e \leq p-1$ and the set of $(p-1)$-tuples $(c_1, c_2, \ldots, c_{p-1})$ of integers modulo p so that

$$\prod_{e=1}^{p-1} (1 + eX)^{\alpha_e} = 1 + c_1 X + c_2 X^2 + \cdots + c_{p-1} X^{p-1} + c_p X^p + \cdots + c_{(p-1)^2} X^{(p-1)^2} \tag{P}$$

as a polynomial in the indeterminate X over $\mathrm{GF}(p)$. If α_e is regarded as an integer modulo p, then this bijection expresses α_e as a polynomial in $c_1, c_2, \ldots, c_{p-1}$. In addition, if $p \leq m \leq (p-1)^2$, then the coefficient c_m in Equation (P) is determined by the n-tuple $(c_1, c_2, \ldots, c_{p-1})$. Specifically, there is a polynomial R_m such that

$$c_m = R_m(c_1, c_2, \ldots, c_{p-1}). \tag{R}$$

Proof. Expanding the product on the left, all the coefficients c_m (and hence the first $p-1$ of them) are determined by $(\alpha_1, \alpha_2, \ldots, \alpha_{p-1})$. We will think of the coefficients as functions $c_m(\alpha_1, \alpha_2, \ldots, \alpha_{p-1})$. The functions c_m are elementary symmetric functions, but with a different (and restricted) domain.

Next we will show that $(c_1, c_2, \ldots, c_{p-1})$ determines $(\alpha_1, \alpha_2, \ldots, \alpha_{p-1})$. We use Newton's identities (5.2.8). When $1 \leq m \leq p-1$, these identities express the power-sum symmetric functions s_m as polynomials with integer coefficients in the elementary symmetric functions. These expressions remain valid when reduced modulo p and applied to s_m and the coefficients $c_m(\alpha_1, \alpha_2, \ldots, \alpha_{p-1})$. Since

$$s_m = \sum_{e=1}^{p-1} \alpha_e e^m,$$

we have

$$
\begin{aligned}
s_1 &= \sum_{e=1}^{p-1} \alpha_e e = Q_1(c_1), \\
&\vdots \\
s_m &= \sum_{e=1}^{p-1} \alpha_e e^m = Q_m(c_1, c_2, \ldots, c_m), \\
&\vdots \\
s_{p-1} &= \sum_{e=1}^{p-1} \alpha_e e^{p-1} = Q_{p-1}(c_1, c_2, , \ldots, c_{p-1}),
\end{aligned}
$$

where $Q_m(c_1, c_2, \ldots, c_m)$ is a polynomial in $c_1, c_2, \ldots, c_m$. These equations form a system of $p-1$ linear equations in the "unknowns" α_e over the field GF(p). The determinant of the coefficients on the left side is a van der Monde determinant. Hence, we can solve the system to obtain the integers α_e modulo p as polynomials in the coefficients $c_1, c_2, \ldots, c_{p-1}$. Since we are assuming that $0 \leq \alpha_e \leq p-1$, the value of α_e modulo p determines α_e as an integer. As we observed at the beginning, $(\alpha_1, \alpha_2, \ldots, \alpha_{p-1})$ determines all the coefficients c_m on the right. Hence, since $(c_1, c_2, \ldots, c_{p-1})$ determines $(\alpha_1, \alpha_2, \ldots, \alpha_{p-1})$, the coefficient c_m is a function of $(c_1, c_2, \ldots, c_{p-1})$. By Lemma 5.7.1, there is a polynomial R_m such that Equation (R) holds. □

Let L_p be the set of positive integers of the form tp^s, where s and t are integers such that $s \geq 0$ and $1 \leq t \leq p-1$; explicitly,

$$L_p = \{1, 2, \ldots, p-1, p, 2p, \ldots, (p-1)p, p^2, 2p^2, \ldots, (p-1)p^2, \ldots\}.$$

5.7.4. Theorem. Every symmetric function $f(x_1, x_2, \ldots, x_n)$ over GF(p) can be expressed as a polynomial in the elementary symmetric functions a_m, where $m \in L_p$.

Proof. By the first fundamental theorem (5.2.6), it suffices to show that every elementary symmetric function $a_m(x_1, x_2, \ldots, x_n)$ can be expressed as a polynomial in $a_j(x_1, x_2, \ldots, x_n)$, $j \in L_p$.

Let $\underline{b} \in \mathrm{GF}(p)^n$. Expand each function $\alpha_e(\underline{b})$ in base p and let $\alpha_{e,k}(\underline{b})$ be the kth p-ary digit, so that

$$\alpha_e(\underline{b}) = \alpha_{e,0}(\underline{b}) + \alpha_{e,1}(\underline{b})p + \alpha_{e,2}(\underline{b})p^2 + \cdots ,$$

with $0 \le \alpha_{e,k}(\underline{b}) \le p-1$. Then for all $\underline{b} \in \mathrm{GF}(p)$,

$$\begin{aligned}
&1 + a_1(\underline{b})X + a_2(\underline{b})X^2 + \cdots + a_m(\underline{b})X^m + \cdots \\
&= \prod_{e=1}^{p-1}(1+eX)^{\alpha_e(\underline{b})} \\
&= \prod_{e=1}^{p-1}(1+eX)^{\alpha_{e,0}(\underline{b})} \prod_{e=1}^{p-1}(1+eX)^{\alpha_{e,1}(\underline{b})p} \prod_{e=1}^{p-1}(1+eX)^{\alpha_{e,2}(\underline{b})p^2} \cdots,
\end{aligned}$$

where a_m are elementary symmetric functions. By the binomial theorem modulo p and Fermat's little theorem,

$$\prod_{e=1}^{p-1}(1+eX)^{\alpha_{e,i}(\underline{b})p^i} = \prod_{e=1}^{p-1}(1+eX^{p^i})^{\alpha_{e,i}(\underline{b})}.$$

The product on the right has the form given in Equation (P) in Lemma 5.7.2, with $Y = X^{p^i}$. Hence, we have

$$\begin{aligned}
\prod_{e=1}^{p-1}(1+eX^{p^i})^{\alpha_{e,i}(\underline{b})} &= 1 + c_1^{(i)}X^{p^i} + c_2^{(i)}X^{2p^i} + \cdots + c_{p-1}^{(i)}X^{(p-1)p^i} \\
&\quad + c_p^{(i)}X^{pp^i} + c_{p+1}^{(i)}X^{(p+1)p^i} + \cdots + c_{(p-1)^2}^{(i)}X^{(p-1)^2p^i}.
\end{aligned}$$

If we define $c_0^{(i)}(\underline{b})$ to be the constant function 1, then

$$a_m(\underline{b}) = \sum_{(d_1,d_2,\ldots,d_r)} c_{d_0}^{(0)}(\alpha_{e,0}(\underline{b}))c_{d_1}^{(1)}(\alpha_{e,1}(\underline{b}))\cdots c_{d_r}^{(r)}(\alpha_{e,r}(\underline{b})), \tag{C}$$

the sum ranging over all sequences $(d_0, d_2, \ldots, d_r)$ such that $d_r \neq 0$, $0 \le d_i \le (p-1)^2$, and

$$d_0 + d_1p + d_2p^2 + \cdots + d_rp^r = m.$$

If $p \le m \le (p-1)^2$, then by Lemma 5.7.3,

$$c_m^{(i)} = R_m(c_1^{(i)}, c_2^{(i)}, \ldots, c_{p-1}^{(i)}). \tag{1}$$

Further, when $m = tp^s$, $1 \le t \le p-1$, then one of the summands in Equation (C) is $c_t^{(i)}$. Thus, we can rewrite Equation (C) in the form

$$c_t^{(i)} = a_{tp^s} - \sum c_{d_0}^{(0)}c_{d_1}^{(1)}\cdots c_{d_r}^{(r)}, \tag{2}$$

the sum now ranging over all sequences $(d_1, d_2, \ldots, d_r)$ satisfying the three earlier conditions and the new condition, either $r < s$, or if $r = s$, $d_r < t$.

We can now apply Equations (1) and (2) repeatedly to write any elementary symmetric function a_m as a polynomial in a_{tp^i}, $tp^i \in L_p$. □

The next theorem is the analog of the second fundamental theorem.

5.7.5. Theorem. Let k be a positive integer and X_s be a set of variables indexed by integers s in $L_p \cap \{1, 2, \ldots, k\}$. Suppose the number n of variables is at least p^k. Then there is no polynomial $P(X_s)$ which is not the zero function in the variables X_s such that the polynomial $P(a_s(x_1, x_2, \ldots, x_n))$ in the variables $x_1, x_2, \ldots, x_n$ is the zero function.

Proof. It suffices to show that when $n \geq p^k$, the set $\{(a_s(\underline{b}))_{s \in L_p \cap \{1,2,\ldots,n\}} : \underline{b} \in \mathrm{GF}(p)^n\}$ is the set of all vectors of dimension $|L_p \cap \{1, 2, \ldots, n\}|$. As in the proof of Theorem 5.7.3, we have

$$\begin{aligned}
&1 + a_1 X + a_2 X^2 + \cdots + a_s X^s + \cdots \\
&\quad = \prod_{i=0}^{\infty} (1 + c_1^{(i)} X^{p^i} + c_2^{(i)} X^{2p^i} + \cdots + c_{p-1}^{(i)} X^{(p-1)p^i} + c_p^{(i)} X^{pp^i} + \cdots \\
&\qquad + c_{(p-1)^2}^{(i)} X^{(p-1)^2 p^i}).
\end{aligned}$$

On equating coefficients, $a_1 = c_1^{(0)},\ a_2 = c_2^{(0)},\ \ldots\ , a_{p-1} = c_{p-1}^{(0)}$,

$$\begin{aligned}
a_p &= c_p^{(1)} + c_p^{(0)},\ a_{2p} = c_2^{(1)} + c_p^{(0)} c_p^{(1)} + c_{2p}^{(0)}, \\
a_{3p} &= c_3^{(1)} + c_p^{(0)} c_2^{(1)} + c_{2p}^{(0)} c_1^{(1)} + c_{3p}^{(0)},\ \ldots .
\end{aligned}$$

In general,

$$a_{tp^i} = c_t^{(i)} + c_{t-1}^{(i)} c_p^{(i-1)} + \cdots + c_1^{(i)} c_{(t-1)p}^{(i-1)} + \text{ other terms},$$

where the other terms involve $c_k^{(j)}$, where $j < i$.

Now suppose that a specific vector $(A_s)_{s \in L_p \cap \{1,2,\ldots,n\}}$ is given. Then, we can use Lemma 5.7.3 to find nonnegative integers $(\alpha_{e,0})$ so that

$$a_1 = c_1^{(0)} = A_1,\ a_2 = c_2^{(0)} = A_2,\ \ldots,\ a_{p-1} = c_{p-1}^{(0)} = A_{p-1}.$$

Next, we substitute $c_j^{(0)} = A_j$ and $a_s = A_s$ into the equations for $a_p, a_{2p}, \ldots, a_{(p-1)p}$, obtaining $p - 1$ equations of the form

$$c_j^{(1)} = C_j, \quad j = p, 2p, \ldots, (p-1)p.$$

Using Lemma 5.7.1, we can determine nonnegative integers $\alpha_{e,1}$ so that $c_j^{(1)} = C_j$, and hence, $a_j = A_j$ for $j = p, 2p, \ldots, (p-1)p$. Continuing, we obtain nonnegative integers $\alpha_{e,0}, \alpha_{e,1}, \ldots, \alpha_{e,m-1}$, so that $a_s = A_s$, $s \in L_p \cap \{1, 2, \ldots, m\}$. The specified vector (A_s) can now be obtained as the vector $(a_s(\underline{b}))$, where $\underline{b}$ is a vector $\mathrm{GF}(p)^n$ satisfying $\alpha_e(\underline{b}) = \alpha_{e,0} + \alpha_{e,1} p + \alpha_{e,2} p^2 + \cdots$. □

The simplest case of the theory discussed in this section is when $p = 2$. In this case, there is only one nonzero element 1 and we write α_i instead of $\alpha_{1,i}(\underline{b})$. Since $(1+X)^{\alpha_i} = 1 + \alpha_i X$, we have

$$\prod_{i=1}^{\infty}(1+X^{2^i})^{\alpha_i} = \prod_{i=1}^{\infty}(1+\alpha_i X^{2^i}) = \sum_{m=0}^{\infty} \alpha_{k_1}\alpha_{k_2}\cdots\alpha_{k_r} X^m,$$

where $m = 2^{k_1} + 2^{k_2} + \cdots + 2^{k_r}$ is the base-2 expansion of m. In this case, the elementary symmetric functions a_i are directly related to the functions α_i, and we do not need to go through the functions c_i. Hence, we conclude that $L_2 = \{2^j : 0 \le j < \infty\}$ and $a_m = a_{k_1}a_{k_2}\cdots a_{k_r}$.

Exercises

5.7.1. (a) Prove *Lucas' theorem*:

$$\binom{n}{m} = \binom{n_0}{m_0}\binom{n_1}{m_1}\binom{n_2}{m_2}\cdots \mod p,$$

where $n = n_0 + n_1 p + n_2 p^2 + \cdots$ and $m = m_0 + m_1 p + m_2 p^2 + \cdots$ are the base-p expansions of n and m.

(b) Prove the case $p = 2$ of Theorem 5.7.4 using Lucas's theorem.

5.7.2. Work out explicitly the case $p = 3$ of Theorem 5.7.3.

5.7.3. *Asymptotic distributions of elementary symmetric functions mod p.*[24] Let $P_n(m, e)$ be the number of n-tuples $\underline{b}$ in $\mathrm{GF}(p)^n$ such that $a_m(\underline{b}) = e$.

(a) Show that $\lim_{n\to\infty} P_n(m, e)$ exists for all m and e.

Let $P(m, e) = \lim_{n\to\infty} P_n(m, e)$.

(b) Show that

$$\left(1 - \frac{1}{p}\right)^h \le 1 - P(m, 0) \le \left(1 - \frac{1}{p^{2(p-1)}}\right)^{\lfloor (h+1)/2 \rfloor},$$

where h is the number of nonzero digits in the base-p expansion of m. Using this, show that $\limsup_{n\to\infty} P(m, 0) = 1$.

5.7.4. (Research problem) Extend Theorems 5.7.4 and 5.7.5 to arbitrary finite fields.

[24] Aberth (1964) and Fine (1950).

5.7.5. Let GF(q) be a finite field of order q and m be a positive integer. Then

$$\sum_{e:\, e\in \mathrm{GF}(q)} e^m = \begin{cases} 0 & \text{if } 1 \le m < q-1, \\ -1 & \text{if } m = q-1. \end{cases}$$

5.7.6. *Permutation polynomials.* [25]

Let GF(q) be a finite field of order q, where $q = p^a$. By Lemmas 5.7.1 and 5.7.2, a function $\mathrm{GF}(q) \to \mathrm{GF}(q)$, $X \mapsto f(X)$ can be written as a polynomial $f(X)$ of degree at most $q-1$. A *permutation polynomial* is a polynomial of degree at most $q-1$ representing a permutation on GF(q).

(a) Show that X^m is a permutation polynomial if and only if m and $q-1$ are relatively prime.

(b) Prove *Hermite's theorem.* A polynomial $f(X)$ in GF(q)[X] of degree at most $q-1$ is a permutation polynomial if and only if the following two conditions hold:

H1. The polynomial $f(X)$ has exactly one zero in GF(q).
H2. If $t < q-2$ and p do not divide t, then the tth power $[f(X)]^t$ is equivalent to a polynomial of degree at most $q-2$ modulo $X^q - X$.

(c) Let $f(X)$ be a polynomial of degree strictly less than m in GF(p^{am})[X] and

$$\bar{f}(X) = f(X^{p^a}).$$

Show that the polynomial $\bar{f}(X)$ is a permutation polynomial if and only if 0 is the only zero of $\bar{f}(X)$ in GF(p^{am}).

(d) The polynomial

$$f(X) = \sum_{i=0}^{m-1} A_{m-i} X^{p^{ai}}$$

is a permutation polynomial in GF(p^{am}) if and only if the determinant

$$\begin{vmatrix} A_1 & A_2 & \dots & A_{m-1} & A_m \\ A_2^{p^a} & A_3^{p^a} & \dots & A_m^{p^a} & A_1^{p^a} \\ \vdots & \vdots & \ddots & & \vdots \\ A_m^{p^{a(m-1)}} & A_1^{p^{a(m-1)}} & \dots & A_{m-2}^{p^{a(m-1)}} & A_{m-1}^{p^{a(m-1)}} \end{vmatrix}$$

is nonzero.

[25] Dickson (1901, chapter 5) and Netto (1892). This exercise gives a glimpse of this area. The deeper theory of permutation polynomials was developed by L. Carlitz and his school.

5.7.7. *The Dickson invariants.*[26]

Let $\mathrm{GL}(n,p)$ be the *general linear group* of $n \times n$ nonsingular matrices with entries in $\mathrm{GF}(p)$. A polynomial f in $\mathrm{GF}(p)[x_1, x_2, \ldots, x_n]$ is a *relative invariant of weight s* if for all matrices (a_{ij}) in $\mathrm{GL}(n,p)$,

$$f(x_1, x_2, \ldots, x_n) = (\det(a_{ij}))^s f\left(\sum_{j=1}^n a_{1j}x_j, \sum_{j=1}^n a_{2j}x_j, \ldots, \sum_{j=1}^n a_{nj}x_j\right).$$

A *Dickson invariant* is a relative invariant of weight 0, that is, an absolute invariant. For an n-tuple $(\alpha_1, \alpha_2, \ldots, \alpha_n)$ of nonnegative integers, let

$$[\alpha_1, \alpha_2, \ldots, \alpha_n] = \det(x_i^{p^{\alpha_j}})_{1 \le i,j \le n}.$$

Such determinants are called *p-alternants*. For example,

$$[3, 0] = \begin{vmatrix} x_1^{p^3} & x_1 \\ x_2^{p^3} & x_2 \end{vmatrix} = x_1x_2(x_1^{p-1} - x_2^{p-1})(x_1^{p^2+p+1} + x_1^{p^2+p}x_2 + \cdots + x_2^{p^2+p+1}).$$

Let (a_{ij}) be a matrix in $\mathrm{GL}(n,p)$. By the binomial theorem modulo p and Fermat's little theorem,

$$\left(\sum_{j=1}^n a_{ij}x_j\right)^{p^\alpha} = \sum_{j=1}^n a_{ij}x_j^{p^\alpha}.$$

Thus,

$$\det\left(\left(\sum_{k=1}^n a_{ik}x_k\right)^{p^{\alpha_j}}\right) = \det(a_{ij})[\alpha_1, \alpha_2, \ldots, \alpha_n].$$

In other words, a p-alternant is a weight-1 relative invariant of $\mathrm{GL}(n,p)$.

(a) Let $\ell(x_1, x_2, \ldots, x_n)$ be a nonzero linear form and $(\alpha_1, \alpha_2, \ldots, \alpha_n)$ be an n-tuple of nonnegative integers. Then ℓ divides $[\alpha_1, \alpha_2, \ldots, \alpha_n]$.

Two linear forms $a_1x_1 + a_2x_2 + \cdots + a_nx_n$ and $b_1x_1 + b_2x_2 + \cdots + b_nx_n$ are *projectively equivalent* if there is a non-zero element c in $\mathrm{GF}(p)$ such that $a_i = cb_i$ for all i. There are $p-1$ linear forms in each equivalence class and $p^{n-1} + p^{n-2} + \cdots + p + 1$ equivalence classes.

(b) Show that

$$[n-1, n-2, \ldots, 2, 1, 0] = C \prod \ell(x_1, x_2, \ldots, x_n),$$

[26] Dickson (1911) and Ore (1933). For modern renditions, see Steinberg (1987).

where the product ranges over linear forms, one from each equivalence class, and C is a nonzero constant.

For example, over GF(2),

$$[2,1,0] = x_1(x_1+x_2)(x_1+x_2+x_3)(x_1+x_3)x_2(x_2+x_3)x_3.$$

The p-alternant $[n-1, n-2, \ldots, 2, 1, 0]$ is the nonzero p-alternant of minimum degree. It is an analog of the van der Monde determinant. From (a) and (b), it follows that $[n-1, n-2, \ldots, 2, 1, 0]$ divides the alternant $[\alpha_1, \alpha_2, \ldots, \alpha_n]$.

For a sequence $\lambda_1, \lambda_2, \ldots, \lambda_n$ of integers such that $\lambda_1 \geq \lambda_2 \geq \cdots \geq \lambda_n \geq 0$, let

$$D_{\lambda_1,\lambda_2,\ldots,\lambda_n}(x_1, x_2, \ldots, x_n) = \frac{[\lambda_1+n-1, \lambda_2+n-2, \ldots, \lambda_{n-1}+1, \lambda_n]}{[n-1, n-2, \ldots, 2, 1, 0]}.$$

The polynomial $D_{\lambda_1,\lambda_2,\ldots,\lambda_n}$ is a Dickson invariant of degree

$$p^{n-1}(p^{\lambda_n}-1) + p^{n-2}(p^{\lambda_{n-1}}-1) + \cdots + p(p^{\lambda_1}-1) + (p^{\lambda_n}-1).$$

Let $a_m(x_1, x_2, \ldots, x_n)$ be defined by

$$\begin{vmatrix} X & x_1 & x_2 & \ldots & x_n \\ X^p & x_1^p & x_2^p & \ldots & x_n^p \\ X^{p^2} & x_1^{p^2} & x_2^{p^2} & \ldots & x_n^{p^2} \\ \vdots & \vdots & & \ddots & \vdots \\ X^{p^n} & x_1^{p^n} & x_2^{p^n} & \ldots & x_n^{p^n} \end{vmatrix}$$
$$= [0, 1, 2, \ldots, n-1](X^{p^n} - a_1 X^{p^{n-1}} + a_2 X^{p^{n-2}} - \cdots \pm a_n X),$$

or equivalently,

$$a_m = D_{1,1,\ldots,1,0,0,\ldots,0} = \frac{[\widehat{n-m}]}{[n-1, n-2, \ldots, 1, 0]},$$

where $\widehat{n-m}$ is the length-n sequence obtained from $0, 1, 2, \ldots, n$ by omitting $n-m$.

(c) Show the following analogs of the first fundamental theorem: every relative invariant is a polynomial in $[n-1, n-2, \ldots, 2, 1, 0], a_1, a_2, \ldots, a_n$. Every Dickson invariant is a polynomial in $a_1, a_2, \ldots, a_n$.

(d) Show that the polynomials $D_{\lambda_n,\lambda_{n-1},\ldots,\lambda_1}$ form a basis for the vector space of Dickson invariants on n variables.

(e) Prove an analog of Chevalley's theorem (see Exercise 5.2.7) for Dickson invariants.

5.7.8. *The ring of Ore polynomials.*[27] Let $\mathbb{F}$ be a field of positive characteristic p with prime field $\mathbb{H}$ (which is a finite field of order p). An *Ore polynomial* over $\mathbb{F}$ is a polynomial of the form

$$a_0x^{p^n} + a_1x^{p^{n-1}} + \cdots + a_{n-1}x^p + a_nx.$$

(a) Show that a polynomial in $\mathbb{F}[x]$ is an Ore polynomial if and only if its zeros form a vector subspace over the prime field $\mathbb{H}$.

(b) Let $f(x)$ and $g(x)$ be Ore polynomials. Show that the composition $f(g(x))$ is an Ore polynomial. Show that the Ore polynomials form a (noncommutative) ring under addition and composition.

5.7.9. *Wronskians.*[28]

Let $\mathcal{X}$ be the set $\{D^kx_j\colon 1 \le j \le n, 0 \le k < \infty\}$ of indeterminates. For example, x_2, D^5x_3, $D^{101}x_7$ are in $\mathcal{X}$ if $n \ge 7$. A *differential polynomial* over an infinite field $\mathbb{F}$ is a polynomial with variables in $\mathcal{X}$. Differential polynomials form an $\mathbb{F}$-algebra $\mathbb{F}_D[x_1, x_2, \ldots, x_n]$. The field of differential rational functions is the field of fractions of $\mathbb{F}_D[x_1, x_2, \ldots, x_n]$. The differential operator D is the linear operator on $\mathbb{F}_D[x_1, x_2, \ldots, x_n]$ defined on the variables in $\mathcal{X}$ by specifying

$$D(D^kx_j) = D^{k+1}x_j$$

for variables and extended to polynomials and rational functions by linearity, the *product rule,* and *quotient rule,*

$$D(fg) = f(Dg) + (Df)g, \quad D\left(\frac{f}{g}\right) = \frac{(Df)g - f(Dg)}{g^2}.$$

The (*generalized*) Wronskians $W[\alpha_1, \alpha_2, \ldots, \alpha_n]$ are the differential polynomials defined by

$$W[\alpha_1, \alpha_2, \ldots, \alpha_n] = \det[D^{\alpha_i}x_j]_{1\le i,j\le n} = \begin{vmatrix} D^{\alpha_1}x_1 & D^{\alpha_1}x_2 & \ldots & D^{\alpha_1}x_n \\ D^{\alpha_2}x_1 & D^{\alpha_2}x_2 & \ldots & D^{\alpha_2}x_n \\ \vdots & \vdots & \ddots & \vdots \\ D^{\alpha_n}x_1 & D^{\alpha_n}x_2 & \ldots & D^{\alpha_n}x_n \end{vmatrix}.$$

The Wronskian (as Wronski defined it) is $W[n-1, n-2, \ldots, 2, 1, 0]$. We will denote this particular Wronskian by W.

To define differential symmetric functions, we need to divide by W. An easy way is to take the subalgebra $\mathcal{S}_n$ in the field of differential rational functions

[27] Ore (1933).
[28] Kung (2000) and Kung and Rota (1984a).

consisting of differential rational functions of the form f/W^k, where f is a differential polynomial. We can now define differential Schur functions $S_{\lambda_1,\lambda_2,\ldots,\lambda_n}$ and differential elementary symmetric functions a_m as in Exercise 5.7.8. Explicitly,

$$S_{\lambda_1,\lambda_2,\ldots,\lambda_n} = \frac{W[\lambda_1+n-1,\lambda_2+n-2,\ldots,\lambda_{n-1}+1,\lambda_n]}{W},$$
$$a_m = S_{1,1,\ldots,1,0,0,\ldots,0} = \frac{W[\widehat{n-m}]}{W}.$$

The general linear group $\mathrm{GL}(n,\mathbb{F})$ acts on $\mathcal{S}_n$ in the following way: if $A=(a_{ij})$ is a matrix in $\mathrm{GL}(n,\mathbb{F})$, then A acts on $\mathcal{S}_n$ as follows: for $0 \le k < \infty$,

$$AD^k x_i = \sum_{j=1}^{n} a_{ij} D^k x_i,$$

on the variables, $AW^{-1} = (\det A)^{-1} W^{-1}$, and the action of A is extended to $\mathcal{S}_n$ by requiring that A acts as an $\mathbb{F}$-algebra homomorphism. A polynomial f in $\mathcal{S}_n$ is a *relative invariant* (*of weight* s) if for any matrix A in $\mathrm{GL}(n,\mathbb{F})$,

$$A(f/W^k) = (\det A)^s (f/W^k).$$

(a) Prove *Appell's theorem*. The ring of absolute invariants (that is, invariants of weight 0) is differentially generated by the differential elementary symmetric functions; explicitly, every absolute differential invariant can be written as $f(a_1,a_2,\ldots,a_n)$, where $f(x_1,x_2,\ldots,x_n)$ is a differential polynomial (and one makes the substitution $D^j x_i \leftarrow D^j a_i$).

It follows from Appell's theorem that a relative invariant of weight s can be expressed as $W^s f(a_1,a_2,\ldots,a_n)$, where f is a differential polynomial.

(b) Show the following analog of the second fundamental theorem. Any relation among $W, a_1, a_2, \ldots, a_n$ can be deduced (algebraically and differentially) from *Abel's identity*: $DW = -a_1 W$.

(c) Develop a comprehensive theory of differential symmetric functions.

(d) Find analogs of Chevalley's theorem (Exercise 5.2.7) for differential symmetric functions.

5.8 Historical Remarks and Further Reading

It is arguable that the theory of symmetric functions began with the quadratic formula. Let x_1 and x_2 be the two zeros of a quadratic polynomial. Then x_1+x_2 and $(x_1-x_2)^2$ are the two simplest symmetric polynomials of the zeros and

we have

$$x_1 + x_2 = a_1, \quad (x_1 - x_2)^2 = a_1^2 - 4a_2.$$

From this, the quadratic formula follows: one might try the same strategy for higher-degree polynomials. For example, for a cubic polynomial, one might try to find three independent algebraic expressions in $x_1 + x_2 + x_3$, $x_1 + \omega x_2 + \omega x_3$, and $x_1 + \omega^2 x_2 + \omega x_3$ (where ω is a cube root of unity), which are symmetric in x_1, x_2, and x_3. This turns out to be reasonably easy to do for cubics, harder for quartics, and impossible for quintics. After Abel and Galois, symmetric functions played a supporting but indispensable role in the "theory of equations" until the emergence of representation theory. In the twentieth century, the theory of symmetric functions is regarded as part of the constructive theory of representations of the symmetric group.

For different views of the theory of symmetric functions, we recommend the following books and survey papers:

P. Hall, The algebra of partitions, in *Proceedings of the 4th Canadian Mathematical Congress, Banff*, pp. 147–59, reprinted in *The Collected Works of Philip Hall*, Oxford University Press, Oxford, 1988.

W. Ledermann, *Introduction to Group Characters*, 2nd edition, Cambridge University Press, Cambridge, 1987.

I.G. MacDonald, *Symmetric Functions and Hall Polynomials*, 2nd edition, Oxford University Press, New York, 1995.

P.A. MacMahon, *Collected Papers, Vol. I, Combinatorics, Vol. 2, Number Theory, Invariants and Applications*, G.E. Andrews, ed., M.I.T. Press, Cambridge, MA, 1978, 1986.

B.E. Sagan, *The Symmetric Group*, 2nd edition, Springer, New York, 2001.

R. Stanley, Theory and application of plane partitions, I and II, *Stud. Appl. Math.* 50 (1971) 167–188, 259–279.

R. Stanley, *Enumerative Combinatorics II*, Cambridge University Press, Cambridge, 1999.

6

Determinants, Matrices, and Polynomials

6.1 Polynomials

Polynomials occur in all of combinatorics (and indeed, in all of mathematics). For example, the following theorem of Isaac Newton underlies many conjectures and theorems in combinatorics.

6.1.1. Newton's inequalities. Let

$$p(x) = x^n + c_1 x^{n-1} + c_2 x^{n-2} + \cdots + c_{n-1} x^{n-1} + c_n,$$

where the coefficients c_i are real numbers. If all the zeros of $p(x)$ are real numbers, then the coefficients c_i satisfy the inequalities:

$$c_{i-1} c_{i+1} \leq \frac{i(n-i)}{(i+1)(n-i+1)} c_i^2.$$

One way to prove Newton's inequalities uses binary forms. A *binary form of degree n* (or an *n-form*) $f(x_0, x_1)$ on the variables x_0 and x_1 is a homogeneous polynomial of degree n in x_0 and x_1; that is,

$$\begin{aligned} f(x_0, x_1) &= \sum_{i=0}^{n} \binom{n}{i} a_i x_0^i x_1^{n-i} \\ &= a_0 x_1^n + \binom{n}{1} a_1 x_0 x_1^{n-1} + \binom{n}{2} a_2 x_0^2 x_1^{n-2} + \cdots + \binom{n}{n-1} a_{n-1} x_0^{n-1} x_1 \\ &\quad + a_n x_0^n. \end{aligned}$$

The numbers a_i are the *normalized coefficients* of the form. Over the real or complex numbers (or any field of characteristic zero), one can simply divide by a binomial coefficient. However, this is a problem over fields of positive characteristic. From an n-form $f(x_0, x_1)$, one obtains polynomials $f(1, x)$ and $f(x, 1)$ of degree at most n. Conversely, if $p(x)$ has degree at most n, then $x_0^n p(x_1/x_0)$ and $x_1^n p(x_0/x_1)$ are n-forms. Like polynomials, an n-form can be factored into n linear or 1-forms over an algebraically closed field. Two nonzero linear forms $a_1x_0 + a_0x_1$ and $b_1x_0 + b_0x_1$ are *(projectively) distinct* if one is not a constant multiple of the other, or equivalently, the 2×2 determinant $a_1b_0 - a_0b_1$ is not zero.

Binary forms allow more possibilities for transformations than polynomials. The next lemma is a good example of this.

6.1.2. Lemma. Let $f(x_0, x_1)$ be an n-form, j and k be nonnegative integers such that $j + k \leq n$, and

$$f_{j,k}(x_0, x_1) = \frac{\partial^{j+k}}{\partial x_0^j \partial x_1^k} f(x_0, x_1).$$

Suppose that all the zeros of the polynomial $f(1, x)$ are real. Then all the zeros of $f_{j,k}(1, x)$ and $f_{j,k}(x, 1)$ are real.

Proof. If λ is a nonzero zero of $f(1, x)$, then $1/\lambda$ is a zero of $f(x, 1)$. If 0 occurs as a zero with multiplicity m in $f(1, x)$, then $f(x, 1)$ has degree $n - m$, and conversely, it follows that all the zeros of $f(1, x)$ are real if and only if all the zeros of $f(x, 1)$ are also real.

The lemma now follows from Rolle's theorem: that if a degree-n polynomial $p(x)$ has n real zeros, then its derivative $p'(x)$ has $n - 1$ real zeros. □

To prove Newton's inequalities, write the polynomial $p(x)$ as the binary form

$$f(x_0, x_1) = \sum_{i=0}^{n} \binom{n}{i} a_i x_0^i x_1^{n-i},$$

where

$$c_i = \binom{n}{i} a_i.$$

Then

$$\frac{2}{n!} \frac{\partial^{n-2}}{\partial x_0^{i-1} \partial x_1^{n-i-1}} = a_{i-1}x_1^2 + 2a_i x_0 x_1 + a_{i+1}x_0^2.$$

By the lemma, the quadratic $a_{i-1} + 2a_i x + a_{i+1}x^2$ has two real zeros. We conclude that its discriminant is nonnegative; that is, $a_i^2 - a_{i-1}a_{i+1} \geq 0$. Newton's inequalities now follow easily.

It is not hard to show that if a sequence $a_0, a_1, a_2, \cdots, a_n$ of real numbers is *logarithmically concave*, that is,

$$a_{i-1}a_{i+1} \leq a_i^2,$$

then it is *unimodal,* that is, there exists an index m such that

$$a_0 \le a_1 \le a_2 \le \cdots \le a_m \text{ and } a_m \ge a_{m+1} \ge \cdots \ge a_{n-1} \ge a_n.$$

Not all polynomials of interest in combinatorics have all zeros real, but many have been conjectured to have unimodal sequences of coefficients. Often, such conjectures seem to be made based on calculations of small cases.

The relation between reality of zeros and unimodality of the coefficient sequence suggests that theorems about location of zeros of polynomials in the complex plane might have combinatorial applications. In the first half of this chapter, we give a very selective exposition of this area. In addition to making the area better known, we hope that combinatorics, in particular, the umbral calculus, might be used in this area. We have deliberately avoided using complex analysis. Thus, our choice of topics is idiosyncratic.[1]

We end this section with a quick introduction to umbral notation for polynomials. As described in Section 4.2, the idea is to represent a sequence $a_0, a_1, a_2, \ldots$, where $a_0 = 1$, by the sequence $\alpha^0, \alpha^1, \alpha^2, \ldots$, or the umbra α. If $p(x)$ is the degree-n polynomial

$$a_0x^n + \binom{n}{1}a_1x^{n-1} + \binom{n}{2}a_2x^{n-2} + \cdots + \binom{n}{n-1}a_{n-1}x + a_n,$$

(with $a_0 \ne 0$), then we can represent $p(x)$ umbrally by

$$a_0(x + \alpha)^n,$$

where the umbra α represents the sequence

$$1, a_1/a_0, a_2/a_0, \ldots, a_n/a_0, 0, 0, \ldots.$$

This works if we can specify the degree a priori. If we cannot, or do not wish to, then it is more convenient to use binary forms. We will represent the n-form $f(x_0, x_1)$ by the nth power

$$(\alpha_1 x_0 + \alpha_0 x_1)^n.$$

To formalize these ideas, let $\mathcal{A}$ be a sufficiently large alphabet. Some of the letters in $\mathcal{A}$ will represent sequences and others will represent coefficients of forms. If a letter τ represents a sequence, then it is associated with one variable τ. If a letter α represents the coefficients of an n-form, then it is associated with two variables α_0 and α_1. Let $\mathbb{F}[\mathcal{A}, x_0, x_1]$ be the algebra of polynomials over the field $\mathbb{F}$ with variables $\tau, \alpha_0, \alpha_1, x_0, x_1$, where τ and α are letters in $\mathcal{A}$. We

[1] For the analytic theory, see the classic treatment in part 5 of Pólya and Szegö (1976). See also Marden (1949). We also omit any discussion of Sturm theory, which can be found in many algebra texts. See, for example, Jacobson (1985).

define the (*umbral*) *evaluation operator* eval : $\mathbb{F}[\mathcal{A}, x_0, x_1] \to \mathbb{F}[x_0, x_1]$ by the following rules:

(1) If τ represents the sequence (t_i), then $\mathrm{eval}(\tau^i) = t_i$.
(2) If α represents (the normalized coefficients of) the n-form $f(x_0, x_1) = \sum_{i=0}^{n} \binom{n}{i} a_i x_0^i x_1^{n-i}$, then

$$\mathrm{eval}(\alpha_0^i, \alpha_1^j) = \begin{cases} a_i & \text{if } j = n - i \\ 0 & \text{otherwise.} \end{cases}$$

(3) If $\sigma^i \cdots \tau^j \alpha_0^{k_0} \alpha_1^{k_1} \cdots \beta_0^{k_0} \beta_1^{k_1}$ is a monomial, where $\sigma, \ldots, \tau$ are distinct letters representing sequences and $\alpha, \ldots, \beta$ are distinct letters representing forms, then

$$\mathrm{eval}(\sigma^i \cdots \tau^j \alpha_0^{k_0} \alpha_1^{k_1} \cdots \beta_0^{l_0} \beta_1^{l_1}) = \mathrm{eval}(\sigma^i) \cdots \mathrm{eval}(\tau^j) \mathrm{eval}(\alpha_0^{k_0} \alpha_1^{k_1}) \cdots \mathrm{eval}(\beta_0^{l_0} \beta_1^{l_1})$$

and extended to all of $\mathbb{F}[\mathcal{A}, x_0, x_1]$ by $\mathbb{F}[x_0, x_1]$-linearity. In terms of the umbral evaluation operator,

$$f(x_0, x_1) = \mathrm{eval}((\alpha_1 x_0 + \alpha_0 x_1)^n).$$

Similarly, any polynomial in normalized coefficients of forms can be umbrally represented. For example, if $g(x_0, x_1) = a_0 x_0^2 + 2a_1 x_0 x_1 + a_2 x_1^2$ and γ and δ are two umbras representing g, then the discriminant of g can be represented using

$$a_1^2 - a_0 a_2 = \mathrm{eval}(\gamma_0 \gamma_1 \delta_0 \delta_1 - \gamma_0^2 \delta_1^2)$$

or

$$a_1^2 - a_0 a_2 = \frac{1}{2} \mathrm{eval}((\gamma_0 \delta_1 - \gamma_1 \delta_0)^2).$$

Many of the theorems in this chapter can be stated neatly using umbral notation, suggesting that combinatorial methods might be used. However, this remains a suggestion at present. The umbral notation is heavily used in nineteenth-century invariant theory.[2]

[2] For a postmodern account, see Kung and Rota (1984b).

Exercises

6.1.1. *Quermassintegrals of convex sets.*[3]

Let A and B be subsets in $\mathbb{R}^n$ and α be a real number. The *Minkowski sum* $A+B$ is the set $\{a+b\colon a \in A,\ b \in B\}$ and αB is the set $\{\alpha b\colon b \in B\}$. If C is a compact convex set in $\mathbb{R}^n$, let $\mathrm{Vol}(C)$ be the volume of C. Let B_n be the unit ball

$$\{(x_1, x_2, \dots, x_n)\colon \sqrt{x_1^2 + x_2^2 + \cdots + x_n^2} \le 1\}.$$

(a) If A and B are compact convex sets, show that $A+B$ is compact and convex.

(b) Let A be a compact convex set in $\mathbb{R}^n$. Show that $\mathrm{Vol}(A + xB_n)$ is a polynomial of degree n in x.

Write

$$\mathrm{Vol}(A + xB_n) = \sum_{i=0}^{n} \binom{n}{i} W_i(A) x^i.$$

Then the normalized coefficients $W_i(A)$ are the *quermassintegrals* of A.

(c) Show that if A is a rectangular two-dimensional rectangle in $\mathbb{R}^2$ with sides a and b, then

$$\mathrm{Vol}(A + xB_2) = ab + 2(a+b)x + \pi x^2.$$

Extend this result to all dimensions.

(d) Show that the quermassintegrals of a convex set A are logarithmically concave; that is, $W_{i-1}(A)W_{i+1}(A) \le W_i^2(A)$, $1 \le i \le n-1$.

We remark that the quermassintegrals are valuations on the ring of subsets generated by the compact convex sets in $\mathbb{R}^n$.

6.1.2. (Research problem) Prove or find counterexamples to the following famous conjectures:

(a) The coefficients of chromatic polynomials of graphs are unimodal.

(b) The Whitney numbers of the second kind of geometric lattices are logarithmically concave.

The second conjecture, made implicitly by Rota in *Matching theory,* was motivated by logarithmic concavity of the quermassintegrals. Most naturally occurring probability distributions are logarithmically concave or unimodal. Are there ways to prove logarithmic concavity or unimodality of sequences in combinatorics from this perspective? For example, if one can generate flats of

[3] The material is part of the theory of Minkowski mixed volumes. See McMullen (1993, pp. 933–988) and Sangwire-Yager (1993, pp. 43–71).

matroids in a randomly additive way, then one might obtain insight into the second conjecture.

6.1.3. *The golden ratio and zeros of chromatic polynomials.*[4]

Let Γ be a triangulation of the 2-sphere with v vertices and $P(\Gamma; x)$ its chromatic polynomial. Show that

$$|P(\Gamma, 1+\tau)| \leq \tau^{5-v},$$

where τ is the golden ratio $\frac{1}{2}(1+\sqrt{5})$. Thus, the chromatic polynomial of a triangulation "tends to" have a zero near τ.

6.1.4. *Three-term recurrences and interlacing of zeros.*[5]

Let $(p_n(x))_{n=0}^{\infty}$ be a sequence of polynomials with real coefficients such that $p_n(x)$ has degree n and for $n \geq 1$,

$$p_{n+1}(x) = (A_n x + B_n) p_n(x) - C_n p_{n-1}(x),$$

where A_n and C_n are positive real numbers and B_n is a real number. Show that $p_n(x)$ has n distinct real zeros.

6.1.5. (Research problem) To use umbral notation for polynomials or forms, as defined in this section, one needs to assume that binomial coefficients are nonzero. Find a *characteristic-free* umbral notation, that is, a (useful) umbral notation valid over an infinite field of any characteristic. Alternately, find umbral notations valid over a field with a given positive characteristic.

6.1.6. *An umbral notation for matrices.*[6]

A rank-1 $n \times n$ matrix can be represented as the product of an $n \times 1$ column $(\alpha_1, \alpha_2, \ldots, \alpha_n)^T$ and a $1 \times n$ row $(a_1, a_2, \ldots, a_n)$. The ij-entry is $\alpha_i a_j$. As in the umbral notation for forms, we can umbrally represent the entries of a general $n \times n$ matrix (a_{ij}) by entries in a rank-1 matrix with *Greco-Roman umbra* (α, a), using the basic rule

$$\text{eval}(\alpha_i a_j) = a_{ij}.$$

(Research problem) Give umbral proofs of theorems in matrix theory.

[4] Tutte (1970).

[5] This is a theorem in the theory of orthogonal polynomials. See Andrews et al. (1999) and Szegö (1975).

[6] Turnbull (1930).

6.2 Apolarity

The set of binary forms of degree n with complex coefficients forms a vector space $\mathbb{C}_n[x_0, x_1]$ over the complex numbers $\mathbb{C}$ of dimension $n+1$. We define the *polar bilinear form* $\{\cdot, \cdot\} : \mathbb{C}_n[x_0, x_1] \times \mathbb{C}_n[x_0, x_1] \to \mathbb{C}$ by

$$\begin{aligned}\{f, g\} &= a_0 b_n - \binom{n}{1} a_1 b_{n-1} + \binom{n}{2} a_2 b_{n-2} - \cdots + (-1)^n a_n b_0 \\ &= \sum_{i=0}^{n} (-1)^i \binom{n}{i} a_i b_{n-i},\end{aligned}$$

where $f(x_0, x_1) = \sum_{i=0}^{n} \binom{n}{i} a_i x_0^i x_1^{n-i}$ and $g(x_0, x_1) = \sum_{i=0}^{n} \binom{n}{i} b_i x_0^i x_1^{n-i}$. Alternately, if α is an umbra representing f and β is an umbra representing g, then

$$\{f, g\} = (\alpha_0 \beta_1 - \beta_0 \alpha_1)^n.$$

Two forms f and g are *apolar* if $\{f, g\} = 0$. We extend the notion of apolarity to forms of degree less than n in the following way: if $m < n$, f is an n-form, and g is an m-form, then g is *apolar* to f if for all $(n-m)$-forms h, $\{f, gh\} = 0$. Since the polar form is bilinear, g is apolar to f if and only if for $0 \le i \le n-m$,

$$\{f(x_0, x_1), g(x_0, x_1) x_0^i x_1^{n-m-i}\} = 0.$$

Let T be the invertible 2×2 matrix

$$\begin{pmatrix} t_{00} & t_{01} \\ t_{10} & t_{11} \end{pmatrix}.$$

Then T defines the *linear change of variables*

$$x_0 = t_{00}\bar{x}_0 + t_{01}\bar{x}_1, \quad x_1 = t_{10}\bar{x}_0 + t_{11}\bar{x}_1.$$

If f is the n-form $\sum_{i=0}^{n} \binom{n}{i} a_i x_0^i x_1^{n-i}$, then T defines the form $\bar{f}(\bar{x}_0, \bar{x}_1)$ by

$$\bar{f}(\bar{x}_0, \bar{x}_1) = \sum_{i=0}^{n} \binom{n}{i} a_i (t_{00}\bar{x}_0 + t_{01}\bar{x}_1)^i (t_{10}\bar{x}_0 + t_{11}\bar{x}_1)^{n-i}.$$

Expanding and regrouping terms and writing

$$\bar{f}(\bar{x}_0, \bar{x}_1) = \sum_{i=0}^{n} \binom{n}{i} \bar{a}_i \bar{x}_0^i \bar{x}_1^{n-i},$$

we can express the coefficients $\bar{a}_i$ of $\bar{f}$ as polynomials in the coefficients a_i of f and the entries t_{ij} of the matrix T. Let $I(f_1, f_2, \ldots, f_s, x_0, x_1)$ be a polynomial in the normalized coefficients a_{ji} of f_j and the variables x_0, x_1. Then T acts on I by

$$TI(f_1, f_2, \ldots, f_s, x_0, x_1) = I(\bar{f}_1, \bar{f}_2, \ldots, \bar{f}_s, \bar{x}_0, \bar{x}_1).$$

The next lemma says that the polar bilinear form is a "relative invariant" under linear changes of variables.

6.2.1. Lemma. If T is an invertible 2×2 matrix and, f and g are n-forms, then

$$(\det T)^n \{f, g\} = \{\bar{f}, \bar{g}\}.$$

The proof of Lemma 6.2.1 depends on a more general fact: that the action of a matrix T almost commutes with umbral evaluation. Let α be an umbra representing f. Then

$$\begin{aligned}\bar{f} &= \mathrm{eval}(\alpha_1(t_{00}\bar{x}_0 + t_{01}\bar{x}_1) + \alpha_0(t_{10}\bar{x}_0 + t_{11}\bar{x}_1))^n \\ &= \mathrm{eval}((t_{00}\alpha_1 + t_{10}\alpha_0)\bar{x}_0 + (t_{01}\alpha_1 + t_{11}\alpha_0)\bar{x}_1))^n.\end{aligned}$$

Thus, $\bar{f}$ is represented umbrally by $\bar{\alpha}_0$ and $\bar{\alpha}_1$, where

$$\bar{\alpha}_0 = t_{11}\alpha_0 + t_{01}\alpha_1 \quad \text{and} \quad \bar{\alpha}_1 = t_{10}\alpha_0 + t_{00}\alpha_1,$$

or in matrix notation,

$$\begin{pmatrix}\bar{\alpha}_0 \\ \bar{\alpha}_1\end{pmatrix} = \begin{pmatrix} t_{11} & t_{01} \\ t_{10} & t_{00}\end{pmatrix}\begin{pmatrix}\alpha_0 \\ \alpha_1\end{pmatrix}.$$

Returning to the polar bilinear form and the proof of Lemma 6.2.1, let α and β be umbras representing f and g. Since $\alpha_0\beta_1 - \beta_0\alpha_1$ is a determinant and hence,

$$\bar{\alpha}_0\bar{\beta}_1 - \bar{\beta}_0\bar{\alpha}_1 = (\det T)(\alpha_0\beta_1 - \beta_0\alpha_1),$$

we conclude that

$$\begin{aligned}\{\bar{f}, \bar{g}\} &= \mathrm{eval}((\bar{\alpha}_0\bar{\beta}_1 - \bar{\beta}_0\bar{\alpha}_1)^n) \\ &= \mathrm{eval}([(\det T)(\alpha_0\beta_1 - \beta_0\alpha_1)]^n) = (\det T)^n\{f, g\}.\end{aligned}$$

This completes the proof of Lemma 6.2.1.

6.2.2. Lemma. If $(\mu x_0 - \nu x_1)^m$ is a factor of the n-form $f(x_0, x_1)$, then $f(x_0, x_1)$ is apolar to $(\mu x_0 - \nu x_1)^{n-m+1}$.

Proof. By Lemma 6.2.1, it suffices to prove the lemma for the form x_0^m. By hypothesis, $f(x_0, x_1) = x_0^m g(x_0, x_1)$, where g is an $(n-m)$-form. Thus, if f has normalized coefficients a_i, then $a_i = 0$ for $0 \leq i \leq m - 1$. On the other hand, all except one of the normalized coefficients of $x_0^{n-i}x_1^i$ are zero. Thus, if $0 \le i \le m-1$, $\{f(x_0, x_1), x_0^{n-i}x_1^i\} = 0$. □

Let f be an n-form. Since the polar form is bilinear, the set $f^\perp$ of all m-forms apolar to f is a subspace in $\mathbb{C}_m[x_0, x_1]$. Our next task is to describe such subspaces. There are two cases, depending on whether $n \ge m$ or $m < n$.

Before stating the theorems, it will be useful to write out explicitly the conditions for two forms to be apolar. Let $m \le n$, $f(x_0, x_1) = \sum_{i=0}^n \binom{n}{i} a_i x_0^i x_1^{n-i}$, and $g(x_0, x_1) = \sum_{i=0}^m \binom{m}{i} b_i x_0^i x_1^{m-i}$. Then f is apolar to g if and only if

$$\{f(x_0, x_1), g(x_0, x_1)x_0^{n-m}\} = 0,\ \{f(x_0, x_1), g(x_0, x_1)x_0^{n-m-1}x_1\} = 0,\ \ldots,$$
$$\{f(x_0, x_1), g(x_0, x_1)x_0 x_1^{n-m-1}\} = 0,\ \{f(x_0, x_1), g(x_0, x_1)x_1^{n-m}\} = 0.$$

Explicitly, g is apolar to f if and only if the following system (∗) of $n - m + 1$ linear equations holds:

$$\begin{aligned}
b_0 a_n - \binom{m}{1} b_1 a_{n-1} + \binom{m}{2} b_2 a_{n-2} - \cdots + (-1)^m b_m a_{n-m} &= 0, \\
b_0 a_{n-1} - \binom{m}{1} b_1 a_{n-2} + \cdots + (-1)^m b_m a_{n-m-1} &= 0, \\
&\ \vdots \\
b_0 a_m - \binom{m}{1} b_1 a_{m-1} + \cdots + (-1)^m b_m a_0 &= 0.
\end{aligned}$$

6.2.3. Theorem. Let $n \ge m$ and g be a nonzero m-form. Then the subspace $g^\perp$ in $\mathbb{C}_n[x_0, x_1]$ has dimension (exactly) m. If

$$g(x_0, x_1) = a(\mu_1 x_0 - \nu_1 x_1)^{m_1}(\mu_2 x_0 - \nu_2 x_1)^{m_2} \cdots (\mu_s x_0 - \nu_s x_1)^{m_s}$$

is a factorization of $g(x_0, x_1)$ into distinct linear forms, then the forms

$$(\mu_i x_0 - \nu_i x_1)^{n-m_i+1} x_0^j x_1^{m_i-j-1}, \quad 0 \le j \le m_i - 1,$$

as i ranges from 1 to s, form a basis for the subspace $g^\perp$. In particular, if g can be factored into n distinct factors, then $g^\perp$ is the subspace of all forms

$$c_1(\mu_1 x_0 - \nu_1 x_1)^n + c_2(\mu_2 x_0 - \nu_2 x_1)^n + \cdots + c_m(\mu_m x_0 - \nu_m x_1)^n,$$

where c_i are constants.

Proof. We can find all n-forms f apolar to the given form g by regarding f as an "unknown" form and solving for the unknowns a_i in (∗). The system (∗) is a triangular system of $n - m + 1$ linear equations in $n + 1$ unknowns. Hence, the subspace $g^\perp$ has dimension exactly m.

By Lemma 6.2.2, the m_i forms $(\mu_i x_0 - \nu_i x_1)^{n-m_i+1} x_0^j x_1^{m_i-j-1}$, where $0 \leq j \leq m_i - 1$, are in $g^\perp$. It is elementary to show they are linearly independent. Hence, the set of all such forms is a basis for $g^\perp$. □

6.2.4. Theorem. Let $n \geq m$ and f be a nonzero n-form. Then the subspace $f^\perp$ in $\mathbb{C}_m[x_0, x_1]$ has dimension at least $2m - n$.

Proof. We can find all m-forms apolar to f by solving the system $(*)$ with $n - m + 1$ linear equations for the unknowns b_i. These equations need not be linearly independent. Hence the dimension of the solution space is at least $(m + 1) - (n - m + 1)$. □

We shall now discuss the special case of a cubic form. Let

$$f(x_0, x_1) = a_0 x_1^3 + 3a_1 x_0 x_1^2 + 3a_2 x_0^2 x_1 + a_3 x_0^3.$$

The subspace $f^\perp$ in $\mathbb{C}_2[x_0, x_1]$ has dimension at least 1. Let $g(x_0, x_1) = b_0 x_1^2 + 2b_1 x_0 x_1 + b_2 x_0^2$. Then g is apolar to f if and only if

$$\begin{aligned} a_0 b_2 + 3a_1 b_1 + 3a_2 b_0 &= 0 \\ 3a_1 b_2 + 3a_2 b_1 + a_3 b_0 &= 0. \end{aligned}$$

Almost all the time, or "generically,"[7] the 2-form $g(x_0, x_1)$ is the product of two distinct linear forms $\mu_1 x_0 - \nu_1 x_1$ and $\mu_2 x_0 - \nu_2 x_1$. In this case,

$$f(x_0, x_1) = c_1(\mu_1 x_0 - \nu_1 x_1)^3 + c_2(\mu_2 x_0 - \nu_2 x_1)^3.$$

In the case the linear forms in the factorization of g are not distinct,

$$f(x_0, x_1) = a(\mu x_0 - \nu x_1)^2(\gamma x_0 - \delta x_1).$$

This gives an algorithm for solving the cubic polynomial by taking square roots and cube roots. Given a cubic polynomial $p(x)$, rewrite it as the 3-form $x_0^3 p(x_1/x_0)$. Then, generically,

$$p(x) = c(x - \xi)^3 + d(x - \eta)^3,$$

where c and d are nonzero. The numbers ξ and η can be found by solving the quadratic polynomial $g(1, x)$, where g is a form apolar to $x_0^3 p(x_1/x_0)$. The

[7] The coefficients of the quadratic g are polynomials in a_0, a_1, a_2, and a_3 and the property of having a double zero is equivalent to the discriminant of g being zero. Regarding a 3-form as a 4-tuple in $\mathbb{C}^4$, the set of 3-forms such that a 2-form apolar to it has a double zero is a union of varieties or zero-sets of polynomials. Thus, g is the product of two distinct linear forms except on a "lower-dimensional set."

zeros $p(x)$ can be found by solving the equation

$$c(x-\xi)^3 = -d(x-\eta)^3$$

or, equivalently,

$$x-\xi = -\omega(c/d)^{1/3}(x-\eta),$$

where ω is a cube root of unity. The nongeneric cases have multiple zeros and can be solved by taking square or cube roots.

We end this section with a famous theorem of Sylvester.[8]

6.2.5. Sylvester's theorem. Almost all forms f of odd degree $2j+1$ can be written as a sum of $j+1$ or fewer $(2j+1)$th powers of linear forms.

Proof. Let $n = 2j+1$ and $m = j+1$. Since $2m-n$ is positive, there exists an m-form g apolar to f. For almost all forms f, g has a factorization into $j+1$ distinct linear forms. The theorem now follows from Theorem 6.2.4. □

Exercises

6.2.1. *The polar covariant.*

Let $n \geq m$. The *polar covariant* is the bilinear function $\{\cdot,\cdot\} : \mathbb{C}_n[x_0,x_1] \times \mathbb{C}_m[x_0,x_1] \to \mathbb{C}_{n-m}[x_0,x_1]$ defined by

$$\{f(x_0,x_1), g(x_0,x_1)\} = \text{eval}((\alpha_0\beta_1 - \alpha_1\beta_0)^m(\alpha_1 x_0 + \alpha_0 x_1)^{n-m}),$$

where f is an n-form represented by the umbra α and g is an m-form represented by β.

(a) Show that f is apolar to g (as defined in the text) if and only if the polar covariant $\{f,g\}$ is identically zero.

(b) Show that if f is an n-form, g is an m_1-form, h is an m_2-form, and $n \geq m_1 + m_2$, then

$$\{f, gh\} = \{\{f,g\},h\}.$$

6.2.2. Show that up to a constant factor, the polar bilinear form is the only bilinear form $B(f,g)$ on $\mathbb{C}_n[x_0,x_1]$ such that for every invertible 2×2 matrix T,

$$B(f,g) = (\det T)^n B(Tf, Tg).$$

[8] Sylvester (1851a, b; 1904–1912).

6.2.3. *The Gundelfinger invariants.*[9]

Let $f(x_0, x_1)$ be an n-form and let $G_k(f)$ be the form defined by

$$G_k(f) = \det\left(\frac{\partial^{2k}}{\partial x_0^{2k-i-j}\partial x_1^{i+j}} f(x_0, x_1)\right)_{0\le i,j\le k}.$$

(a) Let s be the minimum positive integer s such that there exists an s-form apolar to f. Let s' be the minimum positive integer such that $G_s(f)$ is identically zero. Show that $s = s'$.

(b) Let s be as defined in part (a). Show that $f^\perp$ in $\mathbb{C}_s[x_0, x_1]$ is a one-dimensional subspace spanned by

$$\det\begin{pmatrix} a_0 & a_1 & a_2 & \dots & a_s \\ a_1 & a_2 & a_3 & \dots & a_{s+1} \\ \vdots & \vdots & \vdots & & \vdots \\ a_{s-2} & a_{s-1} & a_s & \dots & a_{2s-2} \\ a_{s-1+t} & a_{s+t} & a_{s+t+1} & \dots & a_{2s-1+t} \\ x_0^s & -x_0^{s-1}x_1 & x_0^{s-2}x_1^2 & \dots & (-1)^s x_1^s \end{pmatrix},$$

where t is the minimum exponent such that the coefficient of $x_0^t x_1^{s(n-2s-2)-t}$ in $G_{s-1}(f)$ is nonzero.

(c) Show that $f(x_0, x_1)$ equals an nth power if and only if its *Hessian*

$$\frac{\partial^2 f}{\partial x_0^2}\frac{\partial^2 f}{\partial x_1^2} - \left(\frac{\partial^2 f}{\partial x_0 \partial x_1}\right)^2$$

is zero.

6.3 Grace's Theorem

In this section, we consider polynomials in the complex variable z. Let $p(z)$ and $q(z)$ be polynomials of degree (exactly) n. We extend the polar bilinear form to such pairs of polynomials by

$$\{p(z), q(z)\} = \{x_0^n p(x_1/x_0), x_0^n q(x_1/x_0)\}.$$

In addition, we say that $p(z)$ and $q(z)$ are *apolar* if $\{p(z), q(z)\} = 0$. In this section, we give two proofs of Grace's theorem relating the zeros of two polynomials apolar to each other. For the first proof, we begin with two lemmas: one analytic and the other algebraic.

9 Gundelfinger (1886) and Kung (1986).

6.3.1. Lucas' lemma.[10] Let $p(z)$ be a nonconstant polynomial over the complex field. Then every zero of the derivative $p'(z)$ lies inside the convex hull C of the zeros of $p(z)$.

Proof. If $m > 1$ and ξ is a zero of multiplicity m of $p(z)$, then ξ is a zero of multiplicity $m-1$ of $p'(z)$. Such a zero ξ of $p'(z)$ lies in the convex hull C. Thus, we may assume that all the zeros of $p(z)$ have multiplicity 1. Let $\xi_1, \xi_2, \ldots, \xi_n$ be the zeros of $p(z)$ and let w be a zero of $p'(z)$. Then for all i, $w \neq \xi_i$. By logarithmic differentiation,

$$\frac{p'(z)}{p(z)} = \sum_{i=1}^{n} \frac{1}{z - \xi_i}.$$

Setting $z = w$ and taking the complex conjugate, we obtain

$$0 = \sum_{i=1}^{n} \frac{1}{\overline{w - \xi_i}} = \sum_{i=1}^{n} \frac{w - \xi_i}{|w - \xi_i|^2}.$$

Thus,

$$w = \sum_{i=1}^{n} \lambda_i \xi_i,$$

where

$$\lambda_i = \frac{1}{|w - \xi_i|^2} \Big/ \sum_{i=1}^{n} \frac{1}{|w - \xi_i|^2}.$$

Since $\lambda_i \geq 0$ and $\sum_{i=1}^{n} \lambda_i = 1$, we conclude that w is in the convex hull of the zeros $\xi_1, \xi_2, \ldots, \xi_n$. □

A *linear fractional transformation* ℓ from the extended complex plane $\mathbb{C} \cup \{\infty\}$ to itself is a function of the form

$$\ell(z) = \frac{az + b}{cz + d},$$

where $ad - bc \neq 0$. We shall use the following properties of linear fractional transformations, familiar from a first course in complex analysis.[11] First, linear fractional transformations send circles to circles, where straight lines are regarded as circles of infinite radius. Second, given any two triples z_1, z_2, z_3

[10] Lucas (1879). Lucas used a fact from mechanics due to Gauss. For this reason, this lemma is sometimes attributed to Gauss and Lucas.

[11] See, for example, Alhfors (1966, p. 76).

and z_1', z_2', z_3' of complex numbers, there exists a (unique) linear fractional transformation ℓ such that $\ell(z_1) = z_1'$, $\ell(z_2) = z_2'$, and $\ell(z_3) = z_3'$.

6.3.2. Lemma. The polar bilinear form for two degree-n polynomials is a relative invariant under linear fractional transformations; that is, if

$$\bar{p}(z) = (cz+d)^n p\left(\frac{az+b}{cz+d}\right) \quad \text{and} \quad \bar{q}(z) = (cz+d)^n q\left(\frac{az+b}{cz+d}\right),$$

then

$$\{\bar{p}(z), \bar{q}(z)\} = (ad-bc)^n \{p(z), q(z)\}.$$

Proof. Let $f(x_0, x_1) = x_0^n p(x_1/x_0)$. Then

$$\bar{p}(z) = (cz+d)^n p\left(\frac{az+b}{cz+d}\right) = f(cz+d, az+b).$$

Thus, $\bar{p}(z) = \bar{f}(1, z)$, where $\bar{f}(x_0, x_1) = f(\bar{x}_0, \bar{x}_1)$ under the linear change of variables

$$\bar{x}_0 = dx_0 + cx_1 \quad \text{and} \quad \bar{x}_1 = bx_0 + ax_1.$$

A similar relation holds between $q(z)$ and the form $g(x_0, x_1)$ defined by $g(x_0, x_1) = x_0^n q(x_1/x_0)$. By Lemma 6.2.1,

$$\{\bar{p}, \bar{q}\} = \{\bar{f}, \bar{g}\} = (ad-bc)^n \{f, g\} = (ad-bc)^n \{p, q\}. \qquad \square$$

6.3.3. Grace's theorem.[12] Let $p(z)$ and $q(z)$ be two degree-n polynomials apolar to each other. If the zeros of $p(z)$ are all contained in a disk D, then at least one zero of $q(z)$ is also in D.

Proof. We proceed by induction on the degree n. It is easy to check that $a_0 + a_1 z$ and $b_0 + b_1 z$ are apolar if and only if for some nonzero constant c, $a_0 + a_1 z = c(b_0 + b_1 z)$. Hence, Grace's theorem holds when the degree equals 1.

Now suppose that Grace's theorem holds when $p(z)$ and $q(z)$ have degree less than n. Let $p(z)$ and $q(z)$ be degree-n polynomials, which we may assume to be monic, apolar to each other, $z_1, z_2, \ldots, z_n$ be the zeros of $p(z)$, and $w_1, w_2, \ldots, w_n$ be the zeros of $q(z)$. We may suppose that all the zeros $z_1, z_2, \ldots, z_n$ of $p(z)$ are inside D and one of the zeros, w_n, say, is outside D.

Let

$$e_i(\underline{t}) = e_i(t_1, t_2, \ldots, t_n) = \sum_{\{j_1, j_2, \ldots, j_i\}} t_{j_1} t_{j_2} \cdots t_{j_i},$$

[12] Grace (1902).

where the sum ranges over all i-subsets of $\{1, 2, \ldots, n\}$. The function e_i is the ith elementary symmetric function in the variables $t_1, t_2, \ldots, t_n$. (The notation is different from the notation in Chapter 5.)

If $p(z) = \sum_{i=0}^{n} \binom{n}{i} a_i z^{n-i}$ and $q(z) = \sum_{i=0}^{n} \binom{n}{i} b_i z^{n-i}$, with $a_0 = 1$ and $b_0 = 1$, then

$$(-1)^i e_i(\underline{z}) = \binom{n}{i} a_i, \qquad (-1)^i e_i(\underline{w}) = \binom{n}{i} b_i.$$

Hence,

$$(-1)^n \{p, q\} = e_n(\underline{w}) - \frac{e_{n-1}(\underline{w}) e_1(\underline{z})}{\binom{n}{1}} + \frac{e_{n-2}(\underline{w}) e_2(\underline{z})}{\binom{n}{2}} - \cdots + (-1)^n e_n(\underline{z}).$$

We consider first the generic case, when the zero w_n of $q(z)$ has multiplicity 1. In that case,

$$\begin{aligned} e_n(w_1, w_2, \ldots, w_n) &= w_n e_{n-1}(w_1, w_2, \ldots, w_{n-1}), \\ e_j(w_1, w_2, \ldots, w_n) &= w_n e_{j-1}(w_1, w_2, \ldots, w_{n-1}) + e_j(w_1, w_2, \ldots, w_{n-1}) \end{aligned}$$

and

$$\begin{aligned} 0 &= (-1)^n \{p, q\} \\ &= w_n X(w_1, w_2, \ldots, w_{n-1}, z_1, z_2, \ldots, z_n) \\ &\quad + Y(w_1, w_2, \ldots, w_{n-1}, z_1, z_2, \ldots, z_n), \end{aligned}$$

where

$$\begin{aligned} &X(w_1, \ldots, w_{n-1}, z_1, \ldots, z_n) \\ &\quad = e_{n-1}(w_1, \ldots, w_{n-1}) - \frac{e_{n-2}(w_1, \ldots, w_{n-1}) e_1(\underline{z})}{\binom{n}{1}} \\ &\qquad + \frac{e_{n-3}(w_1, \ldots, w_{n-1}) e_2(\underline{z})}{\binom{n}{2}} - \cdots + (-1)^{n-1} \frac{e_{n-1}(\underline{z})}{\binom{n}{n-1}} \end{aligned}$$

and

$$\begin{aligned} Y(w_1, \ldots, w_{n-1}, z_1, \ldots, z_n) = &- \frac{e_{n-1}(w_1, \ldots, w_{n-1}) e_1(\underline{z})}{\binom{n}{1}} \\ &+ \frac{e_{n-2}(w_1, \ldots, w_{n-1}) e_2(\underline{z})}{\binom{n}{2}} \\ &- \cdots + (-1)^n e_n(\underline{z}). \end{aligned}$$

Thus, we can rewrite the equation $\{p, q\} = 0$ as

$$X(w_1, \ldots, w_{n-1}, z_1, \ldots, z_n) = -\frac{1}{w_n} Y(w_1, \ldots, w_{n-1}, z_1, \ldots, z_n).$$

We will now use the fact that there exists a linear fractional transformation ℓ such that $\ell(D)$ is the unit disk and $\ell(w_n)$ equals ∞; that is to say, $1/\ell(w_n) = 0$. Applying ℓ, we obtain

$$X(\ell(w_1), \ldots, \ell(w_{n-1}), \ell(z_1), \ldots, \ell(z_n)) = 0.$$

The expression X is in fact an evaluation of the polar bilinear form. To see this, consider the derivative $p'(z)$. Then

$$p'(z) = n \sum_{i=0}^{n-1} \binom{n-1}{i} a_i z^{n-1-i}.$$

If $u_1, u_2, \ldots, u_{n-1}$ are the zeros of $p'(z)$, then, for $0 \leq i \leq n-1$,

$$e_i(u_1, u_2, \ldots, u_{n-1}) = (-1)^i \binom{n-1}{i} a_i.$$

It follows that

$$(-1)^i \frac{e_i(z_1, z_2, \ldots, z_n)}{\binom{n}{i}} = a_i = (-1)^i \frac{e_i(u_1, u_2, \ldots, u_{n-1})}{\binom{n-1}{i}}.$$

Therefore,

$$\begin{aligned} &X(w_1, w_2, \ldots, w_{n-1}, z_1, z_2, \ldots, z_n) \\ &= e_{n-1}(w_1, \ldots, w_{n-1}) - \frac{e_{n-2}(w_1, \ldots, w_{n-1}) e_1(u_1, \ldots, u_{n-1})}{\binom{n-1}{1}} \\ &\quad + \frac{e_{n-3}(w_1, \ldots, w_{n-2}) e_2(u_1, \ldots, u_{n-1})}{\binom{n-1}{2}} - \cdots \\ &\quad + (-1)^{n-1} e_{n-1}(u_1, \ldots, u_{n-1}) \\ &= \left\{ p'(z), \prod_{i=1}^{n-1} (z - w_i) \right\}. \end{aligned}$$

Since $X(\ell(w_1), \ldots, \ell(w_{n-1}), \ell(z_1), \ldots, \ell(z_n)) = 0$ and the polar bilinear form is relatively invariant under linear fractional transformations, this implies that the derivative $p'(z)$ is apolar to the polynomial $\prod_{i=1}^{n-1}(z - w_i)$. In addition, under ℓ, the zeros of $p(z)$ are inside the unit disk. By Lucas' lemma, all the transformed zeros of $p'(z)$ are inside the unit disk, and hence, by the inductive hypothesis, at least one of the transformed zeros $\ell(w_1), \ell(w_2), \ldots, \ell(w_{n-1})$ lies inside the unit disk. Reversing the transformation ℓ, we conclude that at least one of the zeros $w_1, w_2, \ldots, w_{n-1}$ lies inside the disk D. This verifies the induction step.

The nongeneric case is similar. When w_n is a zero of multiplicity ν and $w_{n-\nu+1} = w_{n-\nu+2} = \cdots = w_n$, we have

$$e_j(w_1, \ldots, w_n) = w_n^{\nu} e_{j-\nu}(w_1, \ldots, w_{n-\nu}) + \text{ terms with lower powers of } w_n.$$

With these equations, we can proceed in a similar way to prove the induction step. □

6.3.4. Corollary. Let $n \geq m$, $p(z)$ be a degree-n polynomial, and $q(z)$ be a degree-m polynomial. Suppose that the forms $x_0^n p(x_1/x_0)$ and $x_0^m q(x_1/x_0)$ are apolar to each other. Then a disk containing all the zeros of $p(z)$ contains at least one zero of $q(z)$.

Proof. Choose complex numbers $\xi_1, \ldots, \xi_{n-k}$ outside the disk. Since the associated forms are apolar,

$$\{p(z), q(z) \prod_{i=1}^{n-k} (z - \xi_i)\} = 0.$$

We can now apply Grace's theorem to finish the proof. □

We note, for use in the next section, the following variation on Grace's theorem.

6.3.5. Proposition. Let $p(z)$ and $q(z)$ be two degree-n polynomials apolar to each other. If the zeros of $p(z)$ are all contained in a half-plane H, then at least one zero of $q(z)$ is also in H.

This can be proved by imitating the proof of Theorem 6.3.3 or using the fact that one can transform a half-plane into a disk by a linear fractional transformation.

Our next topic is the polar derivative. Let y be the point (y_0, y_1) in $\mathbb{C}^2$. Then $D_{yx} : \mathbb{C}_n[x_0, x_1] \to \mathbb{C}_{n-1}[x_0, x_1]$ is the operator defined by

$$D_{yx} = y_0 \frac{\partial}{\partial x_0} + y_1 \frac{\partial}{\partial x_1}.$$

The operator D_{yx} is a derivation; that is, D_{yx} is linear, $D_{yx}(1) = 0$, and

$$D_{yx}(fg) = D_{yx}(f)g + f D_{yx}(g).$$

The $(n-1)$-form $D_{yx} f(x_0, x_1)$ is the *polar derivative of* $f(x_0, x_1)$ *at the point* y. Explicitly,

$$D_{yx} f(x_0, x_1) = n \sum_{i=0}^{n-1} \binom{n-1}{i} x_0^i x_1^{n-1-i} (y_0 a_{i+1} + y_1 a_i).$$

The operator D_{yx} is independent of the choice of coordinates. As in Section 6.2, let T be the linear change of variables

$$x_0 = t_{00}\bar{x}_0 + t_{01}\bar{x}_1, \quad x_1 = t_{10}\bar{x}_0 + t_{11}\bar{x}_1,$$

defined by the 2×2 matrix $(t_{ij})_{0 \le i,j \le 1}$. By the chain rule,

$$\begin{pmatrix} \partial/\partial x_0 \\ \partial/\partial x_1 \end{pmatrix} = \begin{pmatrix} t_{00} & t_{10} \\ t_{01} & t_{11} \end{pmatrix} \begin{pmatrix} \partial/\partial \bar{x}_0 \\ \partial/\partial \bar{x}_1 \end{pmatrix}.$$

Hence,

$$\begin{aligned} \bar{y}_0 \frac{\partial}{\partial \bar{x}_0} + \bar{y}_1 \frac{\partial}{\partial \bar{x}_1} &= (\bar{y}_0, \bar{y}_1) \begin{pmatrix} \partial/\partial \bar{x}_0 \\ \partial/\partial \bar{x}_1 \end{pmatrix} \\ &= (y_0, y_1) \begin{pmatrix} t_{00} & t_{10} \\ t_{01} & t_{11} \end{pmatrix} \begin{pmatrix} t_{00} & t_{10} \\ t_{01} & t_{11} \end{pmatrix}^{-1} \begin{pmatrix} \partial/\partial x_0 \\ \partial/\partial x_1 \end{pmatrix} \\ &= (y_0, y_1) \begin{pmatrix} \partial/\partial x_0 \\ \partial/\partial x_1 \end{pmatrix} = y_0 \frac{\partial}{\partial x_0} + y_1 \frac{\partial}{\partial x_1}. \end{aligned}$$

6.3.6. Lemma. Let $f(x_0, x_1)$ be an n-form, $g(x_0, x_1)$ be an $(n-1)$-form, and $y = (y_0, y_1)$. Then

$$\{g(x_0, x_1), D_{yx} f(x_0, x_1)\} = n\{(y_1 x_0 - y_0 x_1) g(x_0, x_1), f(x_0, x_1)\}.$$

Proof. The proof is a routine formal calculation, going through the formula

$$D_{yx} f(x_0, x_1) = n \sum_{i=0}^{n-1} \binom{n-1}{i} x_0^i x_1^{n-1-i} (y_0 a_{i+1} + y_1 a_i). \qquad \square$$

If $(cx_0 + dx_1)$ is a linear factor of the form $f(x_0, x_1)$, then we say that $(-d, c)$ is a homogeneous zero of $f(x_0, x_1)$. If $d \neq 0$, then $(-d, c)$ is a homogeneous zero of the form f if and only if $-c/d$ is a zero of the polynomial $f(1, z)$. Two homogeneous zeros $(-d, c)$ and $(-d', c')$ are *equivalent* if $dc' = cd'$. Let K be a subset of the extended complex plane $\mathbb{C} \cup \{\infty\}$. The homogeneous zero $(-d, c)$ is *inside* K if the zero $-c/d \in K$ and *outside* K if $-c/d \notin K$.

6.3.7. Laguerre's theorem. Let $f(x_0, x_1)$ be an n-form and D be a disk in the complex plane. If all the homogeneous zeros of $f(x_0, x_1)$ are inside D and (y_0, y_1) is a point in $\mathbb{C}^2$ such that $y_1/y_0 \notin D$, then all the homogeneous zeros of the polar derivative $D_{yx} f(x_0, x_1)$ are also inside D.

Proof. Since the operator D_{yx} is independent of the choice of coordinates, we can make a linear change of variables and assume that $y = (0, 1)$ and

$D_{yx} = \partial/\partial x_1$. In particular, the polynomial $D_{yx} f(1, z)$ is the derivative $p'(z)$. Since D is convex and all the zeros of $p(z)$ are inside D, we can apply Lucas' lemma to conclude that all the zeros of $p'(z)$, and hence, all the homogeneous zeros of $D_{yx} f(x_0, x_1)$, lie inside D. □

We will prove Grace's theorem in the following equivalent version.

6.3.8. Grace's theorem for forms. Let $f(x_0, x_1)$ and $g(x_0, x_1)$ be two n-forms, apolar to each other. Suppose that all the homogeneous zeros of f are in a disk D. Then at least one homogeneous zero of g is in D.

The proof uses polar derivatives and is similar in spirit to the first proof. Let

$$g(x_0, x_1) = \prod_{i=1}^{n} (c_i x_0 + d_i x_1)$$

and $y_i = (-d_i, c_i)$. If the homogeneous zero $(-d_1, c_1)$ is outside D, then Laguerre's theorem implies that all the homogeneous zeros of the polar derivative $D_{y_1 x} f(x_0, x_1)$ are inside D. By Lemma 6.2.6,

$$\left\{ \prod_{i=2}^{n} (c_i x_0 + d_i x_1), D_{y_1 x} f(x_0, x_1) \right\} = -n\{g(x_0, x_1), f(x_0, x_1)\} = 0,$$

and hence, the $(n-1)$-forms $\prod_{i=2}^{n}(c_i x_0 + d_i x_1)$ and $D_{y_1} f(x_0, x_1)$ are apolar.

If $(-d_2, c_2)$ is outside D, then we repeat the argument. Continuing in this way, either we have, for some $i < n$, $(-d_i, c_i) \in D$ (and the theorem is proved) or the argument is repeated $n-1$ times. In the latter case, the homogeneous zero of the 1-form $D_{y_{n-1}x} \cdots D_{y_2 x} D_{y_1 x} f(x_0, x_1)$ lies inside D and the 1-forms $c_n x_0 + d_n x_1$ and $D_{y_{n-1}x} \cdots D_{y_2 x} D_{y_1 x} f(x_0, x_1)$ are apolar. Recalling that two 1-forms are apolar if and only if one is a constant multiple of the other, we conclude that the homogeneous zero $(-d_n, c_n)$ is in D. □

Exercises

6.3.1. Let $p(z)$ and $q(z)$ be degree-n polynomials apolar to each other, C_1 be a convex set containing all the zeros of $p(z)$, and C_2 be a convex set containing all the zeros of $q(z)$. Show that $C_1 \cap C_2 \neq \emptyset$.

6.3.2. Let $y = (y_0, y_1)$. Show that

$$D_{yx} f(x_0, x_1) = n \,\mathrm{eval}((\alpha_1 y_0 + \alpha_0 y_1)(\alpha_1 x_0 + \alpha_0 x_1)^{n-1}).$$

6.3.3. Let ξ_1 and ξ_2 be two distinct zeros of the polynomial $p(z)$. Show that $\{p'(z), (z-\xi_1)^n - (z-\xi_2)^n\} = 0$. Generalize this result.

6.3.4. *The Lee–Yang circle theorem.*

Let $x_{ij},\ 1 \leq i, j \leq n$, be real numbers such that $|x_{ij}| < 1$ and $x_{ij} = x_{ji}$. Let

$$p(z) = \sum_{S:\, S \subseteq \{1,2,\ldots,n\}} z^{|S|} \prod_{i:\, i \in S,\, j \notin S} x_{ij}.$$

Then all the zeros of $p(z)$ lie on the unit circle.

6.3.5. *Szegö's determinant theorem.*

Let D be a disk in the complex plane, $z_1, z_2, \ldots, z_n$ be n distinct points inside D, and $w_1, w_2, \ldots, w_n$ be n distinct points outside D. Then

$$\det((z_i - w_j)^n)_{1 \leq i,j \leq n} \neq 0.$$

6.4 Multiplier Sequences

A sequence of real numbers $\mu_0, \mu_1, \mu_2, \ldots$ is a *multiplier sequence of the first kind* if $\mu_0 = 1$ and whenever all the zeros of the polynomial

$$a_0 x^n + \binom{n}{1} a_1 x^{n-1} + \cdots + \binom{n}{n-1} a_{n-1} x + a_n$$

are real, then all the zeros of

$$a_0 \mu_n x^n + \binom{n}{1} a_1 \mu_{n-1} x^{n-1} + \cdots + \binom{n}{n-1} a_{n-1} \mu_1 x + a_n \mu_0$$

are real. If we represent the sequence $(\mu_i)_{i=0}^{\infty}$ umbrally as $(\mu^i)_{i=0}^{\infty}$, then the main condition can be restated: whenever all the zeros of $\mathrm{eval}(a_0(x+\alpha)^n)$ are real, then all the zeros of $\mathrm{eval}(a_0(\mu x+\alpha)^n)$ are real. A sequence $(\mu_i)_{i=0}^{\infty}$ is a *multiplier sequence of the second kind* if whenever all the zeros of $a_0(x+\alpha)^n$ are real and positive, then all the zeros of $a_0(\mu x+\alpha)^n$ are real (but not necessarily positive).

For example, if ν is a nonzero real number, then $1, \nu, \nu^2, \nu^3, \ldots$ is obviously a multiplier sequence of the first kind. Less obviously, if w is a positive real number and $w^{(n)}$ is the rising factorial $w(w+1)(w+2)\cdots(w+n-1)$, then

$$1, 1/w^{(1)}, 1/w^{(2)}, 1/w^{(3)}, \ldots$$

is a multiplier sequence of the first kind. Finally, if n is a (fixed) positive integer and $x_{(n)}$ is the falling factorial $x(x-1)(x-2)\cdots(x-n+1)$, then

$$0_{(n)}, 1_{(n)}, 2_{(n)}, 3_{(n)}, \ldots$$

is a multiplier sequence of the first kind. An example of a multiplier sequence of the second kind is

$$\cos(\lambda), \cos(\lambda+\theta), \cos(\lambda+2\theta), \cos(\lambda+3\theta), \ldots.$$

The theory of multiplier sequences is a part of complex analysis, particularly the theory of Laguerre–Pólya classes. Our aim in this section is to present elementary characterizations of multiplier sequences.

Let $p(x)=\sum_{i=0}^{n}\binom{n}{i}a_i x^{n-i}$ and $q(x)=\sum_{i=0}^{n}\binom{n}{i}b_i x^{n-i}$. The *Szegö composition* $p*q(x)$ of $p(x)$ and $q(x)$ is the polynomial

$$\sum_{i=0}^{n}\binom{n}{i}a_i b_i x^{n-i}.$$

Umbrally, the Szegö composition of $\mathrm{eval}(a_0(x+\alpha)^n)$ and $\mathrm{eval}(b_0(x+\beta)^n)$ is the polynomial $\mathrm{eval}(a_0b_0(x+\alpha\beta)^n)$.

6.4.1. Szegö's theorem.[13] Suppose that all the zeros of $p(x)$ are real and all the zeros of $q(x)$ are real and those that are nonzero have the same sign. Then all the zeros of the Szegö composition $p*q(x)$ are real.

The proof is in three steps.

6.4.2. Lemma. Suppose that $h(x)=p*q(x)$ and all the zeros of $p(x)$ lie in a disk D or a half-plane H. Then for every zero ξ of $h(x)$, there exists a complex number κ in D and a zero β of $q(x)$ such that $\xi=-\beta\kappa$.

Proof. Since ξ is a zero of $h(x)$, we have

$$\sum_{i=0}^{n}\binom{n}{i}a_i b_i \xi^{n-i}=0.$$

Let

$$\bar{q}(x)=x^n q(-\xi/x)=\sum_{i=0}^{n}(-1)\binom{n}{i}b_i\xi^i x^{n-i}.$$

Then $p(x)$ and $\bar{q}(x)$ are apolar. Let D be a disk containing all the zeros of $p(x)$. By Grace's theorem (6.3.3), there is a zero κ of $\bar{q}(x)$ in D. However, $-\xi/\kappa$ is a zero of $q(x)$. We conclude that $\xi=-\beta\kappa$, where β is a zero of $q(x)$. The proof for a half-plane H is similar. □

[13] Szegö (1922).

In the proof of the next lemma, we need a simple fact: if C is a convex set in $\mathbb{R}^n$ containing the origin and η be a real number in the interval $[-1, 0]$, then for every $x \in C$, the point $-\eta x$ is also in C. To see this, observe that since C is convex and contains the origin and the point x, then it contains the line segment joining the origin and x. The point $-\eta x$ lies on this line segment and hence is in C.

6.4.3. Schur's lemma.[14] Let C be a convex set containing the origin in the complex plane. If all the zeros of $p(x)$ are in C and all the zeros of $q(x)$ are real numbers in $[-1, 0]$, then all the zeros of $p * q(x)$ are in C.

Proof. We will use the following fact: let C be a compact convex set with a finite number of extreme points. Then C is an intersection of a finite number of half-planes.

Let C be the convex closure of the zeros of $p(x)$ and the origin, $C = \bigcap_{i=1}^{t} H_i$, where H_i are suitable half-planes. Each half-plane H_i occurring in the intersection contains the origin and all the zeros of $p(x) = 0$. Let ξ be an arbitrary zero of $h(x) = p * q(x)$. Then by Szegö's theorem, ξ can be written as $-\eta h$ for some $h \in H_i$ and η is a zero of $q(x)$ in $(-1, 0)$. By the half-plane version of Grace theorem (6.3.5), $\xi \in H_i$. Thus, ξ lies in all the half-planes H_i and hence lies in their intersection C. □

Applying Schur's lemma to the real line in the complex plane, we conclude that if $p(x)$ has all real zeros and $q(x)$ has all zeros in $[-1, 0]$, then $p * q(x)$ has all real zeros.

To finish the proof of Theorem 6.4.1, suppose that all the zeros of $q(x)$ are, say, nonnegative. Let M be the maximum zero and $\hat{q}(x) = q(-xM)$. Then all the zeros of $\hat{q}(x)$ are in $[-1, 0]$. Thus, all the zeros of $p * \hat{q}(x)$, and hence, $p * \hat{q}(-x/M)$, are real. Since $p * \hat{q}(-x/M) = p * q(x)$, this completes the proof of Theorem 6.4.1.

We now use Theorem 6.4.1 to characterize multiplier sequences. The nth *Appell polynomial* $g_n(\underline{\mu}; x)$ of the sequence $(\mu_i)_{i=0}^{\infty}$ is the polynomial defined by

$$g_n(\underline{\mu}; x) = \mu_0 + \binom{n}{1}\mu_1 x + \binom{n}{2}\mu_2 x^2 + \cdots + \binom{n}{n-1}\mu_{n-1}x^{n-1} + \mu_n x^n.$$

6.4.4. Theorem. A sequence $(\mu_i)_{i=0}^{\infty}$ is a multiplier sequence of the second kind if and only if $\mu_0 = 1$ and for all n, $1 \leq n < \infty$, all the zeros of the Appell polynomial $g_n(\underline{\mu}; x)$ are real.

[14] Schur (1914).

Proof. Apply the multiplier sequence $1, \mu_1, \mu_2, \ldots$ to the polynomial $(1-x)^n$. Then the polynomial

$$\mu_0 - \binom{n}{1}\mu_1 x + \binom{n}{2}\mu_2 x^2 - \cdots + (-1)^{n-1}\binom{n}{n-1}\mu_{n-1}x^{n-1} + (-1)^n \mu_n x^n$$

has all real zeros. Changing variables from x to $-x$, we conclude that all the zeros of $g_n(\underline{\mu}; x)$ are real.

Conversely, suppose that all the zeros of $g_n(\underline{\mu}; x)$ are real. If $p(x) = \sum_{i=0}^{n} \binom{n}{i} a_i x^{n-i}$ and all the zeros of $p(x)$ are positive, then, by Theorem 6.4.1, all the zeros of the polynomial

$$\sum_{i=0}^{n} \binom{n}{i} a_i \mu_{n-i} x^{n-i}$$

(obtained by taking the Szegö composition of $p(x)$ and $g_n(\underline{\mu}; x)$) are real. Hence, $(\mu_i)_{i=0}^{\infty}$ is a multiplier sequence of the second kind. □

Characterizing multiplier sequences of the first kind is more complicated and requires two preliminary results.

6.4.5. Lemma. A multiplier sequence (μ_i) of the first kind does not have a subsequence of the form $a, \underbrace{0, 0, \ldots, 0}_{m}, b$, where $a, b \neq 0$ and $m \geq 1$.

Proof. Suppose that for some index j, $\mu_j = a$, $\mu_{j+1} = \mu_{j+2} = \cdots = \mu_{j+m} = 0$, and $\mu_k = b$. Let

$$q(x) = \begin{cases} x^j(1+x)^{k-j} & \text{if } a \text{ and } b \text{ have the same sign,} \\ x^j(1+x)^{k-j-1}(1-x) & \text{if } a \text{ and } b \text{ have opposite signs.} \end{cases}$$

In both cases, all the zeros of $q(x)$ are real. If we apply the sequence (μ_i) as multipliers to $q(x)$, we obtain the polynomial

$$x^j(a + b'x^{k-j}),$$

where $b' = \pm b$, with the sign chosen so that b' has the same sign as a. However, since $k - j > 1$, the polynomial $a + b'x^{k-j}$ has complex (nonreal) zeros. We conclude that (μ_i) is not a multiplier sequence of the first kind. □

Lemma 6.4.5 implies that none of the terms in a multiplier sequence of the first kind equals zero.

6.4.6. Lemma. Let (μ_i) be a multiplier sequence of the first kind. Then either all the terms μ_i have the same sign or the signs are alternating.

Proof. It suffices to show that for $i \geq 0$, μ_i and μ_{i+2} have the same sign. To do this, apply the multipliers (μ_i) to $x^{i+2} - x^i$. Then all the zeros of $x^i(\mu_{i+2}x^2 - \mu_i) = 0$ are real and hence $\mu_i/\mu_{i+2} > 0$. □

6.4.7. Theorem. A sequence $1, \mu_1, \mu_2, \ldots$ is a multiplier sequence of the first kind if and only if all the zeros of the Appell polynomials $g_n(\underline{\mu}; x)$ are real and have the same sign.

Proof. A multiplier sequence of the first kind is also one of the second kind. Hence, by Theorem 6.4.4, all the zeros of the Appell polynomials $g_n(\underline{\mu}; x)$ are real.

By Lemma 6.4.6, the signs of the coefficients of polynomial $g_n(\underline{\mu}; x)$ are either the same or alternating. If μ_i's have the same sign, then $g_n(\underline{\mu}; x)$ cannot have a positive zero. If μ_i's have alternating signs, then $g_n(\underline{\mu}; x)$ cannot have a negative zero.

The converse of the theorem follows from Szegö's theorem (6.4.1). □

Exercises

6.4.1. If the zeros of $p(x)$ lie in the interval $(-a, a)$ and the zeros of $q(x)$ lie in the interval $(-b, 0)$, where $a, b > 0$, then the zeros of $p * q(x)$ lie in $(-ab, ab)$.

6.4.2. If the polynomial $\sum_{k=0}^{n} a_k z^{n-k}$ has all its zeros in unit disk, then so does

$$\sum_{k=0}^{n} \frac{a_k z^{n-k}}{\binom{n}{k}}.$$

In particular, all the zeros of the polynomial

$$\sum_{k=0}^{n} \frac{z^{n-k}}{\binom{n}{k}}$$

lie inside the unit disk.

6.4.3. If $2j+1$ is an odd positive integer, let

$$(2j+1)!! = (2j+1)(2j-1)(2j-3)\cdots 5 \cdot 3 \cdot 1.$$

Show that

$$1, 0, -1!!, 0, 3!!, 0, -5!!, 0, 7!!, 0, -9!!, 0, \ldots$$

is a multiplier sequence of the second kind.

6.4.4. If all the zeros of $p(x) = \sum_{k=0}^{s} a_k x^k$ are real and all the zeros of $q(x) = \sum_{k=0}^{t} b_k x^k$ have the same sign, and $m = \min\{s, t\}$, then all the zeros of the polynomials

$$\sum_{k=0}^{m} a_k b_k x^k \quad \text{and} \quad \sum_{k=0}^{m} k! a_k b_k x^k$$

are real.

6.5 Totally Positive Matrices

The second half of Chapter 6 is about totally positive matrices. We give two families of examples and prove the Perron–Frobenius theorem in this section. In the next section, we introduce compound matrices and prove several determinantal formulas. Using compound matrices, we study three more advanced topics in Sections 6.7–6.9.[15]

We use the following notation and conventions: we usually work over the real field, with occasional excursions into the complex field. In particular, all matrices are assumed to have real entries unless specifically stated otherwise. We use the following notation: $\vec{x}$ is the column vector with coordinates x_i. Thus, if $\vec{x}$ is an n-dimensional vector, then $\vec{x} = (x_1, x_2, \ldots, x_n)^T$ and $\|\vec{x}\| = \sqrt{x_1^2 + x_2^2 + \cdots + x_n^2}$.

A matrix is *positive* if every entry of A is positive. It is *nonnegative* if every entry is nonnegative. Let A be a matrix with rows labeled by the set R and columns labeled by the set C, with a fixed linear order on the sets R and S. If $I \subseteq R$ and $J \subseteq S$, the *submatrix* $A[I|J]$ is the matrix obtained by restricting A to the rows I and columns J, keeping the same order. When $|I| = |J|$, then *minor* (*supported by rows I and columns* J) is the determinant $\det A[I|J]$. For example, if $R = S = \{0, 1, 2, \ldots\}$ and $I = \{1, 6, 9\}$ and $J = \{2, 6, 7\}$, then the minor supported by these sequences is

$$\det \begin{pmatrix} a_{12} & a_{16} & a_{17} \\ a_{62} & a_{66} & a_{67} \\ a_{92} & a_{96} & a_{97} \end{pmatrix}.$$

A matrix is *totally positive* if every minor is positive. It is *totally nonnegative* if every minor is nonnegative. A *line* in a matrix is a row or a column.

[15] The area of total positivity of matrices and operators is a venerable area in probability and functional analysis. An excellent reference is Karlin (1968).

6.5.1. Lemma. The set of totally positive (respectively, totally nonnegative) matrices is closed under the following operations:

(1) Deleting a line.
(2) Multiplying a line by a positive (respectively, nonnegative) constant.
(3) Adding a line to an adjacent line, where a *line* is a row or a column.

Proof. Closure under the first two operations is obvious. When a line is added to an adjacent line, a minor in the new matrix is either unchanged or the sum of two minors in the original matrix. Closure under the third operation follows from this. □

Next we give two examples of totally positive matrices. The first is due to G. Pólya.[16]

6.5.2. Theorem. Let $\alpha_1, \alpha_2, \ldots, \alpha_n$ and $\beta_1, \beta_2, \ldots, \beta_n$ be two strictly increasing sequences of real numbers and M be the matrix

$$(e^{\alpha_i \beta_j})_{1 \le i,j \le n}.$$

Then M is totally positive.

The proof is in four steps. We begin by observing that a submatrix of M has the same form. Thus, to prove the theorem, it suffices to show that $\det M > 0$.

Next, we recall the explicit formula for the van der Monde determinant. Let $x_1, x_2, \ldots, x_n$ be numbers and $V_n(x_1, x_2, \ldots, x_n)$ be the matrix whose ijth entry is x_j^{i-1}. Then

$$\det(V_n) = \det \begin{pmatrix} 1 & 1 & \cdots & 1 \\ x_1 & x_2 & \cdots & x_n \\ \vdots & \vdots & \ddots & \vdots \\ x_1^{n-1} & x_2^{n-1} & \cdots & x_n^{n-1} \end{pmatrix} = \prod_{i,j: 1 \le i < j \le n} (x_j - x_i).$$

(To see this, regard $x_1, x_2, \ldots, x_n$ as variables. Observe that the determinant is zero if x_i is set equal to x_j. Hence for any pair i and j, the difference $x_i - x_j$ divides $V_n(x_1, x_2, \ldots, x_n)$. We conclude that the product $\prod_{i,j: 1 \le i < j \le n} (x_j - x_i)$ divides $V_n(x_1, x_2, \ldots, x_n)$. Comparing degrees and leading coefficients, we obtain the formula.) In particular, if $x_1 < x_2 < \cdots < x_n$, then $\det V_n(x_1, x_2, \ldots, x_n) > 0$.

In the third step, we prove a result of Sylvester.

[16] See Pólya and Szegö (1976, problem v. 76, p. 46).

6.5.3. Sylvester's lemma. Let $\alpha_1, \alpha_2, \cdots, \alpha_n$ be a sequence of distinct real numbers and $c_1, c_2, \ldots, c_n$ be a sequence of real numbers, not all zero. Then the function

$$c_1 x^{\alpha_1} + c_2 x^{\alpha_2} + \cdots + c_n x^{\alpha_n}$$

(defined on the positive real axis) has at most $n-1$ distinct positive real zeros.

Proof. We induct on n, the length of the sequence. If $n = 1$, the lemma holds. We show the induction step by contradiction. Modify the indexing so that $\alpha_1 < \alpha_2 < \cdots < \alpha_n$. Suppose that

$$c_1 x^{\alpha_1} + c_2 x^{\alpha_2} + \cdots + c_n x^{\alpha_n}$$

has n distinct positive real zero. Then the function $f(x)$, defined by

$$f(x) = c_1 + c_2 x^{\alpha_2 - \alpha_1} + \cdots + c_n x^{\alpha_n - \alpha_1},$$

also has n distinct positive roots. By Rolle's theorem in calculus, between any two real zeros of $f(x)$ lies at least one real zero of the derivative $f'(x)$. Hence, $f'(x)$ has at least $n-1$ distinct positive roots; that is,

$$c_2(\alpha_2 - \alpha_1)x^{\alpha_2 - \alpha_1 - 1} + c_3(\alpha_3 - \alpha_2)x^{\alpha_3 - \alpha_2 - 1} + \cdots + c_n(\alpha_n - \alpha_1)x^{\alpha_n - \alpha_1 - 1}$$

has at least $n-1$ distinct positive roots. This contradicts the inductive hypothesis. □

We are now ready to prove Theorem 6.5.2. Let $x_j = e^{\beta_j}$. Then $0 < x_1 < x_2 < \cdots < x_n$ and $M = (x_j^{\alpha_i})$. Consider the system of linear equations in the unknowns $c_1, c_2 \ldots, c_n$:

$$\begin{pmatrix} x_1^{\alpha_1} & x_1^{\alpha_2} & \cdots & x_1^{\alpha_n} \\ \vdots & & \ddots & \vdots \\ x_n^{\alpha_1} & x_n^{\alpha_2} & \cdots & x_n^{\alpha_n} \end{pmatrix} \begin{pmatrix} c_1 \\ \vdots \\ c_n \end{pmatrix} = 0.$$

If $\det(x_j^{\alpha_i}) = 0$, then there exist real numbers $c_1, c_2, \ldots, c_n$, not all zero, such that

$$c_1 x_j^{\alpha_1} + c_2 x_j^{\alpha_2} + \cdots + c_n x_j^{\alpha_n} = 0$$

for distinct positive real numbers $x_1, x_2, \ldots, x_n$, contradicting Lemma 6.5.3.

We have now proved that the determinant is nonzero for a given increasing sequence $x_1, x_2, \ldots, x_n$ of positive real numbers. It remains to determine the sign. Consider $\det(x_j^{\alpha_i})$ as a function on n-tuples $(\alpha_1, \alpha_2, \ldots, \alpha_n)$ of real numbers. Then $\det(x_j^{\alpha_i})$ is nonzero unless for two distinct indices i and j, $\alpha_i = \alpha_j$. Let $\mathcal{A}$ be the union of the hyperplanes $\{(\alpha_1, \alpha_2, \ldots, \alpha_n)\colon \alpha_i = \alpha_j\}$, where i and

j ranges over all pairs of distinct indices. In the complement $\mathbb{R}^n \backslash \mathcal{A}$, any n-tuple $(\alpha_1, \alpha_2, \ldots, \alpha_n)$ such that $\alpha_1 < \alpha_2 < \cdots < \alpha_n$ lies in the same connected component as $(0, 1, 2, \ldots, n-1)$. Since the van der Monde determinant $\det(x_j^{i-1})$ is positive, we conclude that the sign of $\det(x_j^{\alpha_i})$ is also positive. □

If $x_1, x_2, \ldots, x_n$ and $y_1, y_2, \ldots, y_n$ are sequences of numbers such that $x_i + y_j \neq 0$ for every pair of indices i, j, let N be the matrix $(1/(x_i + y_j))_{1 \leq i, j \leq n}$.

6.5.4. Theorem. If x_i and y_j are real numbers such that $x_1 < x_2 < \cdots < x_n$ and $y_1 < y_2 < \cdots < y_n$, then the matrix N is totally positive.

Since the minors of N have the same form, it suffices to show that $\det N > 0$. This follows from a formula of Cauchy.

6.5.5. Cauchy's formula.

$$\det N = \frac{\prod_{i,j:1 \leq i < j \leq n} (x_j - x_i)(y_j - y_i)}{\prod_{i,j:1 \leq i < j \leq n} (x_i + y_j)}.$$

Proof. We induct on n. The formula holds trivially when $n = 1$. To prove the induction step, we perform column and row operations on N. First, subtract column n from columns 1 to $n-1$ in N, obtaining a new matrix N_1. Noting that

$$\frac{1}{x_i + y_j} - \frac{1}{x_i + y_n} = \frac{y_n - y_j}{(x_i + y_j)(x_i + y_n)},$$

every entry in row i of N_1 has the factor $1/(x_i + y_n)$ and when $j < n$, every entry in column j has the factor $y_n - y_j$. Taking out common factors, we obtain

$$\det N = \det N_1 = \frac{\prod_{i=1}^{n-1} (y_n - y_i)}{\prod_{i=1}^{n} (x_i + y_n)} \det N_2,$$

where

$$N_2 = \begin{pmatrix} \frac{1}{x_1+y_1} & \cdots & \frac{1}{x_1+y_{n-1}} & 1 \\ \frac{1}{x_2+y_1} & \cdots & \frac{1}{x_2+y_{n-1}} & 1 \\ \vdots & \ddots & \vdots & \vdots \\ \frac{1}{x_n+y_1} & \cdots & \frac{1}{x_n+y_{n-1}} & 1 \end{pmatrix}.$$

Now subtract row n from rows 1 to $n-1$ in N_2. Then column n becomes $(0, 0, \ldots, 0, 1)^T$, and if $i \neq n$ and $j \neq n$, the ij-entry can be simplified by the identity

$$\frac{1}{x_i + y_j} - \frac{1}{x_n + y_j} = \frac{x_n - x_i}{(x_i + y_j)(x_n + y_j)}.$$

In particular, if $i \neq n$, every entry in row i has the factor $x_n - x_i$, and if $j \neq n$, every entry in column j has the factor $1/(x_n + y_j)$. Taking out common factors and expanding the matrix along the last column, we obtain

$$\det N_2 = \frac{\prod_{i=1}^{n-1}(x_n - x_i)}{\prod_{j=1}^{n}(x_n + y_j)} \det \begin{pmatrix} \frac{1}{x_1+y_1} & \frac{1}{x_1+y_2} & \cdots & \frac{1}{x_1+y_{n-1}} \\ \frac{1}{x_2+y_1} & \frac{1}{x_2+y_2} & \cdots & \frac{1}{x_2+y_{n-1}} \\ \vdots & \vdots & \ddots & \vdots \\ \frac{1}{x_{n-1}+y_1} & \frac{1}{x_{n-1}+y_2} & \cdots & \frac{1}{x_{n-1}+y_{n-1}} \end{pmatrix}.$$

By induction, Cauchy's formula holds for the smaller $(n-1) \times (n-1)$ matrix, completing the proof. □

We end this section with the Perron–Frobenius theorem on matrices with positive entries. We shall use the Brouwer fixed-point theorem in the proof.

6.5.6. The Perron–Frobenius theorem. Let A be an $n \times n$ matrix with positive real entries a_{ij}. Then A has at least one positive real eigenvalue. If the maximum positive eigenvalue is r and λ is another (real or complex) eigenvalue A, then $|\lambda| < r$. The eigenvalue r has multiplicity 1 and there is an associated eigenvector $\vec{u}$ in which all entries are positive.

Proof. We show first that A has a positive eigenvalue. Let

$$S = \{\vec{x} \colon \vec{x} \in \mathbb{R}^n,\ \|\vec{x}\| = 1,\ \text{and } x_i \geq 0 \text{ for all } i\}.$$

Consider the map $\phi : S \to S$ defined by

$$\phi(\vec{x}) = \frac{A\vec{x}}{\|A\vec{x}\|}.$$

The set S is homeomorphic to the closed unit ball B_{n-1} of dimension $n-1$. The Brouwer fixed-point theorem asserts that every continuous function from a closed unit ball to itself has a fixed point. Let $\vec{u}$ be a fixed point of ϕ. Then $\phi(\vec{u}) = \vec{u}$ or

$$A\vec{u} = \|A\vec{u}\|\vec{u}.$$

We conclude that $\vec{u}$ is an eigenvector of A with positive eigenvalue $\|A\vec{u}\|$. Since $\vec{u}$ is in S, all the coordinates u_i of $\vec{u}$ are nonnegative. Since

$$u_i = \frac{1}{r}\sum_{j=1}^{n} a_{ij}u_j,$$

where u_i are not all zero and a_{ij} are positive, we conclude that each u_i is positive.

Let $r = \|A\vec{u}\|$. We will show that (I) r is an eigenvalue of multiplicity 1, and (II) for every other eigenvalue λ of A, $|\lambda| < r$.

To show (I), we need to show that every vector $\vec{v}$ satisfying $A\vec{v} = r\vec{v}$ is a multiple of $\vec{u}$. We argue by contradiction. Suppose that $\vec{v}$ is a vector, not a multiple of $\vec{u}$, such that $A\vec{v} = r\vec{v}$. Let $\vec{w} = \vec{u} + \epsilon\vec{v}$, with ϵ chosen so that every entry of $\vec{w}$ is nonnegative and at least one entry of $\vec{w}$, w_i, say, is zero. However, $A\vec{w} = r\vec{w}$, and hence

$$\sum_{j=1}^{n} a_{ij} w_j = 0.$$

Since the left side is positive, this is a contradiction.

To show (II), note that an eigenvalue of A is also an eigenvalue of the transpose A^T with the same multiplicity. Hence, r is a eigenvalue of A^T of multiplicity 1. Let $\vec{u}'$ be an eigenvector of A^T with eigenvalue r. Let λ be another eigenvalue (not equal to r) of A. Then λ is also an eigenvalue of A^T. Let $\vec{v}$ be an eigenvector of A^T with eigenvalue λ. Suppose that $\vec{v} = (v_1, \ldots, v_n)^T$. Then since $A^T\vec{v} = \lambda\vec{v}$, we have, for $1 \leq i \leq n$,

$$\lambda v_i = \sum_{j=1}^{n} a_{ji} v_j. \tag{V}$$

By the triangle inequality,

$$|\lambda||v_i| \leq \sum_{j=1}^{n} |a_{ji}||v_j| = \sum_{j=1}^{n} a_{ji}|v_j|. \tag{T}$$

On the other hand, for the positive eigenvalue r and its eigenvector $\vec{u}$, we have

$$r u_i = \sum_{j=1}^{n} a_{ij} u_j, \qquad i = 1, 2, \ldots, n.$$

Multiplying the ith equation by $|v_i|$ and summing over i, we obtain

$$\begin{aligned} r\sum_{i=1}^{n} u_i|v_i| &= \sum_{i=1}^{n}\sum_{j=1}^{n} a_{ij}u_j|v_i| \\ &= \sum_{j=1}^{n} u_j \left(\sum_{i=1}^{n} a_{ij}|v_i|\right) \\ &\geq \sum_{j=1}^{n} u_j|\lambda||v_j| \\ &= |\lambda| \sum_{j=1}^{n} u_j|v_j|. \end{aligned}$$

It follows that $|\lambda| \leq r$.

It remains to prove that $|\lambda| = r$ implies $\lambda = r$. Suppose that $|\lambda| = r$. Then

$$\sum_{j=1}^{n} u_j \left(\sum_{i=1}^{n} a_{i,j} |v_i| \right) = \sum_{j=1}^{n} u_j |\lambda| |v_j|.$$

Since $u_j > 0$, inequality (T) implies that

$$\sum_{i=1}^{n} a_{ij} |v_i| = r|v_j|.$$

Switching the roles of i and j in Equation (V) and taking absolute values, we obtain

$$\left| \sum_{i=1}^{n} a_{ij} v_i \right| = r|v_j|.$$

Thus,

$$\left| \sum_{i=1}^{n} a_{ij} v_i \right| = \sum_{i=1}^{n} a_{ij} |v_i|.$$

Using the fact that for complex numbers b and c, $|b + c| = |b| + |c|$ if and only if b and c have the same angle as a vector in the complex plane, we conclude that $v_i = |v_i|\omega$, where ω is a root of unity and (reversing the roles of i and j)

$$r v_i = \sum_{j=1}^{n} a_{ji} v_j.$$

Thus, $\vec{v}$ is an eigenvector of A^T with eigenvalue r. Since r is an eigenvalue of A^T of multiplicity 1, $\vec{v}$ is a constant multiple of an eigenvector of A^T with eigenvalue r. We conclude that $\lambda = r$. □

Exercises

6.5.1. Let $x_1, x_2, \ldots, x_n$ and $y_1, y_2, \ldots, y_n$ be sequences of real numbers such that for all pairs of indices i and j, $x_i y_j \neq 1$. Let P be the matrix $(1/(1 - x_i y_j))_{1 \le i, j \le n}$.

(a) Show that

$$\det P = \frac{\prod_{i,j:1 \le i < j \le n} (x_j - x_i)(y_j - y_i)}{\prod_{i,j:1 \le i < j \le n} (1 - x_i y_j)}.$$

(b) Suppose that x_i, y_j are real numbers such that $x_1 < x_2 < \cdots < x_n$, $y_1 < y_2 < \cdots < y_n$, and $0 < x_i y_j < 1$ for all pairs of indices i and j. Show that the matrix P is totally positive.

6.5.2. Let A be a nonnegative matrix such that 1 is the maximum absolute value of the eigenvalues of A. Show that if λ is an eigenvalue of A with absolute value 1, then λ^2 is also an eigenvalue of A.

6.6 Exterior Algebras and Compound Matrices

In this section, we examine the relation between totally positive matrices and positive matrices using compound matrices. The best way to introduce compound matrices is by exterior algebra. We begin with a brief introduction to exterior algebra.

Let V be an n-dimensional vector space over a field $\mathbb{F}$ of characteristic not equal to 2. The kth *exterior power* $\bigwedge^k(V)$ is the vector space constructed in the following way: if $u_1, u_2, \ldots, u_k$ are k vectors in V, their *exterior product* is the formal expression

$$u_1 \wedge u_2 \wedge \cdots \wedge u_k.$$

An exterior product is also called a *decomposable skew-symmetric tensor.* The exterior power $\bigwedge^k(V)$ is the vector space of all linear combinations of exterior product, modulo the following relations:

$$\begin{aligned} &u_1 \wedge u_2 \wedge \cdots \wedge (au_i + bu_i') \wedge \cdots \wedge u_k \\ &\quad = au_1 \wedge u_2 \wedge \cdots \wedge u_i \wedge \cdots \wedge u_k + bu_1 \wedge u_2 \wedge \cdots \wedge u_i' \wedge \cdots \wedge u_k \end{aligned}$$

and

$$u_1 \wedge \cdots \wedge u_i \wedge \cdots \wedge u_j \wedge \cdots \wedge u_k = -u_1 \wedge \cdots \wedge u_j \wedge \cdots \wedge u_i \wedge \cdots \wedge u_k.$$

The first relation says that the exterior product $\wedge$ is *multilinear.* The second says that it is *skew-symmetric* or *alternating*. Skew-symmetry implies that if an exterior product has a repeated term, then it is zero. To see this, observe that if $u_i = u_j = u$, then

$$u_1 \wedge \cdots \wedge u \wedge \cdots \wedge u \wedge \cdots \wedge u_k = -u_1 \wedge \cdots \wedge u \wedge \cdots \wedge u \wedge \cdots \wedge u_k,$$

and hence, as $\mathbb{F}$ is assumed to have characteristic not equal to 2, the exterior product is zero.

Exterior products look complicated. However, if one chooses a standard ordered basis $e_1, e_2, \ldots, e_n$ for V, then the relations imply that the $\binom{n}{k}$ exterior products

$$e_{i_1} \wedge e_{i_2} \wedge \cdots \wedge e_{i_k},$$

where $i_1 < i_2 < \cdots < i_k$, form a basis for $\bigwedge^k(V)$. We call this basis for $\bigwedge^k(V)$ the *standard* basis. In particular, $\bigwedge^k(V)$ is the $\binom{n}{k}$-dimensional vector space of all linear combinations

$$\sum_{i_1, i_2, \ldots, i_k} a_{i_1 i_2 \ldots i_k} \, e_{i_1} \wedge e_{i_2} \wedge \cdots \wedge e_{i_k}.$$

The zeroth exterior product has dimension 1 and is spanned by the empty product 1. In addition, if $k > n$, any exterior product in the basis vectors e_i contains a repeated term and is hence zero. We will use the following compact notation: if I is a k-subset of $\{1, 2, \ldots, n\}$, then

$$e_I = e_{i_1} \wedge e_{i_2} \wedge \cdots \wedge e_{i_k},$$

where $i_1, i_2, \ldots, i_k$ are the elements of I, arranged in increasing order.

It is useful to put all the exterior powers into one structure. The *exterior algebra* $\bigwedge(V)$ is the $\mathbb{F}$-algebra on the direct sum $\bigoplus_{k=0}^{n} \bigwedge^k(V)$ with multiplication $\wedge$ defined formally on exterior products by

$$\begin{aligned}(u_1 \wedge u_2 \wedge \cdots \wedge u_j) \wedge (v_1 \wedge v_2 \wedge \cdots \wedge v_k)\\ = u_1 \wedge u_2 \wedge \cdots \wedge u_j \wedge v_1 \wedge v_2 \wedge \cdots \wedge v_k\end{aligned}$$

and extended by linearity. The multiplication $\wedge$ is (by definition) associative. The exterior algebra is graded by its summands: if an element is in $\bigwedge^k(V)$, then it has *step* or *grade* k. The exterior algebra has dimension 2^n. Philosophically, the exterior algebra is a "linearization" of the Boolean algebra of all subsets of $\{1, 2, \ldots, n\}$.

The motivation for exterior powers is the following formula: if

$$u_i = a_{i1}e_1 + a_{i2}e_2 + \cdots + a_{in}e_n, \quad 1 \le i \le n,$$

then

$$u_1 \wedge u_2 \wedge \cdots \wedge u_n = (\det A)\, e_1 \wedge e_2 \wedge \cdots \wedge e_n,$$

where A is the matrix $(a_{ij})_{1 \le i,j \le n}$. This formula is purely formal. For example,

$$\begin{aligned}&(a_{11}e_1 + a_{12}e_2)(a_{21}e_1 + a_{22}e_2)\\ &\quad = a_{11}a_{21}\, e_1 \wedge e_1 + a_{11}a_{22}\, e_1 \wedge e_2 + a_{12}a_{21}\, e_2 \wedge e_1 + a_{12}a_{22}\, e_2 \wedge e_2\\ &\quad = (a_{11}a_{22} - a_{12}a_{21})\, e_1 \wedge e_2.\end{aligned}$$

The case $n = 2$ is typical and generalizes to all n. The next result extends this formula.

Recall from Section 6.5 that if I and J are k-subsets of $\{1, 2, \ldots, n\}$, then $A[I|J]$ and $\det A[I|J]$ are the submatrix and minor supported by I and J.

6.6.1. Lemma. Let $k \leq n$, $u_i = a_{i1}e_1 + a_{i2}e_2 + \cdots + a_{in}e_n$, $1 \leq i \leq k$, and A be the $k \times n$ matrix $(a_{ij})_{1\leq i\leq k, 1\leq j\leq n}$. Then

$$u_1 \wedge u_2 \wedge \cdots \wedge u_k = \sum_{J:|J|=k,\, J\subseteq\{1,2,\ldots,n\}} \det A[\{1, 2, \ldots, k\}|J]\; e_J.$$

Proof. By multilinearity,

$$u_1 \wedge u_2 \wedge \cdots \wedge u_k = \sum_{J:|J|=k,\, J\subseteq\{1,2,\ldots,n\}} a_J\, e_J,$$

where

$$a_J = \sum_{\sigma} \operatorname{sign}(\sigma) a_{1,\sigma(1)} a_{2,\sigma(2)} \cdots a_{k,\sigma(k)},$$

where σ ranges over all bijections σ from $\{1, 2, \ldots, k\}$ to J, and if $j_1, j_2, \ldots, j_k$ are the indices in J listed in increasing order, $\operatorname{sign}(\sigma)$ is the sign of the permutation $j_i \mapsto \sigma(i)$. Thus,

$$\begin{aligned} a_J &= \det \begin{pmatrix} a_{1j_1} & a_{1j_2} & \cdots & a_{1j_k} \\ \vdots & & \ddots & \vdots \\ a_{kj_1} & a_{kj_2} & \cdots & a_{kj_k} \end{pmatrix} \\ &= \det A[\{1, 2, \ldots, k\}|J]. \end{aligned}$$ □

6.6.2. Corollary. The vectors $u_1, u_2, \ldots, u_k$ are linearly independent if and only if $u_1 \wedge u_2 \wedge \cdots \wedge u_k \neq 0$.

Proof. Use the fact that if $u_i = a_{i1}e_1 + a_{i2}e_2 + \cdots + a_{in}e_n$, then $u_1, \ldots, u_k$ are linearly independent if and only if at least one $k \times k$ minor of the rectangular matrix $(a_{ij})_{1\leq i\leq k,\, 1\leq j\leq n}$ is nonzero. □

There is also an exterior product construction for linear transformations. Let $A: V \to V$ be a linear transformation. Then A "lifts" to a linear transformation $\wedge^k A: \bigwedge^k(V) \to \bigwedge^k(V)$ defined by

$$(\wedge^k A)(u_1 \wedge u_2 \wedge \cdots \wedge u_k) = Au_1 \wedge Au_2 \wedge \cdots \wedge Au_k$$

and extended by linearity.

Let A be the matrix $(a_{ij})_{1\leq j,k\leq n}$. The kth *compound matrix* of A is the $\binom{n}{k} \times \binom{n}{k}$ matrix indexed by k-subsets of $\{1, 2, \ldots, n\}$ (in some order) with IJ-entry equal to $\det A[I|J]$.

6.6.3. Theorem. Let (a_{ij}) be the matrix of the linear transformation $A : V \to V$ relative to the basis $e_1, e_2, \ldots, e_n$. Then the kth compound matrix $A^{(k)}$ is the matrix of $\wedge^k A : \bigwedge^k(V) \to \bigwedge^k(V)$ relative to the standard basis e_I of $\bigwedge^k(V)$.

Proof. Since $Ae_i = a_{i1}e_1 + a_{i2}e_2 + \cdots + a_{in}e_n$, the argument in the proof of Lemma 6.6.1 gives

$$\begin{aligned}(\wedge^k A)(e_{i_1} \wedge e_{i_2} \wedge \cdots \wedge e_{i_k}) &= Ae_{i_1} \wedge Ae_{i_2} \wedge \cdots \wedge Ae_{i_k} \\ &= \sum_{J:|J|=k,\ J\subseteq\{1,2,\ldots,n\}} \det A[\{i_1, i_2, \ldots, i_k\}|J]\ e_J.\end{aligned}$$

□

The exterior product for linear transformations is 'functorial'; that is, it commutes with composition. Explicitly, if A and B are linear transformations from V to V, then

$$\wedge^k(AB) = (\wedge^k A)(\wedge^k B).$$

This follows immediately from the construction. Putting this in terms of compound matrices, we obtain the following lemma.

6.6.4. Lemma.

$$(AB)^{(k)} = A^{(k)}B^{(k)}.$$

From this lemma, we derive a formula for the determinant of the product of two matrices.

6.6.5. The Binet–Cauchy formula. Let $k \leq n$, F be a $k \times n$ matrix, and G be an $n \times k$ matrix. Then

$$\det FG = \sum_{K:|K|=k,\ K\subseteq\{1,2,\ldots,n\}} \det F[\{1, 2, \ldots, k\}|K] \det G[K|\{1, 2, \ldots, k\}].$$

Proof. Extend F to an $n \times n$ matrix A so that F is the first k rows and G to an $n \times n$ matrix B so that G is the first n columns. Let $I = \{1, 2, \ldots, k\}$. Then $\det FG$ equals the minor of AB supported by I and I. By Lemma 6.6.4, $\det FG$ is the IJ-entry in the product $A^{(k)}B^{(k)}$. Explicitly,

$$\det FG = \sum_{K:|K|=k, K\subseteq\{1,2,\ldots,n\}} \det A[I|K]\ \det B[K|I].$$ □

For the remainder of this section, we shall use the following notation: if $I \subseteq \{1, 2, \ldots, n\}$, then I' is the complement $\{1, 2, \ldots, n\}\backslash I$. We will use also a simple fact about permutations. Let $i_1, i_2, \ldots, i_k$ be the elements of a k-subset

I, listed in increasing order. Define sum(I) and sign(I) by

$$\begin{aligned}\text{sum}(I) &= i_1 + i_2 + \cdots + i_k,\\ \text{sign}(I) &= (-1)^{\sum_{j=1}^{k}(i_j - j)} = (-1)^{\binom{k}{2}+\text{sum}(I)}.\end{aligned}$$

Let σ_I be the permutation of $\{1, 2, \ldots, n\}$ obtained by putting the elements of I before those of I', keeping the same order internally in I and I'. Then

$$\text{sign}(\sigma_I) = \text{sign}(I).$$

To see this, note that we can rearrange $i_1, i_2, \ldots, i_k, i'_1, i'_2 \ldots, i'_{n-k}$ into $1, 2, \ldots, n$ by moving i_j, which is in the jth position, to the i_j position. This takes $i_j - j$ adjacent transpositions.

6.6.6. The multiple Laplace expansion. Let A be an $n \times n$ matrix, I be a k-subset of $\{1, 2, \ldots, n\}$, and I' be its complement. Then

$$\det A = \sum_{J:|J|=k,\, J \in \{1,2,\ldots,n\}} \text{sign}(I)\text{sign}(J) \det A[I|J] \det A[I'|J'].$$

Proof. Let $i_1, i_2, \ldots, i_k$ and $i_{k+1}, i_{k+2}, \ldots, i_n$ be the elements of I and its complement I', listed in increasing order. Let $A = (a_{ij})$ and $\vec{u}_i = \sum_{j=1}^{n} a_{ij} e_j$. Using Lemma 6.6.1, we calculate $\det A$ in two different ways:

$$\begin{aligned}&(\det A)\, e_1 \wedge e_2 \wedge \cdots \wedge e_n\\ &= u_1 \wedge u_2 \wedge \cdots \wedge u_n\\ &= \text{sign}(I)(u_{i_1} \wedge u_{i_2} \wedge \cdots \wedge u_{i_k}) \wedge (u_{i_{k+1}} \wedge u_{i_{k+2}} \wedge \cdots \wedge u_{i_n})\\ &= \text{sign}(I)\left(\sum_{J:|J|=k} \det A[I|J] e_J\right)\left(\sum_{K:|K|=n-k} \det A[I'|K] e_K\right)\\ &= \text{sign}(I)\left(\sum_{J:|J|=k} \text{sign}(J) \det A[I|J] \det A[I'|J']\right) e_1 \wedge e_2 \wedge \cdots \wedge e_n.\end{aligned}$$

The last step is obtained by multiplying out the product and using the fact that by skew-symmetry, $e_J \wedge e_K = 0$ if $J \cap K \neq \emptyset$, and hence, because $|J| + |K| = n$, $e_J \wedge e_K$ if and only if K equals the complement J'. □

6.6.7. Jacobi's formula for compound matrices. Let $k \le n$, A be an $n \times n$ matrix, $A^{(k)} = (b_{IJ})_{I,J:|I|=|J|=k}$, and C be the $\binom{n}{k} \times \binom{n}{k}$ matrix indexed by k-subsets of $\{1, 2, \ldots, n\}$ with IJ-entry c_{IJ} given by

$$c_{IJ} = (-1)^{\text{sum}(I)+\text{sum}(J)} \det A[J'|I'].$$

Then

$$BC = \det(A)I,$$

where I is the $\binom{n}{k} \times \binom{n}{k}$ identity matrix.

Proof. We calculate the entries of the product BC. Let I be a k-subset of $\{1, 2, \ldots, n\}$. Then, the diagonal II-entry of BC equals

$$\sum_{J:|J|=k} b_{IJ}c_{JI}.$$

By the Laplace expansion,

$$\begin{aligned}\sum_{J:|J|=k} b_{IJ}c_{JI} &= \sum_{J:\,|J|=k} \det A[I|J](-1)^{\mathrm{sum}(I)+\mathrm{sum}(J)} \det A[I'|J'] \\ &= \sum_{J:\,|J|=k} \mathrm{sign}(I)\mathrm{sign}(J) \det A[I|J] \det A[I'|J'] \\ &= \det(A).\end{aligned}$$

Next, let I and J be two distinct k-subsets of $\{1, 2, \ldots, n\}$. Then

$$\sum_{K:|K|=k} b_{IK}c_{KJ} = \sum_{K:|K|=k} \mathrm{sign}(I)\mathrm{sign}(J) \det A[I|K] \det A[J'|K'].$$

By the Laplace expansion, the last sum equals $\det D$, where D is the matrix obtained by putting the rows in I on top of the rows in J', keeping the same order in I and J'. Since $I \neq J$, $I \cap J' \neq \emptyset$, and hence, D has two equal rows. We conclude that $\det D = 0$ and all the off-diagonal entries of the product BC are zero. □

We shall now apply our results to positive and totally positive matrices. We need one last definition. If A is the matrix (a_{ij}), let $\tilde{A}$ be the matrix $(\tilde{a}_{i,j})$, where $\tilde{a}_{ij} = (-1)^{i+j}a_{ij}$.

6.6.8. Theorem.

(a) An $n \times n$ matrix A is totally positive if and only if for all k, $0 \leq k \leq n$, the kth compound matrix $A^{(k)}$ is positive.

(b) If A and B are totally positive, then AB is totally positive.

(c) If A is totally positive, then $\tilde{A}^{-1}$ is totally positive.

The analog for Theorem 6.6.8 holds for total nonnegative matrices.

Proof. Part (a) is immediate from the definition of a compound matrix. Part (b) follows from the fact that products of positive matrices are positive.

To show part (c), let B be the kth compound matrix $\tilde{A}^{(k)}$ of $\tilde{A}$ and C be the $\binom{n}{k}$ by $\binom{n}{k}$ matrix indexed by k-subsets of $\{1, 2, \ldots, n\}$ with IJ-entry

$$(-1)^{\mathrm{sum}(I)+\mathrm{sum}(J)} \det \tilde{A}[J'|I'].$$

Then by Jacobi's formula, $BC = \det(\tilde{A})I$. Since $\tilde{a}_{ij} = (-1)^{i+j}a_{i,j}$,

$$\begin{aligned}\det \tilde{A}[I|J] &= \det((-1)^{i+j}a_{ij})_{i\in I,\, j\in J}\\ &= (-1)^{\mathrm{sum}(I)+\mathrm{sum}(J)} \det A[I|J].\end{aligned}$$

Noting that $\mathrm{sum}(I) + \mathrm{sum}(I') = \binom{n}{2}$, we have

$$\begin{aligned}c_{IJ} &= (-1)^{\mathrm{sum}(I)+\mathrm{sum}(J)} \det \tilde{A}(J'|I')\\ &= (-1)^{\mathrm{sum}(I)+\mathrm{sum}(J)+\mathrm{sum}(I')+\mathrm{sum}(J')} \det A(J'|I')\\ &= \det A(J'|I').\end{aligned}$$

Since A is totally positive, we conclude that C is positive. Further, $C = (\det A)B^{-1}$ and $\det(A) > 0$. Hence, B^{-1} is positive and by Lemma 6.6.4,

$$B^{-1} = (\tilde{A}^{(k)})^{-1} = (\tilde{A}^{-1})^{(k)}.$$

We conclude that all the compound matrices of $\tilde{A}^{-1}$ are positive and, by part (a), $\tilde{A}^{-1}$ is totally positive. □

We end this section with a description of the eigenvalues of compound matrices. We will assume that we are working over an algebraically closed field $\mathbb{F}$ of characteristic not equal to 2.

6.6.9. Lemma. Let A be a diagonalizable $n \times n$ matrix and $\lambda_1, \lambda_2, \ldots, \lambda_n$ be the eigenvalues of A with corresponding eigenvectors $\vec{u}_1, \vec{u}_2, \ldots, \vec{u}_n$. If I is a k-subset of $\{1, 2, \ldots, n\}$ with elements $i_1, i_2, \ldots, i_k$ listed in increasing order, let $\lambda_I = \lambda_{i_1}\lambda_{i_2}\cdots\lambda_{i_k}$ and

$$\vec{u}_I = \vec{u}_{i_1} \wedge \vec{u}_{i_2} \wedge \cdots \wedge \vec{u}_{i_k}.$$

Then for all k-subsets I of $\{1, 2, \ldots, n\}$, λ_I is an eigenvalue with corresponding eigenvector $\vec{u}_I$ of the kth compound matrix $A^{(k)}$ and such eigenvalues are all the eigenvalues of $A^{(k)}$.

Proof. Observe that

$$\begin{aligned}(\wedge^k A)\vec{u}_I &= (\wedge^k A)\vec{u}_{i_1} \wedge \vec{u}_{i_2} \wedge \cdots \wedge \vec{u}_{i_k}\\ &= A\vec{u}_{i_1} \wedge A\vec{u}_{i_2} \wedge \cdots \wedge A\vec{u}_{i_k}\\ &= \lambda_{i_1}\lambda_{i_2}\cdots\lambda_{i_k}\vec{u}_{i_1} \wedge \vec{u}_{i_2} \wedge \cdots \wedge \vec{u}_{i_k}\\ &= \lambda_I\vec{u}_I.\end{aligned}$$

Since the exterior products $\vec{u}_I$ are $\binom{n}{k}$ linearly independent vectors, the lemma follows. □

The preceding statement about eigenvalues of $A^{(k)}$ holds even when A is not diagonalizable. In such cases, one uses the Jordan canonical form. We shall not need the more general case and can omit the details with good conscience.

Exercises

6.6.1. From Lemma 6.6.9, it follows that if A is diagonalizable,

$$\det A^{(k)} = (\det A)^{\binom{n-1}{k-1}}.$$

Find a direct proof, not using the assumption that A is diagonalizable.

6.6.2. *The Orlik–Solomon algebra of a geometric lattice.*[17]

Let L be a geometric lattice and M be the matroid on the set S of the atoms of L. A subset C of S is a *circuit* if $\mathrm{rk}(C) = |C| - 1$ and $\mathrm{rk}(C') = |C'|$ for every proper subset C' of C. Choose a linear order on S. Let V be the vector space of all formal linear combinations $\sum_{x:x\in S} a_x x$ and $\bigwedge(V)$ the exterior algebra of V. If C is a circuit with elements $x_1, x_2, \ldots, x_m$ arranged in increasing order, let

$$\partial C = \sum_{i=1}^{m} (-1)^{m-1} x_1 \wedge x_2 \wedge \cdots \wedge x_{i-1} \wedge x_{i+1} \wedge \cdots \wedge x_m.$$

Let I be the ideal in $\bigwedge(V)$ generated by the elements ∂C, where C ranges over all circuits of M. The *Orlik–Solomon algebra* is the quotient $\bigwedge(V)/I$. Show that $\mathcal{O}(L)$ is graded by step and the subspace of step k has dimension equal to

$$(-1)^k \left(\sum_{Y:\,Y\in L,\mathrm{rk}(Y)=k} \mu(\hat{0}, Y) \right),$$

where μ is the Möbius function of the lattice L.

6.6.3. Let $U, U_1, U_2, \ldots, U_k$ be independent identically distributed random variables and $f_1, f_2, \ldots, f_k, g_1, g_2, \ldots, g_k$ be functions.

(a) Show that

$$\frac{1}{k!} E[\det(f_i(U_j)) \det(g_i(U_j))] = \det(E[f_i(U) g_j(U)]),$$

where E is expectation.

(b) Use part (a) to give another proof of the Binet–Cauchy formula (6.6.5).

[17] Orlik and Solomon (1980).

6.7 Eigenvalues of Totally Positive Matrices

In this section, we discuss two theorems of Gantmacher and Krein[18] about the eigenvalues of a totally positive matrix.

6.7.1. Theorem. The eigenvalues of a totally positive matrix are distinct positive real numbers.

Proof. Let A be a totally positive $n \times n$ matrix. Then for $1 \leq k \leq n$, the kth compound matrix $A^{(k)}$ is positive. By the Perron–Frobenius theorem, the matrix A has a unique positive eigenvalue of largest absolute value. Let $\lambda_1, \lambda_2, \ldots, \lambda_n$ be the eigenvalues of A, labeled so that λ_1 is the maximum positive eigenvalue and

$$\lambda_1 > |\lambda_2| \geq \cdots \geq |\lambda_n|.$$

Consider the second compound matrix $A^{(2)}$. The eigenvalue of $A^{(2)}$ with the largest absolute value is $\lambda_1\lambda_2$. By the Perron–Frobenius theorem, $\lambda_1\lambda_2$ is real and positive. Hence, λ_2 is a positive real number. Repeating this argument, using induction and the fact that the eigenvalue having maximum absolute value of $A^{(k)}$ is $\lambda_1\lambda_2\cdots\lambda_k$, we conclude that λ_k is real and positive for all k, $1 \leq k \leq n$. □

The next lemma is an easy consequence of Lemma 6.6.9 and the Perron–Frobenius theorem.

6.7.2. Lemma. Let A be a totally positive $n \times n$ matrix with eigenvalues $\lambda_1, \lambda_2, \ldots, \lambda_n$, listed in decreasing order, with corresponding eigenvectors $\vec{u}_1, \vec{u}_2, \ldots, \vec{u}_n$. Then for each k-subset I of $\{1, 2, \ldots, n\}$, λ_I is an eigenvalue of $A^{(k)}$ with eigenvector $\vec{u}_I$. The eigenvector $\vec{u}_{\{1,2,\ldots,k\}}$ of the maximum eigenvalue $\lambda_{\{1,2,\ldots,k\}}$ has positive coefficients when expanded in the standard basis e_I.

In the remainder of this section, we shall always work with real vectors. Let $\vec{u} = (u_1, u_2, \ldots, u_n)$. A *sign change* in $\vec{u}$ is a pair of indices i and j such that $i < j \leq n$, $u_k = 0$ for all indices k such that $i < k < j$ and $u_i u_j < 0$. The *weak variation* $\mathrm{Var}^-(\vec{u})$ is the number of sign changes in $\vec{u}$. For example,

$$\mathrm{Var}^-(-2, 2, 0, 0, 1, -2, 0, -3, 0, -1, 0) = 2.$$

The *strong variation* $\mathrm{Var}^+(\vec{u})$ is defined by

$$\mathrm{Var}^+(\vec{u}) = \max_{\vec{w}} \mathrm{Var}^-(\vec{w}),$$

[18] Gantmacher and Krein (1937).

where $\vec{w}$ ranges over all vectors obtained from $\vec{u}$ by replacing some zero coordinate by a real (nonzero) coordinate. For example,

$$\mathrm{Var}^+(-2, 2, 0, 0, 1, -2, 0, -3, 0, 1, 0) = 8,$$

partly because, for example,

$$\mathrm{Var}^-(-2, 2, -1, 0, 1, -2, 1, -3, 0, 1, -1) = 8.$$

From the definitions, $\mathrm{Var}^-(\underline{u}) \leq \mathrm{Var}^+(\underline{u})$.

Let $\vec{u}^{\,\mathrm{alt}}$ be the vector $(u_1, -u_2, \ldots, (-1)^{i-1}u_i, \ldots, (-1)^{n-1}u_n)$.

6.7.3. Lemma.

$$\mathrm{Var}^-(\vec{u}) + \mathrm{Var}^+(\vec{u}^{\,\mathrm{alt}}) = n - 1.$$

The proof consists of an induction with a number of easy cases. It is left to the reader.

6.7.4. Theorem. Let A be a totally positive matrix with eigenvalues $\lambda_1, \lambda_2, \ldots, \lambda_n$, listed in decreasing order, and $\vec{u}_i$ be an eigenvector associated with the eigenvalue λ_i. If $1 \leq p \leq q \leq n$, let $\vec{u}$ be a nonzero real linear combination

$$c_p\vec{u}_p + c_{p+1}\vec{u}_{p+1} + \cdots + c_q\vec{u}_q$$

of the eigenvectors $\vec{u}_p, \vec{u}_{p+1}, \ldots, \vec{u}_q$. Then,

$$p - 1 \leq \mathrm{Var}^-(\vec{u}) \leq \mathrm{Var}^+(\vec{u}) \leq q - 1.$$

In particular,

$$\mathrm{Var}^-(\vec{u}_i) = \mathrm{Var}^+(\vec{u}_i) = i - 1.$$

Proof. Note first that the eigenvalue λ_i is real and of multiplicity 1, and the eigenvector $\vec{u}_i$ is real and determined up to a constant. Let $\vec{u}_i = (u_{i1}, u_{i2}, \ldots, u_{in})$.

We prove first that $\mathrm{Var}^+(\vec{u}) \leq q - 1$. This is immediate for $q = n$ and so we may assume that $q < n$. We proceed by contradiction. Suppose that there are coefficients $c_p, c_{p+1}, \ldots, c_q$, such that $\mathrm{Var}^+(\vec{u}) \geq q$. Let $\vec{u} = (u_1, u_2, \ldots, u_n)$. Then, changing all the signs of the coefficients c_j if necessary, we can find a subsequence $u_{i_1}, u_{i_2}, \ldots, u_{i_{q+1}}$ such that for all j, $1 \leq j \leq q + 1$, $(-1)^{j-1}u_{i_j} \geq 0$. Let U be the $(q + 1) \times (q + 1)$ matrix

$$\begin{pmatrix} u_{1,i_1} & u_{2,i_1} & \cdots & u_{q,i_1} & u_{i_1} \\ \vdots & & \ddots & \vdots & \vdots \\ u_{1,i_{q+1}} & u_{2,i_{q+1}} & \cdots & u_{q,i_{q+1}} & u_{i_{q+1}} \end{pmatrix}.$$

Since the last column is a linear combination of columns p to q, $\det U = 0$. Expanding $\det U$ along the last column, we obtain

$$\sum_{j=1}^{q+1}(-1)^{j-1}u_{i_j} \det U[\{1,2,\ldots,j-1,j+1,\ldots,q+1\}|\{1,2,\ldots,q\}] = 0. \tag{D}$$

The minor $\det U[\{1,2,\ldots,j-1,j+1,\ldots,q+1\}|\{1,2,\ldots,q\}]$ is the coefficient of e_J, where $J = \{i_1, i_2, \ldots, i_{j-1}, i_{j+1}, \ldots, i_{q+1}\}$ when we expand $\vec{u}_1 \wedge \vec{u}_2 \wedge \cdots \wedge \vec{u}_q$ as a linear combination of the standard basis vectors e_I in $\bigwedge^q(V)$. By Lemma 6.7.2, all the minors are positive. Hence, since $(-1)^{j-1}u_{i_j} \geq 0$, Equation (D) holds only if $u_{i_j} = 0$ for all j, $1 \leq j \leq q+1$. In particular, the first q columns of the matrix U are linearly dependent.

Finally, construct another $(q+1) \times (q+1)$ matrix V by taking the first q columns of U and adding, as the last column, the column vector

$$(u_{q+1,i_1}, u_{q+1,i_2}, \ldots, u_{q+1,i_{q+1}})^T$$

formed by restricting $\vec{u}_{q+1}$ to the coordinates $i_1, i_2, \ldots, i_{q+1}$. Then $\det V = 0$. However, $\det V$ is the coefficient of $e_{\{i_1,i_2,\ldots,i_{q+1}\}}$ in the expansion of $\vec{u}_1 \wedge \cdots \wedge \vec{u}_q \wedge \vec{u}_{q+1}$ in the standard basis of $\bigwedge^{q+1}(V)$. By Lemma 6.7.2, this coefficient is positive and we have reached a contradiction. We have thus proved the upper bound $\mathrm{Var}^+(\vec{u}) \leq q-1$.

Next, we show the lower bound $p - 1 \leq \mathrm{Var}^-(\vec{u})$ by modifying the matrices and vectors so that we can apply the upper bound. Let $A = (a_{ij})_{1\leq i,j\leq n}$. Recall that $\tilde{A} = ((-1)^{i+j}a_{i,j})_{1\leq i,j\leq n}$. Then $\tilde{A} = SAS$, where S is the $n \times n$ diagonal matrix, with diagonal equal to $(1, -1, 1, -1, \ldots, (-1)^n)$. In addition, $\vec{u}^{\,\mathrm{alt}} = S\vec{u}$. Finally, note that $S^2 = I$.

By Theorem 6.6.8, $\tilde{A}^{-1}$ is totally positive. Since

$$\tilde{A}\vec{u}_i^{\,\mathrm{alt}} = SAS(S\vec{u}_i) = SA\vec{u}_i = \lambda_i \vec{u}_i^{\,\mathrm{alt}},$$

the eigenvalues of $\tilde{A}$ are the positive real numbers, $\lambda_1, \lambda_2, \ldots, \lambda_n$, listed, as usual, in decreasing order. Hence the eigenvalues of $\tilde{A}^{-1}$ are $\lambda_n^{-1}, \lambda_{n-1}^{-1}, \ldots, \lambda_2^{-1}, \lambda_1^{-1}$, with corresponding eigenvectors $\vec{u}_n^{\,\mathrm{alt}}, \vec{u}_{n-1}^{\,\mathrm{alt}}, \ldots, \vec{u}_2^{\,\mathrm{alt}}, \vec{u}_1^{\,\mathrm{alt}}$. Applying the upper bound to the matrix $\tilde{A}^{-1}$ and the vector $\vec{u}^{\,\mathrm{alt}}$, we conclude that

$$\mathrm{Var}^+(\vec{u}^{\,\mathrm{alt}}) \leq n - p.$$

To finish, observe that by Lemma 6.6.3,

$$\mathrm{Var}^+(\vec{u}^{\,\mathrm{alt}}) + \mathrm{Var}^-(\vec{u}) = n - 1.$$

Hence,

$$\mathrm{Var}^-(\vec{u}) = n - 1 - \mathrm{Var}^+(\vec{u}^{\,\mathrm{alt}}) \geq p - 1.$$

This completes the proof of Theorem 6.7.4. □

6.8 Variation-Decreasing Matrices

An $n \times m$ matrix A with real entries is *variation-decreasing* if for all nonzero vectors $\vec{x}$ in $\mathbb{R}^m$,

$$\mathrm{Var}^-(A\vec{x}) \leq \mathrm{Var}^-(\vec{x}).$$

The main objective of this section is to prove a result of Motzkin[19] that a totally nonnegative matrix is variation-decreasing.

6.8.1. Theorem. Let A be an $n \times m$ totally nonnegative matrix. Then for every vector $\vec{x}$ in $\mathbb{R}^m$,

$$\mathrm{Var}^-(A\vec{x}) \leq \mathrm{rank}(A) - 1.$$

Proof. If $\vec{y} \in \mathbb{R}^n$, then $\mathrm{Var}^-(\vec{y}) \leq n - 1$. Hence, if $\mathrm{rank}(A) = n$, then the inequality holds trivially. Thus, we may assume that $\mathrm{rank}(A) < n$.

Let $r = \mathrm{rank}(A)$. We will prove the theorem by induction on r. If $r = 1$, then every row of A is a nonnegative multiple of a fixed row vector $\vec{s}$ with nonnegative coordinates. Let $c_i\vec{s}$ be the ith row of A. If $\vec{x} \in \mathbb{R}^m$, then $A\vec{x} = (c_1 d, c_2 d, \ldots, c_n d)^T$, where d equals the dot product $\vec{s} \cdot \vec{x}$. The coordinates of $A\vec{x}$ are all nonnegative or nonpositive, depending on the sign of d. Thus, $\mathrm{Var}^-(A\vec{x}) = 0$ and the theorem holds.

We shall prove the induction step by contradiction. Let A be a totally nonnegative matrix of rank r. Suppose there exists a vector $\vec{x}$ such that $\mathrm{Var}^-(A\vec{x}) = r$. If $A\vec{x} = (y_1, y_2, \ldots, y_n)^T$, then there exist increasing indices $i_0, i_1, \ldots, i_r$ such that the coordinates $y_{i_0}, y_{i_1}, \ldots, y_{i_r}$ are all nonzero and alternate in sign. Let $\vec{y} = (y_{i_0}, y_{i_1}, \ldots, y_{i_r})^T$ and $A' = A[\{i_0, i_1, \ldots, i_r\}|\{1, 2, \ldots, m\}]$. Since A' is a submatrix of A, $\mathrm{rank}(A') \leq r$. In fact, $\mathrm{rank}(A') = r$. Otherwise, by induction,

$$r = \mathrm{Var}^-(\vec{y}) = \mathrm{Var}^-(A'\vec{x}) \leq \mathrm{rank}(A') - 1 < r - 1,$$

a contradiction. Choose r linearly independent column vectors, $\vec{u}_1, \vec{u}_2, \ldots, \vec{u}_r$, from A' and put them side by side to form the $(r+1) \times r$ matrix M. The vectors $\vec{u}_i$ form a basis of the column space of A'. Since $\vec{y}$ lies in the column

[19] Motzkin (n.d.).

space, the $(r+1) \times (r+1)$ matrix formed by adding $\vec{y}$ as the last column to M has zero determinant. Expanding this determinant along the last column $\vec{y}$, we obtain

$$\sum_{j=0}^{r} (-1)^{r+j} \det M_j y_{i_j} = 0, \tag{E}$$

where M_j is the submatrix of the matrix formed by deleting row j from M. Since $\det M_j$ is a minor of A, $\det M_j \geq 0$. Also, y_{i_j} alternate in sign. Thus, $(-1)^{r+j} y_{i_j}$ are nonzero real numbers, all having the same sign and Equation (E) can hold only if $\det M_j = 0$ for all j, contradicting our earlier choice that the columns $\vec{u}_i$ are linearly independent. This verifies the inductive step. □

6.8.3. Motzkin's theorem. A totally nonnegative matrix is variation-decreasing.

Proof. Let A be an $n \times m$ matrix with columns indexed by $\{1, 2 \ldots, m\}$. We need to prove that for any nonzero vector $\vec{x}$ in $\mathbb{R}^m$,

$$\mathrm{Var}^-(A\vec{x}) \leq \mathrm{Var}^-(\vec{x}).$$

If any coordinate of $\vec{x}$ is zero, then we may remove that coordinate and the corresponding column of A without changing the weak variation. Thus, we may assume that none of the coordinates of $\vec{x}$ is zero. Let $q = \mathrm{Var}^-(\vec{x})$. We can replace $\vec{x}$ by $-\vec{x}$ without changing the weak variation. Thus, we can partition the index set $\{1, 2, \ldots, m\}$ into $q+1$ into intervals

$$\{1, 2, \ldots, i_0\}, \ \{i_0 + 1, i_0 + 2, \ldots, i_1\}, \quad \ldots, \quad \{i_{q-1} + 1, i_{q-1} + 2, \ldots, m\}$$

such that $x_1, x_2, \ldots, x_{i_0}$ have positive sign, $x_{i_0+1}, x_{i_0+2}, \ldots, x_{i_1}$ have negative sign, and so on. We also set $i_{-1} = 0$ and $i_q = m$.

Next, we construct an $n \times (q+1)$ matrix C from A by first multiplying column i by the positive number $|x_i|$ and then adding up all the columns indexed by an interval. Explicitly, if $C = (c_{ij})_{1 \leq i \leq n, 0 \leq j \leq q}$, then

$$c_{ij} = \sum_{k=i_{j-1}+1}^{i_j} a_{ik} |x_k|.$$

By Lemma 6.5.1, C is totally nonnegative.

Let $\vec{s} = (1, -1, 1, -1, \ldots, (-1)^n)$. Then,

$$\sum_{j=0}^{q} (-1)^j c_{ij} = \sum_{j=0}^{q} \sum_{k=i_{j-1}+1}^{i_j} a_{ik}|x_k|(-1)^j$$
$$= \sum_{k=1}^{m} a_{ik}x_k.$$

From this, we conclude that

$$C\vec{s} = A\vec{x}.$$

But C, being an $n \times (q+1)$ matrix, has rank at most $q+1$. Hence, by Theorem 6.8.1,

$$\mathrm{Var}^-(A\vec{x}) = \mathrm{Var}^-(C\vec{s}) \leq \mathrm{rank}(C) - 1 \leq q = \mathrm{Var}^-(\vec{x}). \qquad \square$$

Theorem 6.8.2 is not the strongest possible. Motzkin has characterized variation-decreasing matrices.

6.8.3. Theorem. A matrix A is variation-decreasing if and only if the following two conditions are satisfied:

(a) If $k < \mathrm{rank}(A)$, then all $k \times k$ minors have the same sign.

(b) If $k = \mathrm{rank}(A)$ and I_0 is a given set of rows of size k, then the minors $\det A[I_0|J]$, where J is a k-subset of columns, have the same sign.

Exercises

6.8.1. Prove Theorem 6.8.3.

6.8.2. Let A be a totally nonnegative $n \times n$ matrix with rows and columns indexed by $\{1, 2, \ldots, n\}$, $I \subseteq \{1, 2, \ldots, n\}$, and $I^c = \{1, 2, \ldots, n\} \backslash I$. Show that

$$\det(A) \leq \det A[I|I] \det A[I^c|I^c].$$

6.8.3. Let A be a totally nonnegative $n \times n$ matrix. Then A^p is totally positive for some positive integer p if and only if $\det(A) > 0$ and for all i, $1 \leq i \leq n-1$, $a_{i,i+1} > 0$ and $a_{i+1,i} > 0$.

6.9 Pólya Frequency Sequences

Let $\underline{a}$ be an infinite sequence $a_0, a_1, a_2, \ldots$ of real numbers. The Δ-*matrix* $\Delta(\underline{a})$ of the sequence $\underline{a}$ is the infinite matrix

$$\Delta(\alpha) = \begin{pmatrix} a_0 & a_1 & a_2 & a_3 & \cdots \\ 0 & a_0 & a_1 & a_2 & \cdots \\ 0 & 0 & a_0 & a_1 & \cdots \\ 0 & 0 & 0 & a_0 & \cdots \\ \vdots & \vdots & \vdots & \vdots & \ddots \end{pmatrix}.$$

The sequence $\underline{a}$ is a *Pólya frequency sequence* or *totally positive sequence* if the first term a_0 is nonzero and $\Delta(\underline{a})$ is totally nonnegative.[20]

Pólya frequency sequences can be characterized by their generating functions. The (*ordinary*) *generating function* of the sequence $\underline{a}$ is the formal power series $f(\underline{a}; z)$ defined by

$$f(\underline{a}; z) = a_0 + a_1 z + a_2 z^2 + \cdots = \sum_{m=0}^{\infty} a_m z^m.$$

Conversely, the *sequence* of the formal power series $f(z) = \sum_{m=0}^{\infty} b_m z^m$ is $b_0, b_1, b_2, \ldots$.

6.9.1. Edrei's theorem.[21] A sequence $\underline{a}$ is a Pólya frequency sequence if and only if there exist a nonnegative real number γ and two sequences $\alpha_1, \alpha_2, \alpha_3, \ldots$ and $\beta_1, \beta_2, \beta_3, \ldots$ of nonnegative real numbers such that the sums $\sum_{i=1}^{\infty} \alpha_i$ and $\sum_{i=1}^{\infty} \beta_i$ converge to finite real numbers and

$$f(\underline{a}; z) = e^{\gamma z} \frac{\prod_{i=1}^{\infty}(1 + \alpha_i z)}{\prod_{j=1}^{\infty}(1 - \beta_j z)}.$$

We will prove the easy half of Edrei's theorem: that if $f(\underline{a}; z)$ is of the form given, then $\underline{a}$ is a Pólya frequency sequence. The converse is much more difficult. We begin with methods for constructing new Pólya frequency sequences

[20] Much work has been done on Pólya frequency sequences. See Brenti (1989) for a combinatorial perspective.

[21] Edrei (1952). In the paper Aissen et al. (1952), the authors proved Theorem 6.9.1, with the weaker condition that

$$f(\underline{a}; z) = e^{h(z)} \frac{\prod_{i=1}^{\infty}(1 + \alpha_i z)}{\prod_{j=1}^{\infty}(1 - \beta_j z)},$$

where $e^{h(z)}$ is an entire function and the generating function of a Pólya frequency sequence. Edrei showed, using Nevanlinna theory, that we can take $h(z) = \gamma z$.

from existing Pólya frequency sequences. The first result gives algebraic constructions.

6.9.2. Lemma. Let $f(\underline{a}; z)$ and $f(\underline{b}; z)$ be the generating functions of Pólya frequency sequences $\underline{a}$ and $\underline{b}$. Then the power series $f(\underline{a}; z)f(\underline{b}; z)$ and $1/f(\underline{a}; -z)$ are generating functions of Pólya frequency sequences.

Proof. The fact underlying this lemma is the isomorphism between the algebra of formal power series and the algebra of Δ-matrices (See Exercise 4.1.3). Let $\underline{c}$ be the sequence of the product $f(\underline{a}; z)f(\underline{b}; z)$. Then by an easy formal calculation,

$$\Delta(\underline{c}) = \Delta(\underline{a})\Delta(\underline{b}).$$

Since $\Delta(\underline{a})$ and $\Delta(\underline{b})$ are totally nonnegative, $\Delta(\underline{a})\Delta(\underline{b})$ is totally nonnegative by Theorem 6.6.8. Hence $\underline{c}$ is a Pólya frequency sequence.

Since $a_0 \neq 0$, the matrix $\Delta(\underline{a})$ is invertible and its inverse is also a Δ-matrix. By Theorem 6.6.9, $\tilde{A}^{-1}$, where $\tilde{A}$ is the Δ-matrix for the sequence $a_0, -a_1, a_2, -a_3, \ldots$, is totally non-negative. But

$$f(a_0, -a_1, a_2, \ldots; z) = a_0 - a_1 z + a_2 z^2 - \cdots = f(\underline{a}; -z).$$

Hence, $1/f(\underline{a}; -z)$ is the generating function for the Pólya frequency sequence whose Δ-matrix is $\tilde{A}^{-1}$. □

The next result gives an analytic construction.

6.9.3. Lemma. Suppose that for $0 \leq j < \infty$, the sequences $a_{0j}, a_{1j}, a_{2j}, \ldots$ are Pólya frequency sequences, and for each i, $0 \leq i < \infty$, the limits $\lim_{j\to\infty} a_{ij}$ exist. Let

$$a_i = \lim_{j\to\infty} a_{ij}.$$

Then the sequence $a_0, a_1, a_2, \ldots$ is a Pólya frequency sequence.

Proof. To prove the first part, observe that the determinant of a given finite submatrix is continuous in the entries. Thus, a sequence of nonnegative minors converges to a nonnegative minor. □

The next proposition shows how to construct all Pólya frequency sequences starting with the simplest ones.

6.9.4. Proposition.

(a) If $\alpha > 0$, then the sequence $1, \alpha, 0, 0, \ldots$ is a Pólya frequency sequence.

(b) If $\alpha_i > 0$, then the product

$$(1+\alpha_1 z)(1+\alpha_2 z)\cdots(1+\alpha_n z)$$

is the generating function of a Pólya frequency sequence.

(c) In particular, if $1, \alpha_1, \alpha_2, \alpha_3, \ldots$ is an infinite sequence of nonnegative real numbers such that the sum $\sum_{i=0}^{\infty} \alpha_i$ converges, then

$$\prod_{i=1}^{\infty}(1+\alpha_i z),$$

is the generating function of a Pólya frequency sequence.

(d) Let γ be a positive real number. Then $e^{\gamma z}$ is the generating function of a Pólya frequency sequence.

(e) Let β be a positive real number. Then

$$\frac{1}{1-\beta z}$$

is the generating function of a Pólya frequency sequence.

(f) If $1, \beta_1, \beta_2, \beta_3, \ldots$ is an infinite sequence of nonnegative real numbers such that the sum $\sum_{i=0}^{\infty} \beta_i$ converges, then

$$\prod_{i=1}^{\infty} \frac{1}{1-\beta_i z}$$

is the generating function of a Pólya frequency sequence.

Proof. The Δ-matrix A of $1, \alpha, 0, 0, 0$ is the matrix having diagonal entries 1, superdiagonal entries α, and all other entries zero. Using this, the proof of (a) is an easy induction.

By (a), $1+\alpha z$ is the generating function of a Pólya frequency sequence. Part (b) follows from Lemma 6.9.2.

From calculus, if the infinite sum $\sum_{i=0}^{\infty} \alpha_i$ converges, then the infinite product $\prod_{i\geq 1}(1+\alpha_i z)$ converges to a power series $f(z)$ and the coefficients of the finite partial products

$$\prod_{i=1}^{n}(1+\alpha_i z)$$

converge to the coefficients of $f(z)$. Since the coefficients of each partial product form a Pólya frequency sequence, Lemma 6.9.3 implies that the coefficients of $f(z)$ form a Pólya frequency sequence.

To prove (d), note that for every positive integer j, $(1 + \gamma z/j)^j$ is the generating function of a Pólya frequency sequence. Hence, as

$$\lim_{j\to\infty}\left(1+\frac{\gamma z}{j}\right)^j = e^{\gamma z},$$

$e^{\gamma z}$ is the generating function of a Pólya frequency sequence.

Finally, let $A = \Delta(1, \beta, \beta^2, \beta^3, \ldots)$. Let I and J be subsets of $\{0, 1, 2, \ldots\}$ and let $i_1, i_2, \ldots, i_k$ and $j_1, j_2, \ldots, j_k$ be their elements in increasing order. Since the ij-entry in A equals β^{j-i} if $j \geq i$ and 0 otherwise,

$$\det A[I|J] = \beta^{\sum_{k=1}^{n}(j_k - i_k)} \det B[I|J],$$

where B is the matrix with all entries equal to 1 on or above the diagonal and 0 below the diagonal. It is easy to check that $\det B[I|J] = 1$ if and only if $I = J$. We conclude that $\Delta(1, \beta, \beta^2, \beta^3, \ldots)$ is totally nonnegative.

Part (f) can be proved by the same continuity argument used to prove (d).

□

Proposition 6.9.4 completes the proof of the easy half of Edrei's theorem. Edrei theorem has the following corollary, which unites the two themes, roots of polynomials and positivity of matrices, of this chapter.

6.9.5. Corollary. All the zeros of the polynomial $1 + a_1 z + a_2 z^2 + \cdots + a_n z^n$ are real negative numbers if and only if the sequence $1, a_1, a_2, \ldots, a_n, 0, 0, \ldots$ is a Pólya frequency sequence.

One implication of the corollary is a special case of a theorem of Schoenberg.

Let k be a positive integer. A sequence $\underline{a}$ is *k-positive* if all the minors of size k or less in the matrix $\Delta(\underline{a})$ are nonnegative. For example, a sequence $\underline{a}$ is 1-positive if and only if all its terms are nonnegative. A sequence $a_0, a_1, \ldots, a_n$ is 2-positive if and only if if $a_i > 0$ on an finite or infinite interval $\{i\colon s < i < t\}$ of integers (so that s may be $-\infty$ and t may be ∞) and zero elsewhere, and the subsequence of positive terms is logarithmically concave; that is, $a_{i+1}a_{i-1} \leq a_i^2$ for $s < i < t$. A characterization of k-positivity is not known. The best result is the following necessary condition. We shall be working with complex zeros and $i = \sqrt{-1}$.

6.9.7. Schoenberg's theorem.[22] Let $1, a_1, \ldots, a_n, 0, 0, \ldots$ be a k-positive sequence and

$$f(z) = 1 + a_1 z + a_2 z^2 + \cdots + a_n z^n.$$

[22] Schoenberg (1952).

Then $f(z)$ has no zeros z in the sector

$$\{z\colon |\arg z| < k\pi/(n+k-1)\}.$$

To prove Schoenberg's theorem, we need the following lemma.

6.9.8. Lemma. Let A be an $m \times n$ matrix and B be a $p \times q$ matrix. If A and B are totally nonnegative, then the $(m+p) \times (n+q-1)$ matrix

$$A \oslash B = \begin{pmatrix} a_{1,1} & \cdots & a_{1,n-1} & a_{1,n} & 0 & \cdots & 0 \\ \vdots & \ddots & \vdots & \vdots & \vdots & \ddots & \vdots \\ a_{m,1} & \cdots & a_{m,n-1} & a_{m,n} & 0 & \cdots & 0 \\ 0 & \cdots & 0 & b_{1,1} & b_{1,2} & \cdots & b_{1,q} \\ \vdots & \ddots & \vdots & \vdots & \vdots & \ddots & \vdots \\ 0 & \cdots & 0 & b_{p,1} & b_{p,2} & \cdots & b_{p,q} \end{pmatrix}$$

is totally nonnegative.

Proof. If A and B are totally nonnegative, then so is the matrix

$$\begin{pmatrix} A & 0 \\ 0 & B \end{pmatrix}.$$

The matrix $A \oslash B$ can be obtained by adding column $n+1$ to column n and then deleting column $n+1$. By Lemma 6.5.1, $A \oslash B$ is totally nonnegative. □

Let A be the $k \times (n+k)$ matrix defined by

$$A = \begin{pmatrix} a_0 & a_1 & a_2 & \cdots & a_n & 0 & 0 & \cdots & \\ 0 & a_0 & a_1 & a_2 & \cdots & a_n & 0 & \cdots & \\ & \ddots & \ddots & \ddots & \ddots & & \ddots & & \\ 0 & \cdots & 0 & a_0 & a_1 & a_2 & \cdots & a_n & \end{pmatrix}.$$

The assumption that the sequence $a_0, a_1, a_2, \ldots, a_n$ is k-positive implies that the matrix A is totally nonnegative.

Let z be a zero of the polynomial $f(z)$. If z is real and negative, then the theorem holds. Since $a_i \geq 0$ for all i, z is not a nonnegative real number. Hence, we may assume that

$$z = re^{i\theta}, \quad r > 0 \text{ and } 0 < |\theta| < \pi.$$

We need to show that $|\theta| \geq k\pi/(n+k-1)$.

For any positive integer N, let M_N be the matrix

$$\underbrace{A \oslash A \oslash \cdots \oslash A}_{N \text{ copies}}.$$

In particular, M_N has size $s \times t$, where $s = Nk$ and $t = N(m+k) - N + 1$. By Lemma 6.9.8, M_N is totally nonnegative, and by Motzkin's theorem (6.8.3), it is variation-decreasing. In addition, since its rows are independent, M_N has rank Nk.

Let $\vec{x}$ be the vector whose jth coordinate x_j is the imaginary part of z^j; that is,

$$x_j = \mathrm{Im}(z^j) = r^j \sin j\theta.$$

The number v of sign changes in the sequence $x_0, x_1, x_2, \ldots, x_n$ can be easily determined. Indeed, the arguments of $x_0, x_1, x_2, \ldots$ increase stepwise by θ and v equals the number of times the sequence $x_0, x_1, x_2, \ldots$ crosses the real axis from the upper to the lower half-plane or the other way around. Hence,

$$\mathrm{Var}^-((x_0, x_1, \ldots, x_n)) \le \lfloor n\theta/\pi \rfloor.$$

This may be an inequality because $n\theta$ may be a multiple of π.

On the other hand, $f(z) = 0$ implies that $z^i + a_1 z^{i+1} + a_2 z^{i+2} + \cdots + a_n z^{i+n} = 0$ for all nonnegative integers i. Taking the imaginary part, we deduce that for all positive integers N greater than n,

$$M_N(x_0, x_1, \ldots, x_{t-1})^T = 0.$$

Since the rows of M_N are linearly independent, we can perturb the coordinates x_i by small values to new values x_i' in such a way that $M_N(x_0', x_1', \ldots, x_{t-1}')^T$ is a vector whose coordinates are all nonzero and alternate in sign. Thus,

$$\mathrm{Var}^-(M_N(x_0', x_1', \ldots, x_{t-1}')) = Nk - 1.$$

On the other hand, provided that the differences $|x_i' - x_i|$ are sufficiently small,

$$\mathrm{Var}^-((x_0', x_1', \ldots, x_{t-1}')) \le \mathrm{Var}^-(\vec{x}) + 1.$$

As M_N is variation-decreasing,

$$Nk - 1 = \mathrm{Var}^-(M_N(x_0', x_1', \ldots, x_{t-1}')) \le \mathrm{Var}^-((x_0', x_1', \ldots, x_{t-1}')).$$

Combining all the inequalities, we have

$$Nk - 1 \le \mathrm{Var}^-(\vec{x}) + 1 \le \left\lfloor \frac{N(n+k-1)\theta}{\pi} \right\rfloor + 1$$

and, in particular,

$$Nk - 1 \leq \frac{N(n+k-1)\theta}{\pi} + 1.$$

This inequality holds for all integers N greater than n. Dividing by N and letting N go to infinity, we finally obtain

$$k \leq \frac{(n+k-1)\theta}{\pi},$$

an inequality equivalent to the inequality asserted. □

Exercises

6.9.1. Define the *central Gaussian coefficients* $[n, \nu; \theta]$ by $[n, 0; \theta] = 1$ and for $\nu \geq 1$,

$$[n, \nu; \theta] = \frac{\sin n\theta \, \sin(n-1)\theta \, \cdots \, \sin(n-\nu+1)\theta}{\sin \nu\theta \, \sin(\nu-1)\theta \, \cdots \, \sin\theta}.$$

Let

$$P_n(z) = \sum_{\nu=0}^{n} [n, \nu; \theta] z^\nu.$$

(a) Show that $P_n(z) = \prod_{\nu=0}^{n-1} (z + e^{i(n-1-2\nu)\theta})$.

(b) Show that

$$\det([n, m+s-t; \theta])_{0 \leq s,t \leq \nu}] = \frac{[n, m; \theta][n+1, m; \theta][n+\nu-1, m; \theta]}{[m, m; \theta][m+1, m; \theta][m+\nu-1, m; \theta]}.$$

(c) Show that the sequence $[n, 0; \theta], [n, 1; \theta], [n, 2; \theta], \ldots, [n, n; \theta], 0, 0, \ldots$ is k-positive if $0 \leq \theta \leq \pi/(m+k-1)$.

(d) Show that if $\theta = \pi/(n+k-1)$, then $P_n(z)$ has a zero on the half-line $\{z\colon \arg z = \pi/(n+k-1)\}$ and another zero on $\{z\colon \arg z = -\pi/(n+k-1)\}$.

(e) Using (a) and (d), show that the upper bound of $\pi/(n-k+1)$ in Schoenberg's theorem is sharp.

6.9.2. (Research problem) Find an algebraic proof, perhaps using Witt vectors, of Edrei's theorem.

Selected Solutions

Chapter 1

1.1.1. Both inequalities can be proved given a little patience. One can also use Whitman's algorithm, described in Exercise 1.3.8.

1.1.2. In the diamond M_5,

$$x \wedge (y \vee z) = x \wedge M = x, \quad (x \wedge y) \vee (x \wedge z) = m.$$

In the pentagon N_5,

$$y \wedge (x \vee z) = y \wedge M = y, \quad (y \wedge x) \vee (y \wedge z) = x \wedge z = m.$$

Hence, the two sublattices give counterexamples to the distributive axioms. To prove the converse, suppose that there are elements x, y, z in L such that $x \wedge (y \vee z) \neq (x \wedge y) \vee (x \wedge z)$ and $x > z$. Then $m = x \wedge y$, $M = y \vee z$, $a = (x \wedge y) \vee z$, $b = x \wedge (y \vee z)$, and $c = y$ form a sublattice isomorphic to the pentagon. To prove closure under meets and joins, use Exercise 1.1.1.

We may now assume that $(*)$ for all x, y, z such that $x \geq z$, $x \wedge (y \vee z) = (x \wedge y) \vee (x \wedge z)$. Let x, y, z satisfy $x \wedge (y \vee z) \neq (x \wedge y) \vee (x \wedge z)$. From x, y, z, define

$$\begin{aligned} m &= (x \wedge y) \vee (y \wedge z) \vee (z \wedge x), \\ a &= (x \wedge y) \vee [(x \vee y) \wedge z], \\ b &= (y \wedge z) \vee [(y \vee z) \wedge x], \\ c &= (z \wedge x) \vee [(z \vee x) \wedge y], \\ M &= (x \vee y) \wedge (y \vee z) \wedge (z \vee x). \end{aligned}$$

By $(*)$, we can assume $x \not\geq z$ to show that these five elements are distinct and closed under meets and joins.

1.1.3. This is easy if one is prepared to use Exercise 1.1.2.

1.1.5. Observe that $[x, \hat{0}, z] = x \wedge z$ and $[x, \hat{1}, z] = x \vee z$. A study of n-ary operations on Boolean algebras can be found in Post (1921), reprinted in Davis (1994).

1.1.7. Use the interpretation $a|b = a^c \wedge b^c$, so that, for example, $a|a = a^c$ and Axiom Sh1 says that $(a^c)^c = a$. There is a complementary version, the "double stroke," defined by $a\|b = a^c \vee b^c$. It might be amusing to write down a set of axioms for Boolean algebras using $\|$.

1.2.1. For each element x, let $I(x)$ be the intersection of all open sets containing x. Thus, $I(x)$ is the minimum open set containing x. Then $x \leq y$ if and only if $y \in I(x)$ defines a partial order. Conversely, the collection of unions and intersections of the principal ideals $I(x)$, $x \in P$, is a T_0-topology.

1.2.2. (a) Let L_i be the linear extension

$$A_1, A_2, \ldots, A_{i-1}, A_{i+1}, \ldots, A_n, B_i, A_i, B_1, B_2, \ldots, B_{i-1}, B_{i+1}, \ldots, B_n.$$

Then $S_n = \bigcap L_i$, and hence the order dimension of S_n is at most n.

Now suppose that $S_n = L_1 \cap L_2 \cap \cdots \cap L_s$. Since A_i and B_i are incomparable, there is at least one index j, $1 \leq j \leq s$, such that $B_i < A_i$ in L_j. For each index i, $1 \leq i \leq n$, let $H(i)$ be the (nonempty) set $\{j: B_i < A_i \text{ in } L_j\}$. The sets $H(i)$ are pairwise disjoint. To prove this, suppose that $t \in H(i) \cap H(j)$, where $i \neq j$. Then

$$B_i <_{L_t} A_i <_{S_n} B_j <_{L_t} A_j <_{L_t} B_i,$$

a contradiction. It follows that $s \geq n$.

1.2.3. (a) Let $L_1, L_2, \ldots, L_n$ be linear extensions of $P \times Q$ intersecting to $P \times Q$. Then those extensions L_i such that $(\hat{0}_P, \hat{1}_Q) < (\hat{1}_P, \hat{0}_Q)$, suitably restricted, give a set of extensions of P intersecting to P.

(b) This requires an intricate argument. An exposition is in Trotter (1992, p. 41).

1.2.4. (b) Observe that if $f : P \to \langle n \rangle$ is an order-preserving map, then $f(x) \leq f(y)$, $f(y) \leq f(x)$, or $f(x) = f(y)$.

1.2.5. (e) We will show that Q satisfies the finite basis property. Let F be a filter and let a be an element in F. Then $F \backslash F(a)$ is a filter in $Q \backslash F(a)$, and by hypothesis, there is a finite basis B for $F \backslash F(a)$ in $Q \backslash F(a)$. The set $B \cup \{a\}$ is a finite basis for F in Q.

(f) We will show that the infinite-nondecreasing-subsequence condition holds in $P \times Q$. Let $((p_i, q_i))_{1 \leq i < \infty}$ be an infinite sequence in $P \times Q$. Then there exists an infinite nondecreasing subsequence $(p_i)_{i\in I}$ in $(p_i)_{1\le i<\infty}$ and an infinite nondecreasing subsequence $(q_i)_{i\in J}$ in the infinite sequence $(q_i)_{i\in I}$. The subsequence $((p_i, q_i))_{i\in J}$ is an infinite nondecreasing subsequence of $((p_i, q_i))_{1\le i<\infty}$.

(g) We use a Gröbner basis argument. (An excellent introduction to Gröbner bases can be found in Cox et al., 1992, chapter 2.)

The lexicographic order Lex is the following linear extension of the n-fold product $\mathbb{N} \times \mathbb{N} \times \cdots \times \mathbb{N}$: $(a_1, a_2, \ldots, a_n) <_{\mathrm{Lex}} (b_1, b_2, \ldots, b_n)$ if for some index i, $a_1 = b_1, a_2 = b_2, \ldots, a_{i-1} = b_{i-1}$, and $a_i < b_i$. Order the monomials in $\mathbb{F}[x_1, x_2, \ldots, x_n]$ by

$$x_1^{a_1} x_2^{a_2} \cdots x_n^{a_n} \le_{\mathrm{Lex}} x_1^{b_1} x_2^{b_2} \cdots x_n^{b_n} \text{ whenever}$$
$$(a_1, a_2, \ldots, a_n) \le_{\mathrm{Lex}} (b_1, b_2, \ldots, b_n).$$

(Other orderings of monomials will also work.) The *leading monomial* $\mathrm{LM}(f)$ of a polynomial $f(x_1, x_2, \ldots, x_n)$ is the maximum monomial occurring in f with a nonzero coefficient. If I is an ideal, let $\mathrm{LM}(I)$ be the monomial ideal generated by all the monomials $\mathrm{LM}(f)$, where f is a polynomial in I. By Dickson's lemma, there exist finitely many monomials $\mathrm{LM}(g_1), \mathrm{LM}(g_2), \ldots, \mathrm{LM}(g_t)$ generating $\mathrm{LM}(I)$. The last step is to show that I is generated by $g_1, g_2, \ldots, g_t$. Suppose that $f \in I$. Then there are polynomials a_i such that

$$\mathrm{LM}(f) = \sum_{i=1}^{t} a_i \mathrm{LM}(g_i).$$

Then

$$\mathrm{LM}\left(f - \sum_{i=1}^{t} a_i g_i\right) <_{\mathrm{Lex}} \mathrm{LM}(f).$$

Repeating this, we obtain polynomials b_i such that

$$\mathrm{LM}\left(f - \sum_{i=1}^{t} b_i g_i\right) <_{\mathrm{Lex}} \mathrm{LM}(g_i)$$

for every g_i. Since the "remainder" $f - \sum_{i=1}^{t} b_i g_i$ is in the ideal I, this implies that it is equal to zero and $f = \sum_{i=1}^{t} b_i g_i$.

(h) The function

$$\mathrm{Seq}(Q) \to \mathrm{Set}(Q), \ (a_1, a_2, \ldots, a_m) \mapsto \{a_1, a_2, \ldots, a_m\}$$

is an order-preserving map. Thus, by (d), it suffices to show that $\mathrm{Seq}(Q)$ is a well-quasi-order.

The following proof is similar to the proof in Haines (1968). We will write a finite sequence in Q as a word $x_1 x_2 \cdots x_m$, where $x_i \in Q$.

We shall do a double induction: an outer induction on Q and an inner induction on Seq(Q). The inner induction requires yet another induction argument. By the ascending chain condition for filters, we may do (finite) induction on filters for the outer induction. The theorem holds if Q is a one-element (well)-quasi-order and the induction can start. For the induction step, suppose that Q is a well-quasi-order such that Seq($Q \backslash F(a)$) is well-quasi-ordered for every element a in Q.

For the inner induction, we use the induction principle in (e). It suffices to show that for every sequence $x_1 x_2 \cdots x_m$ in Seq(Q), the complement Seq(Q)$\backslash F(x_1 x_2 \cdots x_m)$ satisfies the infinite-non-decreasing-subsequence condition. Let $(w_i)_{1 \leq i < \infty}$ be an infinite sequence of words in Seq(Q)$\backslash F(x_1 x_2 \cdots x_m)$. We argue by induction on the length m. If $m = 1$, then $x_1 \in Q$ and $(w_i)_{1 \leq i < \infty}$ is a sequence in Seq($Q \backslash F(x_1)$). Hence, by the outer induction, $(w_i)_{1 \leq i < \infty}$ has an infinite nondecreasing subsequence.

Returning to the inner induction, consider an infinite sequence $(w_i)_{1 \leq i < \infty}$ of words in Seq(Q)$\backslash F(x_1 x_2 \cdots x_m)$. We may assume that for $1 \leq i \leq m$, Seq(Q)$\backslash F(x_i)$ are well-quasi-orders. Further, we may assume that every word w_i in the sequence is contained in Seq(Q)$\backslash F(x_1 x_2 \cdots x_m)$, but not in Seq($Q$)$\backslash F(x_1 x_2 \cdots x_{m-1})$. (To see why we may assume this, suppose there are an infinite number of words w_i, $i \in J$, such that $x_1 x_2 \cdots x_{m-1} \not\leq w_i$. Then $(w_i)_{i \in J}$ is an infinite sequence in Seq(Q)$\backslash F(x_1 x_2 \cdots x_{m-1})$. By the inner induction, $(w_i)_{i \in J}$ has an infinite nondecreasing subsequence and we are done. Hence, there are only a finite number of words w_i in Seq(Q)$\backslash F(x_1 x_2 \ldots x_{m-1})$. These words can be deleted from $(w_i)_{1 \leq i < \infty}$.)

Thus, for every index i, $x_1 x_2 \cdots x_{m-1} \leq w_i$; that is,

$$w_i = w_{i1} x_{i1} w_{i2} x_{i2} \cdots w_{i,m-1} x_{i,m-1} w_{im},$$

where for $1 \leq j \leq m-1$, $x_j \leq x_{ij}$, and w_{ij} is a subword of w_i such that $x_j \not\leq w_{ij}$.

To finish the proof, we construct an infinite nondecreasing subsequence in $(w_i)_{1 \leq i < \infty}$. To start, the sequence $(w_{i1})_{1 \leq i < \infty}$ is an infinite sequence in Seq($Q \backslash F(x_1)$) and hence contains an infinite subsequence $(w_{i1})_{i \in J_0}$. Next, consider the infinite subsequence $(x_{i1})_{i \in J_0}$, which is an infinite sequence in Q. Thus, it has an infinite nondecreasing subsequence $(x_{i1})_{i \in J_1}$, where $J_1 \subseteq J_0$. Continuing, we obtain an infinite nondecreasing subsequence $(w_{i2})_{i \in J_2}$, where $J_2 \subseteq J_1$, and then an infinite nondecreasing subsequence $(x_{i2})_{i \in J_3}$, where $J_3 \subseteq J_2$. Continuing this, we obtain an infinite nondecreasing subsequence $(w_{im})_{i \in J_{2m}}$, where $J_0 \supseteq J_1 \supseteq J_2 \supseteq \cdots \supseteq J_{2m}$. The last index set gives an infinite nondecreasing subsequence $(w_i)_{i \in J_{2m}}$ in $(w_i)_{1 \leq i < \infty}$. This completes the induction and the theorem follows.

This proof generalizes to the more general theorem of Higman alluded to in the text. A different proof can be found in the paper of Higman.

(i) We use an ingenious argument of Nash-Williams. Suppose that there is a *bad sequence* in $\mathcal{T}$, that is, an infinite sequence (T_i) of trees such that for any two indices i and j with $i < j$, $T_i \not\leq T_j$. Choose a tree T_1 such that T_1 is the first term of a bad sequence and $|V(T_1)|$, the number of vertices in T_1, is as small as possible. Next choose T_2 such that T_1, and T_2 are the first two terms of a bad sequence and $|V(T_2)|$ is as small as possible. Continue this to obtain an infinite bad sequence $(T_i)_{1 \leq i < \infty}$.

When the root is removed from a rooted tree T, one obtains a set of disjoint subtrees. These subtrees can be rooted at the successor to the original root. The rooted subtrees obtained in this way are the *principal branches* of T. Let B_i be the set of principal branches of T_i and let $B = \bigcup_{i=1}^{\infty} B_i$.

We will show that B is well-quasi-ordered. Suppose $(R_i)_{1 \leq i < \infty}$ is an infinite sequence in B. Let k be the minimum integer such that one of the terms, R_h, say, is in B_k. Then, for all i, $|V(R_i)| < |V(T_k)|$, and for $1 \leq j \leq k-1$ and all i, $T_j \not\leq R_i$. By the choice of (T_i), the sequence $T_1, T_2, \ldots, T_{k-1}, R_h, R_{h+1}, R_{h+2}, \ldots$ is not a bad sequence. Since $T_i \not\leq T_j$ (obviously) and $T_i \not\leq R_j$ (otherwise, $R_j \leq T_l$, where T_l is a tree in which R_j occurs as a principal branch, and hence, $T_i \leq T_l$), we have, for some i and j, $i < j$ and $R_i \leq R_j$.

By Higman's lemma, Set(B) is a well-quasi-order. Thus, looking at the infinite sequence $(B_i)_{1 \leq i < \infty}$, we conclude that there exist i and j, $i < j$ and $B_i \leq B_j$. It follows that $T_i \leq T_j$, a contradiction. We conclude that $\mathcal{T}$ is a well-quasi-order.

(j,k) The proof is based on a difficult graph-theoretic result: if a graph does not contain a complete graph K_m as a minor, then it admits a treelike decomposition into small pieces. This is proved in a series of papers, the crowning paper being Robertson and Seymour (2004). The matroid minor project is an ongoing project (in 2008) to extend the graph minor theorem to matroid minors. See, for example, Geelen et al. (2006).

1.3.1. If L is complete, let

$$a = \bigvee \{b \colon b \in L \text{ and } b \leq f(b)\}.$$

Then $f(a) = a$.

To prove the converse, suppose L is not complete. Then using Zorn's lemma, one can construct a descending chain C with no infimum in which every ascending chain has a supremum. Let $C_* = \{y\colon y \leq x \text{ for all } x \text{ in } C\}$. Using Zorn's lemma again, there exists an ascending chain D in which every descending chain has an infimum. The sets C_* and D may be empty. By construction, there is no element x in L such that $c \geq x \geq d$ for all pairs $c \in C$ and $d \in D$. If $x \in L$, let

$$C(x) = \{c\colon c \in C, x \not\leq c\} \quad \text{and} \quad D(x) = \{d\colon d \in D, x \not\geq d\}.$$

At least one of the sets $C(x)$ or $D(x)$ is nonempty . Define a function $f : L \to L$ by setting $f(x)$ to be the supremum of $C(x)$ if $C(x) \neq \emptyset$ and the infimum of $D(x)$ if $C(x) = \emptyset$. For each x, either $x \not\leq f(x)$ or $x \not\geq f(x)$. Thus, f has no fixed point. A case analysis shows that f is order-preserving.

1.3.2. The supremum of a subset A equals

$$\inf\{x\colon x \geq a \text{ for all } a \text{ in } A\}.$$

The set of upper bounds of A is nonempty since it contains $\hat{1}$.

1.3.4. (a) If $j \in J(L)$, let $m(j) = \min\{x\colon x \in C, x \geq j\}$. Then $m : J(L) \to C\backslash\{\hat{0}\}$ is a bijection. To show injectivity, suppose $m(h) = m(j)$ for two join-irreducibles. Let y be the element in C covered by $m(h)$. Then $h \vee y = m(h) = m(j) = j \vee y$ and, by distributivity, $h = h \wedge (j \vee y) = (h \wedge j) \vee (h \wedge y)$. This implies that $h \leq h \wedge j$ or $h \leq h \wedge y$. If the latter occurs, then $h \leq y$, contradicting $m(h) > y$. Hence, $h \leq j$. Reversing the roles of h and j, we also have $j \leq h$, allowing us to conclude that $h = j$ and m is injective.

(b) Let j be a join-irreducible, j_* be the unique element covered by j, M be the set of meet-irreducibles containing j, and M_* be the set of meet-irreducibles containing j_*. Then $M_*\backslash M$ contains a single element $m(j)$.

1.3.7. (b) Let $x' = j_2 \vee j_3 \vee \cdots \vee j_m$. Then $x = x' \vee j_1$. By irredundancy, $x > x'$; hence, x is a join-irreducible in $\{y\colon y \geq x'\}$. Let x_* be the element covered by x in $\{y\colon y \geq x'\}$. Suppose for all i, $1 \leq i \leq l$, $x' \vee h_i \neq x$. Then $x' \vee h_i \leq x_* < x$. It follows that $h_1 \vee h_2 \vee \cdots \vee h_l \leq x_* < x$, a contradiction. We conclude that $x' \vee h_i = x$ for some i, as required.

1.3.8. (a) Define $\phi(x_i) = a_i$ and extend ϕ so that ϕ is a lattice homomorphism.

(c) Let $\alpha_1(x_1, x_2, \ldots, x_m) = \beta_1(x_1, x_2, \ldots, x_m)$ and $\alpha_2(x_1, x_2, \ldots, x_n) = \beta_2(x_1, x_2, \ldots, x_n)$ be two identities. Then, it is not hard to show that

$$\begin{aligned}\alpha_1(x_1, x_2, \ldots, x_m) \wedge \alpha_2(x_{m+1}, x_{n+2}, \ldots, x_{m+n})\\ = \beta_1(x_1, x_2, \ldots, x_m) \wedge \beta_2(x_{m+1}, x_{m+2}, \ldots, x_{m+n})\end{aligned}$$

holds in a lattice L if and only if $\alpha_1 = \beta_1$ and $\alpha_2 = \beta_2$ both hold in L.

1.3.10. (c) If y_1 and y_2 are complements of y, then $y_2 \wedge y = \hat{0}$, and hence, by Pierce's property, $y_2 \leq y_1$. Similarly, $y_1 \leq y_2$. Altogether, $y_1 = y_2$.

(d) By (c), L has unique complements. For an element z in L, let z^c be the unique complement of z. Taking complements is order-reversing. (To see this, suppose that $x \leq y$ but $y^c \not\leq x^c$ and then by Pierce's property, $y^c \wedge y \geq y^c \wedge x \neq \hat{0}$, a contradiction.) It follows that $(x^c)^c = x$.

Next, we show that L satisfies De Morgan's law; that is, $(x \vee y)^c = x^c \wedge y^c$. Since $x \vee y \geq x$ and complementation reverses order, $(x \vee y)^c \leq x^c$. Hence, $(x \vee y)^c \leq x^c \wedge y^c$. In addition, $x^c \wedge y^c \leq x^c$ and $(x^c \wedge y^c)^c \geq x$. Hence, $(x^c \wedge y^c)^c \geq x \vee y$ and, taking complements, $x^c \wedge y^c \leq (x \vee y)^c$.

Finally, we show that L is distributive. By Exercise 1.1.1, it suffices to prove

$$(x \wedge y) \vee (x \wedge z) \geq x \wedge (y \vee z).$$

We argue by contradiction. Let $u = (x \wedge y) \vee (x \wedge z)$ and $v = x \wedge (y \vee z)$. Suppose $u < v$. Then by Pierce's property applied to u^c, $u^c \wedge v \neq \hat{0}$. Let $t = u^c \wedge v$. By the lattice axioms, $t \wedge y = \hat{0}$ and $t \wedge z = \hat{0}$. Using Pierce's property twice, $t \leq y^c$ and $t \leq z^c$, and hence, $t \leq y^c \wedge z^c$. By De Morgan's law, proved earlier, $t \leq (y \vee z)^c$. However, $t \leq v = x \wedge (y \vee z) \leq y \vee z$. Hence, $t \leq (y \vee z)^c \wedge (y \vee z) = \hat{0}$, a contradiction.

(e) If L is modular and has unique complements, then it is easy to prove that $x \wedge y = \hat{0}$ implies $x \leq y^c$ and $x < y$ implies $x \vee (x^c \wedge y) = y$. By Exercise 1.1.2, if L is not distributive, it has a sublattice isomorphic to the diamond, and using the two preliminary results, we can obtain a contradiction.

(f) This exercise follows easily from McLaughlin's theorem (see Exercise 3.4.2).

(h) We sketch Dilworth's proof. (Shorter proofs have been found.) Begin with any lattice L_0. Construct a free lattice L_1 with a unary operator with L_0 as sublattice. If the unary operator $\cdot^*$ is complementation, then it is reflexive; that is, $a^{**} = a$. Selecting a suitable sublattice L_2 of L_1, obtain a free lattice with a reflexive unary operator containing L_0 as a sublattice. Take a homomorphic image L_3 of L_2 so that the reflexive unary operator becomes complementation. This will produce a unique complement for each element.

1.3.11. (c) The maximum is $[\hat{0} \to \hat{0}]$. To show one of the distributive axioms, suppose that $a \wedge (b \vee c) \leq x$. Then $b \vee c \leq [a \to x]$; that is, $b \leq [a \to x]$ and $c \leq [a \to x]$. These inequalities imply that $a \wedge b \leq x$ and $a \wedge c \leq x$, which in turn, imply, that $(a \wedge b) \vee (a \wedge c) \leq x$.

1.4.1. This is a *philosophical* problem. R. Dedekind used this property as the definition of finiteness.

1.4.5. (a) Since $\Pi(\mathcal{B}, S)$ has a maximum, it suffices to show that meets exist. Let $\mathcal{A}_1$ and $\mathcal{A}_2$ be antichains partitioning $\mathcal{B}$ and $\mathcal{A}_1 \wedge \mathcal{A}_2$ be the collection of maximal subsets in the order-ideal $I(\mathcal{A}_1) \cap I(\mathcal{A}_2)$. The antichain $\mathcal{A}_1 \wedge \mathcal{A}_2$ is the meet of $\mathcal{A}_1$ and $\mathcal{A}_2$ in the partial order of all antichains in 2^S.

We will show that $\mathcal{A}_1 \wedge \mathcal{A}_2$ partitions $\mathcal{B}$. Let $B \in \mathcal{B}$ and A_i be the unique subset in $\mathcal{A}_i$ containing B. Suppose that $B \subseteq C$ and $C \in I(\mathcal{A}_1) \cap I(\mathcal{A}_2)$. Since $C \in I(\mathcal{A}_1)$, there exists $C_1 \in \mathcal{A}_1$ such that $C \subseteq C_1$. Since $B \subseteq C \subseteq C_1$, this implies $C_1 = A_1$. Similarly, $C \subseteq A_2$. Hence, $C \subseteq A_1 \cap A_2$. However, $A_1 \cap A_2 \in I(\mathcal{A}_1) \cap I(\mathcal{A}_2)$ and $B \subseteq A_1 \cap A_2$. Hence, $A_1 \cap A_2$ is the unique maximal subset in $I(\mathcal{A}_1) \cap I(\mathcal{A}_2)$ containing B.

(c) Let σ be an n-partition not equal to the minimum, the n-partition of all n-element subsets of S, and σ' be the subcollection of subsets in σ with at least $n + 1$ elements. Let M be a maximal subset such that $|M| \geq n$ and $|M \cap T| \leq n$ for every $T \in \sigma$ and τ be the n-partition consisting of the set M and all n-element subsets in the complement $S \backslash M$. Then it is not hard to show that τ is a complement of σ.

(d) Let S be a sufficiently large set of points. Represent each point p_i of L by a line ℓ_i with a sufficiently large number of points in S, so that $\ell_i \cap \ell_j = \emptyset$ if $p_i \neq p_j$. We can now add points and 3-point lines so that line-closure simulates joins in L. For example, suppose $p_5 \leq p_1 \vee p_2 \vee p_3 \vee p_4$ and $p_5 \not\leq p_i \vee p_j \vee p_k$, $i, j, k \in \{1, 2, 3, 4\}$. We add the new points y_i^α and put them on the following 3-point lines, where the upper index α always ranges from 1 to 4:

$\{x_1^\alpha, x_2^\alpha, y_2^\alpha\}$, where $x_1^\alpha \in \ell_1$, $x_2^\alpha \in \ell_2$,
$\{y_2^\alpha, x_3^\alpha, y_3^\alpha\}$, where $x_3^\alpha \in \ell_3$
$\{y_3^\alpha, x_4^\alpha, y_4^\alpha\}$, where $x_4^\alpha \in \ell_4$,
$\{y_4^1, y_4^2, x_{12}\}$, $\{y_4^3, y_4^4, x_{34}\}$, where $x_{12}, x_{34} \in \ell_5$.

These 3-point lines would put ℓ_5 is the line-closure of $\ell_1, \ell_2, \ell_3, \ell_4$ but would not introduce any other closure relations.

(d) This requires a complicated bookkeeping argument (see Hartmanis, 1956).

1.4.7. Recall from group theory that $|HK| = |H||K|/|H \cap K|$.

1.5.1 (c) Observe that $R \circ R^{-1}$ and $R^{-1} \circ R$ are symmetric and transitive.

1.5.4. Let P be a partial order on $\{1, 2, \ldots, n\}$. Let i be the minimal element in P with the smallest label, $F_i = \{x\colon x > i\}$, and $P\backslash\{i\}$ the partial order on $\{1, 2, \ldots, i-1, i+1, \ldots, n\}$. Then $P \mapsto (i, P\backslash i, F_i)$ is an injection. We conclude that

$$q(n) \leq n2^{n-1}q(n-1).$$

The upper bound now follows by induction. To prove the lower bound, observe that a relation $R : S \to X$ gives a partial order (usually of rank 2). Just put S "below" X.

Chapter 2

2.1.1. (a) There are many elementary proofs. One way is to find a "rook tour" of the chessboard where one steps from one square to an adjacent square, going through every square exactly once and ending at the starting square.

(b) The answer is "almost always" and the exceptions can be described precisely.

2.1.2. If π is a permutation of $\{1, 2, \ldots, n\}$, let

$$\Delta(\pi) = \sum_{i=1}^{n} d(a_i, b_{\pi(i)}),$$

where $d(a, b)$ is the distance between the points a and b. Consider a permutation π_0 at which $\Delta(\pi_0)$ is minimum.

2.2.2. Let $R_1, R_2, \ldots, R_i$ be the right cosets and $L_1, L_2, \ldots, L_i$ be the left cosets of H in G. Define the relation $I : \{R_1, R_2, \ldots, R_i\} \to \{L_1, L_2, \ldots, L_i\}$ by the following condition:

$$(R_j, L_k) \in I \text{ whenever } R_j \cap L_k \neq \emptyset.$$

Using the fact that the collections $\{R_j\}$ and $\{L_k\}$ of cosets partition G, show that I satisfies the Hall condition.

2.3.1. One way is to show that a bipartite graph minimal under edge-deletion with respect to having a partial matching of size τ is a graph consisting of τ disjoint edges and isolated vertices. This proof appeared in Lovász (1975).

2.3.2. Use induction and the matrix submodular inequality in Section 2.8.

2.3.3. It is easier to prove a more general result first. Let $p(X_1, X_2, \ldots, X_n)$ be a *multilinear* polynomial; that is, each variable X_i occurs at most once in each monomial in $p(X_1, X_2, \ldots, X_n)$. Then $p = p_1 p_2$ if and only if there is a partition of $\{1, 2, \ldots, n\}$ into two subsets A and B such that p_1 is a multilinear polynomial in the variables X_i, $i \in A$, and p_2 is a multilinear polynomial in the variables X_i, $i \in B$.

2.4.1. Parts (b) and (c) are difficult theorems due to J. Edmonds and G.-C. Rota. Their proofs require some matroid theory. An account can be found in the books by Crapo and Rota (1970), and Oxley (1992). The theory of submodular functions and their polyhedra can be viewed as a generalization of matching theory. A key paper is Edmonds (1970).

2.4.6. Use the Binet–Cauchy theorem.

2.6.4. II. (a) An extreme point in a convex polyhedron in real n^2-dimensional space is determined by n^2 linearly independent equations derived from the constraints. There are at most $2n - 1$ linearly independent equations derived from the row and column sum constraints. Thus, at least $(n-1)^2$ equations of the form $d_{ij} = 0$, derived from the nonnegativity constraints, are necessary.

2.6.4. III. Since $\|C\| = \|UCU^T\|$ for any unitary matrix U,

$$\|A - B\| = \|A_0 - UB_0U^T\|,$$

where A_0 is the diagonal matrix $\text{diag}(\alpha_1, \alpha_2, \ldots, \alpha_n)$, B_0 is the diagonal matrix $\text{diag}(\beta_1, \beta_2, \ldots, \beta_n)$, and U is an appropriate unitary matrix.

2.6.4. IV. Use the fact that a linear function attains its minimum on a closed bounded convex set at an extreme point. The linear program can be interpreted as an assignment problem. Suppose that there are n contractors and n projects and the ith contractor quotes the (nonnegative) amount c_{ij} for doing the jth project. Since an integer solution exists, the minimum cost can be attained by giving one project to one contractor, even if it is allowable to divide one project between more than one contractor.

2.6.4. V. It is easy to show an upper bound of $n^2 - n + 1$ from the first proof of Birkhoff's theorem. The stated bound can be proved by first showing that the dimension of the convex set of doubly stochastic matrices is exactly $(n-1)^2$ and then applying Carathéodory's theorem: if C is a convex subset of $\mathbb{R}^d$, then every point in C is a convex combination of at most $d + 1$ extreme points. No constructive proof is known.

2.6.5. Show that a 3×3 product of transfers cannot have 2 or 3 zero entries and construct a 3×3 doubly stochastic matrix with 2 zero entries.

2.6.7. Observe that if C is doubly substochastic, then the augmented matrix

$$\begin{pmatrix} C & I - H_r \\ I - H_c & C^T \end{pmatrix},$$

where H_r is the diagonal matrix with diagonal entries equal to the row sums of C and H_c is the diagonal matrix with diagonal entries equal to the column sums of C, is doubly stochastic.

2.6.10. (a) The result follows from, say, the first proof of Theorem 2.6.1. However, the following argument is interesting. Suppose that D has a positive off-diagonal entry but (*) fails. Consider the matrix D_t defined by

$$D_t = \frac{D + tI}{1 + t}.$$

When t is a nonnegative real number, D_t is a nonidentity doubly stochastic matrix. Since D fails to satisfy (*), D_t also fails to satisfy (*). It follows that

$$\det D_t = (1 + t)^{-n}(d_{11} + t)(d_{22} + t) \cdots (d_{nn} + t)$$

and

$$\det(D + tI) = (d_{11} + t)(d_{22} + t) \cdots (d_{nn} + t).$$

Since 1 is always an eigenvalue of a doubly stochastic matrix, we conclude that one of the diagonal entries of D equals 1. Remove the row and column containing the diagonal entry 1 to obtain a smaller matrix. By induction, the smaller matrix is the identity matrix and we obtain a contradiction.

(b) There are two ways to prove this result. The easier way is to use Theorem 2.6.1. The harder way is to prove it independently using part (a) and then use it to obtain another proof of Theorem 2.6.1.

2.6.11. A reasonable starting conjecture is that the extreme points of $\mathcal{S}_n$ are the *symmetrized permutation matrices,* that is, matrices $\frac{1}{2}(P + P^t)$, where P is a permutation matrix. Using the marriage theorem, it is not hard to show that $\mathcal{S}_n$ is the convex closure of the set of symmetrized permutation matrices. However, not all symmetrized permutation matrices are extreme points. This is due to the following lemma: if P is the permutation matrix associated with the even cycle $(a_1, a_2, \ldots, a_{2m-1}, a_{2m})$ of length $2m$, where $m > 1$, then $\frac{1}{2}(P + P^t)$ is the convex combination

$$\alpha[\tfrac{1}{2}(Q_1 + Q_1^t)] + (1 - \alpha)[\tfrac{1}{2}(Q_2 + Q_2^t)],$$

where Q_1 (respectively, Q_2) is the matrix associated with the product of m transpositions $(a_1, a_2)(a_3, a_4) \cdots (a_{2m-1}, a_{2m})$ (respectively, $(a_{2m}, a_1)(a_2, a_3)$ $(a_4, a_5) \cdots (a_{2m-2}, a_{2m-1})$). In particular, the symmetrized permutation matrix of a permutation whose cycle decomposition contains an even cycle of length greater than 2 is not an extreme point. It is not hard to prove that the other symmetrized permutation matrices are extreme points. Thus, the extreme points of $\mathcal{S}_n$ are the matrices $\frac{1}{2}(P + P^t)$, where P is a matrix associated with a permutation whose cycle decomposition has no even cycle of size greater than 2.

2.6.12. This requires a complicated argument (see Hardy et al., 1952, p. 47).

2.6.13. Use the theorem from the theory of Laplace transforms that $f(x)$ is completely monotonic if and only if

$$f(x) = \int_0^\infty e^{-xt} d\mu(t),$$

where μ is a (nonnegative) measure. First, reduce to the case $m = n$. Since

$$\begin{aligned}\log f(x) &= \sum_{i=1}^{n} [\log(x + z_i) - \log(x + p_i)]\\ &= \int_0^\infty \sum_{i=1}^{n} e^{-xt}[e^{-p_i t} - e^{-z_i t}]\frac{dt}{t},\end{aligned}$$

it suffices to show that for all $t \geq 0$,

$$\sum_{i=1}^{n} [e^{-p_i t} - e^{-z_i t}] \geq 0.$$

This can be done using the analysis argument in the sufficiency proof of Muirhead's inequality.

2.6.14. (a) See Section 6.1.

(b) By (a), $(a_0 a_2)(a_1 a_3)^2 (a_2 a_4)^3 \cdots (a_{k-1} a_{k+1})^k \leq a_1^2 a_2^4 a_3^6 \cdots a_k^{2k}$.

(c) This is a difficult theorem and requires some knowledge of Sturm and Artin-Schreier theory.

2.6.16. (b) Expand the right side formally and observe that every monomial in the permanent expansion occurs in the expansion.

2.6.16. (c) To prove the inequality, observe that every term in

$$\sum_{\pi} \prod_{i=1}^{n} a_{i,\pi(i)} \quad \text{and} \quad \sum_{\pi} \prod_{i=1}^{n} b_{i,\pi(i)}$$

also occurs in

$$\sum_{\pi} \prod_{i=1}^{n} (a_{i,\pi(i)} + b_{i,\pi(i)}).$$

By Exercise 2.6.4V, an $n \times n$ doubly stochastic matrix D is a convex combination of $n^2 - n + 1$ permutation matrices. Hence,

$$\operatorname{per}(D) \geq \sum_{i=1}^{k} \operatorname{per}(\lambda_i P_i) = \sum_{i=1}^{k} \lambda_i^n,$$

where $\sum_{i=1}^{k} \lambda_i = 1,\ 0 < \lambda_i \leq 1,$ and $k \leq n^2 - n + 1$. By elementary arguments, the minimum of $\sum_{i=1}^{k} \lambda_i^n$ is attained when the numbers are all equal. Hence,

$$\operatorname{per}(D) \geq \sum_{i=1}^{k} (1/k)^n = k^{-(n-1)} \geq (n^2 - n + 1)^{-(n-1)}.$$

2.6.17. This was proved by D.I. Falikman and G.P. Egorychev (see, for example, the survey paper, Egorychev 1996).

2.7.2. There is an elementary induction argument (see Brualdi and Ryser, 1991, p. 172).

2.7.4. Prove the more general fact that if v be a valuation on S taking real values, then v^+, defined by

$$v^+(A) = (v(A))^+ ,$$

is a supermodular function; that is,

$$v^+(A) + v^+(B)) \leq v^+(A \cup B) + v^+(A \cap B).$$

Hence, $-v^+$ is a submodular function.

2.7.6. Related work can be found in Converse and Katz (1975).

2.8.1. From the matrix $M[T|S]$, construct the *representation matrix* $[\,I \mid M\,]$, where I is the identity matrix with rows and columns indexed by T. Then

$$\operatorname{rank}(B, A) = \operatorname{rk}(A \cup (T \backslash B)),$$

where the rank on the right is the rank of the column vectors in $A \cup (T \backslash B)$. The bimatroid inequality now follows easily from the matroid submodular inequality.

2.8.2. Let A, B, C, D be subspaces. Then by Grassmann's equality (Section 2.4),

$$\begin{aligned}
&\dim(A \cap B \cap C) \\
&\quad = \dim(A \cap B) + \dim(C) - \dim((A \cap B) \vee C) \\
&\quad \geq \dim(A \cap B) + \dim(C) - \dim((A \vee C) \cap (B \vee C)) \\
&\quad = \dim(A \cap B) + \dim(C) - \dim(A \vee C) - \dim(B \vee C) + \dim(A \vee B \vee C)
\end{aligned}$$

and

$$\begin{aligned}
&\dim(A \cap B \cap C \cap D) \\
&\quad = \dim(A \cap B \cap C) + \dim(A \cap B \cap D) - \dim((A \cap B \cap C) \vee (A \cap B \cap D)) \\
&\quad \geq \dim(A \cap B \cap C) + \dim(A \cap B \cap D) - \dim(A \cap B).
\end{aligned}$$

Combining two instances of the first inequality (for A, B, C and A, B, D) and the second inequality,

$$\begin{aligned}
&\dim(A \cap B \cap C \cap D) \\
&\quad \geq [\dim(C) - \dim(A \vee C) - \dim(B \vee C) + \dim(A \vee B \vee C)] \\
&\qquad + [\dim(D) - \dim(A \vee D) - \dim(B \vee D) + \dim(A \vee B \vee D)] \\
&\qquad + \dim(A \cap B).
\end{aligned}$$

We shall use Grassmann's equality to eliminate the two terms in this inequality involving meets in the following way:

$$\dim(A \cap B) = \dim(A) + \dim(B) - \dim(A \vee B)$$

and

$$\dim(C) + \dim(D) - \dim(C \vee D) = \dim(C \cap D) \geq \dim(A \cap B \cap C \cap D).$$

Hence,

$$\begin{aligned}
&\dim(C) + \dim(D) - \dim(C \vee D) \\
&\quad \geq [\dim(C) - \dim(A \vee C) - \dim(B \vee C) + \dim(A \vee B \vee C)] \\
&\qquad + [\dim(D) - \dim(A \vee D) - \dim(B \vee D) + \dim(A \vee B \vee D)] \\
&\qquad + \dim(A) + \dim(B) - \dim(A \vee B).
\end{aligned}$$

This is a dimension inequality among subspaces involving only $\vee$. It can be converted into a rank inequality on subsets involving $\cup$ using the observation that in a vector space,

$$\operatorname{rk}(X \cup Y) = \dim(\overline{X} \vee \overline{Y}),$$

where $\overline{Z}$ is the subspace spanned by Z. Thus, if we take $A = \overline{X_1}$, $B = \overline{X_2}$, $C = \overline{X_3}$, $D = \overline{X_4}$ and group terms with the same sign, we obtain Ingleton's inequality. For independence from the submodular inequality, use the nonrepresentable Vámos matroid on p. 511 of Oxley (1992).

Chapter 3

3.1.1 Suppose σ is proper. Then PM3 implies that $\mathcal{I}(\sigma)$ maps any function f in $\mathcal{I}(Q)$ to $\mathcal{I}(P)$. By PM1, $\mathcal{I}$ maps the identity δ of $\mathcal{I}(Q)$ to the identity of $\mathcal{I}(P)$. Finally, observe that assuming PM1, condition PM3 is equivalent to PM3′: if $\sigma(x_1) \leq \sigma(x_2)$ and y is in the interval $[\sigma(x_1), \sigma(x_2)]$, then there is a unique x such that $x \in [x_1, x_2]$ and $\sigma(x) = y$. Hence,

$$\begin{aligned}\mathcal{I}(\sigma)(f * g)(x_1, x_2) &= \sum_{y: y \in [\sigma(x_1), \sigma(x_2)]} f(\sigma(x_1), y) g(y, \sigma(x_2)) \\ &= \sum_{x: x \in [x_1, x_2]} f(\sigma(x_1), \sigma(x)) g(\sigma(x), \sigma(x_2)) \\ &= [\mathcal{I}(\sigma)(f) * \mathcal{I}(\sigma)(g)](x_1, x_2).\end{aligned}$$

Thus, $\mathcal{I}(\sigma)$ is an $\mathbb{A}$-algebra homomorphism.

Conversely, let $\sigma : P \to Q$ be a function such that $\mathcal{I}(\sigma) : \mathcal{I}(Q) \to \mathcal{I}(P)$, defined by Equation (C), is an $\mathbb{A}$-algebra homomorphism. Since

$$\delta(\sigma(x_1), \sigma(x_2)) = \mathcal{I}(\sigma)(\delta)(x_1, x_2) = \delta(x_1, x_2),$$

σ is injective. Similarly, applying Equation (C) to the zeta function, σ satisfies PM2. Finally, suppose $y_1 = \sigma(x_1)$, $y_2 = \sigma(x_2)$, and $y_1 \leq y_2$. Let y be a fixed element in $[y_1, y_2]$. Then

$$\begin{aligned}&\sum_{x: x \in [x_1, x_2]} \epsilon_{y_1, y}(y_1, \sigma(x)) \epsilon_{y, y_2}(\sigma(x), y_2) \\ &\quad = \sum_{x: x \in [x_1, x_2]} \mathcal{I}(\sigma)(\epsilon_{y_1, y})(x_1, x) \mathcal{I}(\sigma)(\epsilon_{y, y_2})(x, x_2) \\ &\quad = [\mathcal{I}(\sigma)(\epsilon_{y_1, y}) * \mathcal{I}(\sigma)(\epsilon_{y, y_2})](x_1, x_2) \\ &\quad = \mathcal{I}(\sigma)(\epsilon_{y_1, y} * \epsilon_{y, y_2})(x_1, x_2) \\ &\quad = [\epsilon_{y_1, y} * \epsilon_{y, y_2}](y_1, y_2) \\ &\quad = 1,\end{aligned}$$

and hence PM3 holds.

3.1.2. (a) Since ideals are closed under taking linear combinations, the functions described are in J. Conversely, suppose $f \in J$ and $f(x, y) \neq 0$. Then it follows from

$$\epsilon_{xx} * f * \epsilon_{yy} = f(x, y)\epsilon_{xy}$$

that $\epsilon_{xy} \in J$.

(b) Since

$$\epsilon_{xu} * f * \epsilon_{vy} = f(u, v)\epsilon_{xy},$$

for all $f \in J$, $f(u, v) = 0$ for all $f \in J$ if $\epsilon_{xy} \notin J$.

(c) Part (b) gives an order-reversing function $J \mapsto \text{zero}(J)$ from ideals to order-ideals. If Z is an order-ideal, then define

$$\text{ideal}(Z) = \{f\colon f(x, y) = 0 \text{ whenever } [x, y] \in Z\}.$$

It is routine to check (by writing out the convolution) that J is a (two-sided) ideal. The proof can be completed by observing that the map $Z \mapsto \text{ideal}(Z)$ is an order-reversing function and $J \mapsto \text{ideal}(\text{zero}(J))$ is the identity function.

3.1.3 (a) Consider the elementary matrix functions ϵ_{tt} in $\mathcal{I}(P)$. Then $\epsilon_{tt} * \epsilon_{tt} = \epsilon_{tt}$ and $\epsilon_{ss} * \epsilon_{tt} = 0$ if $s \neq t$. When t ranges over all the elements of P, the functions ϵ_{tt} are n pairwise orthogonal idempotents in $\mathcal{I}(P)$. Hence, IA1 is a necessary condition.

Next, recall that a function f is in the Jacobson radical if and only if $\delta - f * g$ is invertible for every g. Since an upper-triangular matrix is invertible if and only if all diagonal entries are nonzero, f is in the Jacobson radical if and only if all the diagonal entries in f *are* zero. Hence the quotient algebra $\mathcal{I}(P)/J(\mathcal{I}(P))$ is isomorphic to the (commutative) algebra of diagonal matrices. This shows that IA2 is necessary.

To prove the converse, let $e_1, e_2, \ldots, e_n$ be n pairwise orthogonal idempotents in $\mathcal{I}$. Then, by definition, $e_i^2 = e_i$ and $e_i e_j = e_j e_i = 0$. In particular, the idempotents e_i are commuting matrices and we may use the linear algebra theorem that commuting matrices are simultaneously diagonalizable. Change basis so that e_i is the diagonal matrix with all entries zero except for the ii-entry, which equals 1. Relative to this basis, let e_{ij} be the matrix with all entries zero except for the ij-entry, which equals 1.

Define a partial order on the set $\{1, 2, \ldots, n\}$ by specifying that $i \leq j$ if $e_{ij} \in \mathcal{I}$. Since $e_i = e_{ii}$, the relation is reflexive. Transitivity follows from $e_{ij}e_{jk} = e_{ik}$ and the fact that $\mathcal{I}$ is closed under matrix multiplication. To show antisymmetry, suppose that both e_{ij} and e_{ji} are in $\mathcal{I}$. Since $\mathcal{I}/J(\mathcal{I})$ is commutative, the commutator $e_{ij}e_{ji} - e_{ji}e_{ij}$ is in $J(\mathcal{I})$, but the commutator equals $e_i - e_j$. Hence, $\delta - (e_i - e_j)$ is invertible and diagonal, $e_i - e_j = 0$, and $i = j$.

To finish, check that the function $f \mapsto \sum_{i,j=1}^{n} f(i,j)e_{ij}$, $\mathcal{I}(P) \to \mathcal{I}$ is an $\mathbb{A}$-algebra isomorphism.

(b) This follows easily from (a).

3.1.7. Use the elementary matrix functions.

3.1.8. (a) Observe that from a chain C of length $i-2$, we can construct exactly $\binom{n-2}{i-2}$ multichains whose underlying set is C.

(e) Since $Z([c,d];n) = \zeta^n(c,d)$, we have

$$\begin{aligned}\sum_{x:\hat{0}\le x\le\varphi(y)} Z([\hat{0},x];n)\zeta_Q(x,\varphi(y)) &= \sum_{x:\hat{0}\le x\le\varphi(y)} \zeta_Q{}^n(\hat{0},x)\zeta_Q(x,\varphi(y)) \\ &= \zeta_Q{}^{n+1}(\hat{0},\varphi(y)) \\ &= Z([\hat{0},\varphi(y)];n+1).\end{aligned}$$

Since φ, ψ form a Galois coconnection, $x \le \psi(y)$ if and only if $\varphi(x) \le y$. Hence,

$$\sum_{a:a\le y}\left[\sum_{x:\psi(x)=a} Z([\hat{0},x];n)\right] = Z([\hat{0},\varphi(y)];n+1).$$

The theorem now follows from Möbius inversion.

(f) Observe that $Z(P_E;n)$ is the number of multichains of P for which every element is ordered relative to E or, equivalently, strictly ordered relative to some subset D containing E. Hence, by inclusion–exclusion, the number of multichains for which every element is strictly ordered relative to $\emptyset$ is

$$\sum_{E:E\subseteq A} (-1)^{|E|} Z(P_E;n).$$

Next, consider a multichain with underlying chain C. Extend C to a maximal chain C^+. Since A is a cutset, $A \cap C^+ \neq \emptyset$ and each element in the original multichain is ordered relative to the nonempty subset $A \cap C^+$ of A. Thus, no multichain is strictly ordered relative to $\emptyset$ and the alternating sum is zero. We can now finish the proof by observing that $Z(P_\emptyset;n) = Z(P;n)$.

(g) From (f), we obtain

$$\mu_P(\hat{0},\hat{1}) = \sum_{E:E\in A,\, E\neq\emptyset} (-1)^{|E|+1}\mu_{P_E}(\hat{0},\hat{1}).$$

Since P_E is the union of a principal filter and a principal ideal, Theorem 3.1.7 implies that $\mu_{P_E}(\hat{0},\hat{1})$ equals 0 if $P_E \neq \{\hat{0},\hat{1}\}$ and equals -1 if $P_E = \{\hat{0},\hat{1}\}$.

3.1.9. (a) Regard the entries x_{ij} as indeterminates. If $f : \{1, 2, \ldots, n\} \to \{1, 2, \ldots, n\}$, let

$$\mathrm{Mon}(f) = x_{1,f(1)} x_{2,f(2)} \cdots x_{n,f(n)}.$$

Then

$$\mathrm{per}\ X = \sum \{\mathrm{Mon}(f)\colon\ f(\{1, 2, \ldots, n\}) = \{1, 2, \ldots, n\}\}$$

and

$$\prod_{i=1}^{n} \sum_{j:\, j \in E} x_{ij} = \sum \{\mathrm{Mon}(f)\colon\ f(\{1, 2, \ldots, n\}) \subseteq E\}.$$

Ryser's formula now follows from inclusion–exclusion.

(b) By means of a Gray code or Hamiltonian cycle on the cube, the terms in the sum can be calculated one after another, so that we have an algorithm to compute the permanent using $(2^n - 1)(n - 1)$ multiplications, $(2^n - 2)(n + 1)$ additions, and $n + 1$ space (see Knuth, 1969, pp. 467, 640).

(c) Apply Ryser's formula to the $n \times n$ matrix with the top m rows equal to the $m \times n$ matrix X and the bottom $n - m$ rows all 1s.

3.1.10 (b) Observe that if an automorphism α fixes a spanning s-tuple (componentwise), then it is the identity. Hence, every s-tuple is in an orbit of size $|\mathrm{Aut}(G)|$.

(c) This was the motivating application of Theorem 3.1.7. Alternatively, use the fact that $\Phi(G)$ is the set of "nongenerators" of G.

(d) Use the theorem that if $|H|$ and $|K|$ are relatively prime, then $L(H \times K) = L(H) \times L(K)$, where the right-hand product is Cartesian product of lattices.

(e) An s-tuple $(g_1, g_2, \ldots, g_s)$ spanning G gives an s-tuple $(Ng_1, Ng_2, \ldots, Ng_s)$ spanning G/N. Thus, it suffices to show that the number of s-tuples $(x_1, x_2, \ldots, x_s)$ spanning G such that $x_i \in Ng_i$ equals $\phi(G \downarrow N; s)$.

This follows from the following lemma: *let $(Ng_1, Ng_2, \ldots, Ng_s)$ be an s-tuple of cosets and H a subgroup of G. Let $C(H; s)$ be the number of s-tuples $(x_1, x_2, \ldots, x_s)$ such that $x_i \in Ng_i$ and $(x_1, x_2, \ldots, x_s)$ spans a subgroup of H. Then*

$$C(H, s) = \begin{cases} |H \cap N|^s & \text{if for all } i,\ H \cap Ng_i \neq \emptyset, \\ 0 & \text{otherwise.} \end{cases}$$

To prove the lemma, observe that an s-tuple $(x_1, x_2, \ldots, x_s)$ such that $x_i \in Ng_i$ spans a subgroup contained in H if and only if for every i, $x_i \in H \cap Ng_i$.

Therefore,

$$C(H;s) = |H \cap Ng_1||H \cap Ng_2| \cdots |H \cap Ng_s|.$$

It is easy to show that $|H \cap Ng| = |H \cap N|$. The lemma now follows.

To finish, we apply Möbius inversion to conclude that the number of s-tuples $(x_1, x_2, \ldots, x_s)$ such that $x_i \in Ng_i$ and $(x_1, x_2, \ldots, x_s)$ spans G equals

$$\sum_{H:H \in L(G)} \mu(H, G) C(H; s). \qquad (*)$$

The final step is to use the fact that $C(H; s) = 0$ unless for all i, $Ng_i \cap H \neq \emptyset$. Since $(Ng_1, Ng_2, \ldots, Ng_s)$ spans G/N, this condition implies that $H/N = G/N$; that is, $NH = G$. Thus, the sum $(*)$ can be restricted to those subgroups such that $NH = G$. Hence, by the lemma,

$$C(H;s) = \sum_{H:H \in L(G) \text{ and } NH=G} \mu(H, G)|N \cap H|^s = \phi(G \downarrow H; s).$$

(f) If G is a finite p-group G, then the quotient $G/\Phi(G)$ is an elementary Abelian group, that is, a direct sum of cyclic groups of order p. Thus, it suffices to consider a direct sum $\bigoplus_r Z_p$ of r copies of the cyclic group of order p. Let N be the subgroup $\{0\} \oplus \{0\} \cdots \oplus \{0\} \oplus Z_p$. Then $\bigoplus_r Z_p/N \cong \bigoplus_{r-1} Z_p$, and thinking of $\bigoplus_r Z_p$ as a vector space over the finite field GF(p), a subgroup (or subspace) H satisfies $HN = \bigoplus_r Z_p$ if either $H = \bigoplus_r Z_p$ (and $H \cap N = N$) or H is one of p^{r-1} subspaces intersecting N at $\{0\}$. Hence, by Gaschütz's theorem,

$$\phi\left(\bigoplus_r Z_p; s\right) = \phi\left(\bigoplus_{r-1} Z_p; s\right)(p^s - p^{r-1}).$$

Since $\phi(Z_p; s) = p^s - 1$, we obtain, by induction,

$$\phi\left(\bigoplus_r Z_p; s\right) = \prod_{i=0}^{r-1}(p^s - p^i).$$

Finally, if $|G| = p^m$ and $\Phi(G) = p^{m-r}$, then

$$\phi(G; s) = p^{m-r} \prod_{i=0}^{r-1}(p^s - p^{r-1}).$$

(h) Let D_{2n} be the dihedral group of order $2n$. We shall think of D_{2n} as the group of symmetries of the regular n-gon, so that D_{2n} is generated by the rotation ρ by $2\pi/n$ and a reflection. Observe first that if n has prime factorization $p_1^{a_1} p_2^{a_2} \cdots p_r^{a_r}$, then the Frattini subgroup is the group of rotations generated by the rotation $\rho^{p_1 p_2 \cdots p_r}$ and $G/\Phi(G)$ is isomorphic to $D_{2p_1 p_2 \cdots p_r}$. Thus,

$$\phi(D_{2n}; s) = \left(\frac{n}{p_1^{a_1-1} p_2^{a_2-1} \cdots p_r^{a_r-1}} \right)^s \phi(D_{2p_1 p_2 \cdots p_n}; s).$$

It suffices to consider the case when n is square-free.

To calculate $D_{2p_1 p_2 \cdots p_r}$, we construct the subgroup lattice: the subgroups of D_{2n} are either dihedral groups or cyclic groups and it is not difficult to deduce the lattice structure. The final answer is

$$\begin{aligned}
\phi(D_{2p_1 p_2 \cdots p_r}) &= \sum_A (-1)^{r-|A|} p_{A^c} (2p_A)^s \; - \; \sum_A (-1)^{r-|A|} p_{A^c} (p_A)^s \\
&= (2^s - 1) \sum_A (-1)^{r-|A|} p_{A^c} p_A^s \\
&= (2^s - 1) \prod_{i=1}^{r} (p_i^s - p_i),
\end{aligned}$$

where the sums are over all subsets A in $\{1, 2, \ldots, r\}$, $A^c = \{1, 2, \ldots, r\} \backslash A$, $p_A = \prod_{i:i \in A} p_i$, and $p_{A^c} = \prod_{i:i \notin A} p_i$. This exercise can also be done using Gaschütz's theorem.

(i) One needs to determine the subgroup lattice. This is not too hard and was done in Hall's 1936 paper. A related problem is to compute $\mu(\hat{0}, \hat{1})$ for subgroup lattices of celebrity groups. See, for example, Downs (1991) and Shareshian (1997).

(j) Use the classification theorem for finite simple groups.

3.1.11. (a) Define $f(x, y)$ by one of two equivalent conditions:

$$F(x, y) = \sum_{z: z \leq x} f(z, y) \quad \text{or} \quad f(x, y) = \sum_{z: z \leq x} F(z, y) \mu(z, y).$$

Let Z be the matrix of the zeta function of P and H be the upper-triangular matrix with x, y-entry $f(x, y)\zeta(x, y)$. Since determinants are multiplicative, we have

$$\det(Z^T H) = (\det Z)(\det H) = \prod_{x: x \in P} f(x, x).$$

On the other hand, the xy-entry in the matrix product $Z^T H$ equals

$$\begin{aligned}
\sum_{z: z \in P} \zeta(z, x) f(z, y) \zeta(z, y) &= \sum_{z: z \leq x \text{ and } z \leq y} f(z, y) \\
&= \sum_{z: z \leq x \wedge y} f(z, y) \\
&= F(x \wedge y, y).
\end{aligned}$$

We conclude that

$$\det[F(x \wedge y, y)] = \prod_{x:x \in P} f(x,x) = \prod_{x:x\in P} \left(\sum_{z:z \le x} F(z,x)\mu(z,x) \right).$$

The essential idea in this proof was given in Pólya and Szegö (1976, pp. 117, 309).

(c) Apply the Lindström–Wilf formula to the set $\{1, 2, \ldots, n\}$ ordered by divisibility (so that the meet is the greatest common divisor) with $F(i, j) = i$ and note that $\phi(i) = \sum_{k:k\,|\,i} \mu(i/k)k$.

(d) Let $F(x, y) = \delta(\hat{0}, x)$. Then

$$\det[\delta(\hat{0}, x \wedge y)]_{x,y \in P} = \prod_{x:x\in P} \mu(\hat{0}, x) \neq 0.$$

The matrix $[\delta(\hat{0}, x \wedge y)]_{x,y\in P}$ is a (0-1)-matrix with xy-entry equal to 1 if and only if $x \wedge y = \hat{0}$. Since its determinant is nonzero, there is at least one nonzero term in the determinantal expansion. The permutation of that term yields the permutation required.

(e) Apply the Lindström–Wilf formula to a direct product of chains with b elements.

3.1.12. There are $n!$ terms in the permanent, each corresponding to a permutation of $\{1, 2, \ldots, n\}$. Observe that a permutation gives a nonzero term if and only if it is the identity or its cycle decomposition consists of fixed points and one cycle of length greater than 1, and that cycle has the form $(\hat{0}, x_2, x_3, \ldots, x_l)$, where $\hat{0} < x_2 < x_3 < \cdots < x_l$. The permanent formula follows immediately. The determinant formula follows from Philip Hall's theorem (3.1.11).

3.1.13. The proof is a variation on the proof in Exercise 3.1.11. Let Z be the matrix of its zeta function of L and D be the diagonal matrix with xx-entry $\mu(\hat{0}, x)$. Then $Z^T DZ = C$, where

$$C = [\delta(\hat{0}, x \wedge y)]_{x,y\in L},$$

the (0-1)-matrix with entry 1 if $x \wedge y = \hat{0}$ and 0 otherwise. If $\mu(\hat{0}, x) \neq 0$ for every $x \in L$, then C is invertible with inverse C^{-1} equal to $MD^{-1}M^T$, where $M = Z^{-1}$. Explicitly, the xy-entry of C^{-1} is

$$\sum_{z:z \ge x \vee y} \frac{\mu(x,z)\mu(y,z)}{\mu(\hat{0}, x)}$$

and the inversion formula follows.

3.1.14. Consider the following matrices with rows and columns indexed by $L: Z = [\zeta(x, y)]$, D_0 the diagonal matrix with xx-entry $\mu(\hat{0}, x)$, and D_1 the diagonal matrix with xx-entry $\mu(x, \hat{1})$. Then, as in the solution to Exercise 3.1.11(a),

$$\begin{aligned} Z^T D_0 Z &= [\delta(\hat{0}, x \wedge y)], \\ Z D_1 Z^T &= [\delta(\hat{1}, x \vee y)]. \end{aligned}$$

Consider the product $Z D_1 Z^T D_0 Z$. Then writing the product as $(Z D_1 Z^T) D_0 Z$, the xy-entry is

$$\sum_{z: x \vee z = \hat{1}} \mu(\hat{0}, z)\zeta(z, y). \tag{D}$$

Since $x \vee z \leq x \vee y$ if $z \leq y$, the sum is empty, and hence the xy-entry equals 0, unless $x \vee y = \hat{1}$. Similarly, writing the product as $(Z D_1 Z^T) D_0 Z$, the xy-entry is 0 unless $y \wedge x = \hat{0}$. Hence, the xy-entry $Z D_1 Z^T D_0 Z$ is zero unless x and y are complements. To finish the proof, since the determinant is multiplicative, $\det Z D_1 Z^T D_0 Z \neq 0$. Any permutation yielding a nonzero term in the expansion of the determinant yields a complementing permutation. This proof is due to R.M. Wilson (unpublished), recounted in Stanley (1997, p. 185).

Dowling's proof begins with the matrix whose xy-entry is the sum (D). Using the methods in Theorem 3.1.12, he proves that the xy-entry is 0 unless x and y are complements. He also inverts the matrix explicitly, thus proving that it is nonsingular. Disregarding technicalities, Dowling's proof is the same as Wilson's. Finally, note that the sum (D) equals $\mu_f(\hat{0}, y)$, where $f(u) = u \vee x$ in the proof of Theorem 3.1.13. Thus, Wilson's argument offers an alternate way to finish the proof of that theorem.

3.1.15. The interval $[A, B]$ in $L(\mathcal{C})$ is isomorphic to $L(\mathcal{C}')$, where

$$\mathcal{C}' = \{I \backslash (I \cap A): I \in \mathcal{C},\ I \subseteq B\}.$$

Thus, it suffices to prove part (b) for $\mu(\hat{0}, X)$. Further, by Theorem 3.1.9, if X is not a join of atoms, then $\mu(\hat{0}, X) = 0$. Thus, we may assume that X is a join of atoms. In turn, this allows us to assume that $\mathcal{C}$ is an antichain. Removing points in gaps, we can also assume that $I_1 \cup I_2 \cup \cdots \cup I_n = \{1, 2, \ldots, M\}$ for some integer M. Let $\mathcal{C} = \{I_1, I_2, \ldots, I_n\}$, where the intervals are indexed so that the right (and hence left) endpoint is increasing.

Let $X_0 = \emptyset$ and $X_k = I_1 \cup I_2 \cup \cdots \cup I_k$. By the indexing, $X_{k-1} \subseteq X_k$ and the sets X_k grow to the right. By Weisner's theorem (3.1.5) applied to the fixed element I_k,

$$\mu(\emptyset, X_k) = - \sum_{X: X \cup I_k = X_k,\, X \neq X_k} \mu(\emptyset, X).$$

Now observe that $X \cup I_k = X_k$ if and only if $X = X_j$, $j \leq k$, and I_j *touches* I_k; that is, $I_j \cup I_k$ is an interval. Hence,

$$\mu(\emptyset, X_k) = -[\mu(\emptyset, X_{k-1}) + \mu(\emptyset, X_{k-2}) + \cdots + \mu(\emptyset, X_{k-m(k)})],$$

where $m(k)$ is the minimum index j such that I_j touches I_k. Since $\mu(\emptyset, X_1) = -1$, it follows by induction that $\mu(\emptyset, X_k) = -1$, 0, or 1.

For related results and generalizations, see, for example, Kahn (1987).

3.2.1. (a) Represent the term a_i as the element (a_i, i) in the Cartesian product of two chains.

(b) The *depth* of an element a in P is the maximum length of a chain with bottom a (so that, for example, a maximal element has depth zero). The subsets P_i of elements in P of depth i are pairwise disjoint and their union is P. If P has no chain of length $m+1$, then P_i is empty if $i \geq m+1$. Since

$$|P_0| + |P_1| + \cdots + |P_m| \;=\; |P| \;\geq\; mn + 1,$$

there is a subset P_j, say, such that $|P_j| \geq m+1$ (by the pigeon-hole principle). Since the subsets P_i are all antichains, we have found an antichain of size $m+1$.

3.2.5. (a) Let C be a chain in P, y an element in P, and $C \parallel y$ the set of elements in C incomparable with y. The relation $P \cup \{(x, y) \colon x \in C \parallel y\}$ has no directed cycles, and hence its transitive closure is a partial order. Taking a linear extension of the new partial order, we obtain a linear extension $L(C)$ of P in which $(*)$ $x < y$ for every pair x and y, such that $x \in C \parallel y$. Now let $C_1, C_2, \ldots, C_{w(P)}$ be a chain partition of P, and for each chain C_i, let $L(C_i)$ be a linear extension satisfying $(*)$. If x and y are incomparable in P and x is in the chain C_i, then $x < y$ in $L(C_i)$; on the other hand, $y < x$ in $L(C_j)$, where C_j is the chain containing y. Hence, x and y are incomparable in the intersection $\bigcap L(C_i)$. Since $\bigcap L(C_i)$ is an extension of P, we conclude that $P = \bigcap L(C_i)$ and P has order dimension at most its width $w(P)$.

(b) Choose a chain partition $C_1, C_2, \ldots, C_{w(P \backslash A)}$ of $P \backslash A$ and let $L(C_i)$ be the linear extensions in the hint to (a). Add to this a linear extension that puts all the elements in $P \backslash A$ below all the elements of A. These $w+1$ extensions intersect to P.

(c) If A is contained in a larger antichain A', then $w(P \backslash A') \leq w(P \backslash A)$. Thus, we need to prove the inequality only for a maximal antichain A'. If $P = A'$, then $\dim(P) = 2$. Since $A \subset P$, the inequality holds. Otherwise, let $C_1, C_2, \ldots, C_w$

be a chain partition of $P \backslash A'$. We construct two linear extensions from each chain C_i in the partition: one is the extension $L(C_i)$ defined earlier, and the other is its contrary, an extension in which $x > y$ for every pair x and y such that $x \in C \parallel y$. We can also construct an extension such that every element in the ideal $I(A')$ not in A' is less than every element in A', and in turn, every element in the filter $F(A')$ not in A' is above every element in A'. These $2w + 1$ extensions intersect to P.

(d) Dilworth's inequality is tight for an antichain. Examples for the inequalities in (b) and (c) require more work (see Trotter, 1975 or 1992, p. 27).

3.2.6. (e) By part (b), the canonical partition gives a join-preserving injection γ of $\mathcal{A}_k(P)$ into $\mathcal{A}(P)^k$. Hence, the interval $[x, x^*]$ in $\mathcal{A}_k(P)$ is a join-sublattice of $[\gamma(x), \gamma(x^*)]$ and $\gamma(x^*)$ is the join of the elements covering $\gamma(x)$. It follows that $[x, x^*]$ and $[\gamma(x), \gamma(x^*)]$ are isomorphic. To finish, observe that $\mathcal{A}(P)^k$ is a distributive lattice, and hence the interval $[\gamma(x), \gamma(x^*)]$, being atomic, is a Boolean algebra.

Let Q be the partially ordered set on $\{a_1, a_2, a_3, a_4\}$ with $a_1 < a_2$, and $a_3 < a_2$, $a_3 < a_4$. Then $\mathcal{A}_2(Q)$ is not distributive. A nonmodular example is harder (see Example 2.25 in Greene and Kleitman, 1976).

(h) The argument for the case $k = 1$ given in Corollary 3.2.6 generalizes with a little work.

(i) There are several proofs of this. The proof of Greene and Kleitman is somewhat complicated. Hoffman and Schwartz gave a proof using linear programming. Frank's proof uses network flows. Saks' proof starts from the observation that an antichain in the Cartesian product $P \times \langle k \rangle$, where $\langle k \rangle$ is a chain with k elements, gives a k-family in P of the same size.

3.3.1. The collection $\mathcal{C}$ cannot contain both a subset and its complement.

3.3.5. Use the *idea* in the construction given in Theorem 3.3.4. Just for this solution, define a *chain* to be a subset C of sums $\sum \epsilon_i \alpha_i$, such that $s_1, s_2 \in C$ implies $\|s_1 - s_2\| \geq 2$.

We partition the sums into chains inductively. We shall use the following notation: $C + y = \{x + y \colon x \in C\}$. If $n = 1$, then we put the two sums $+\alpha_1$ and $-\alpha_1$ into the same chain C. Suppose a chain partition has been constructed for $\alpha_1, \ldots, \alpha_{n-1}$. If C is a chain in the existing chain partition, then $C + \alpha_n$ and $C - \alpha_n$ are chains.

Next, let β be the vector in $C - \alpha_n$ such that the dot product $(\beta, -\alpha_n)$ is maximum. Then it is easily checked that $\|\beta - \gamma\| \geq 2$ for every $\gamma \in C + \alpha_n$. Put the chains $(C + \alpha_n) \cup \{\beta\}$ and $(C - \alpha_n) \backslash \{\beta\}$ into the new chain partition.

In this way, we partition the sums into $\binom{n}{\lfloor n/2 \rfloor}$ chains. Since a unit ball intersects a chain at most one vector, the required result follows.

3.3.6. Suppose P satisfies the LYM inequality. Let $\mathcal{B} \subseteq \mathcal{W}_i$. Then $\mathcal{B} \cup (\mathcal{W}_{i-1} \backslash I(\mathcal{B}))$ is an antichain. Applying the LYM inequality, we obtain

$$\frac{|\mathcal{B}|}{W_i} + \frac{W_{i-1} - |I(\mathcal{B})|}{W_{i-1}} \leq 1.$$

This simplifies to the normalized matching property.

Next, we use the normalized matching property to construct a regular chain cover. To do this, we construct a ranked partially order set P^{++} such that $|\mathcal{W}_i(P^{++})| = W$, where $W = W_0(P)W_1(P) \cdots W_N(P)$, the product of all the Whitney numbers of P. This is done by taking $\prod_{j:\, j \neq i} W_i$ copies of $\mathcal{W}_i$ and a copy of the element x is less than a copy of y if $x < y$. Using normalized matching property and the marriage theorem, there is a matching between $\mathcal{W}_{i-1}(P^{++})$ and $\mathcal{W}_i(P^{++})$. Splicing these matchings together, we obtain a chain partition of P^{++} and, on identifying copies, a regular chain cover of P.

Finally, suppose P has a regular chain cover $\mathcal{C}$. Then a rank-i element occurs in exactly $|\mathcal{C}|/W_i$ chains. If $\mathcal{A}$ is an antichain, then its rank-i elements occur in $|\mathcal{A} \cap \mathcal{W}_i||\mathcal{C}|/W_i$ chains. Since $\mathcal{A}$ is an antichain, a chain cannot contain two distinct elements. Hence,

$$\sum_{i=0}^{N} \frac{|\mathcal{A} \cap \mathcal{W}_i|}{W_i} |\mathcal{C}| \;\leq\; |\mathcal{C}|.$$

Dividing by $|\mathcal{C}|$, we obtain the LYM inequality.

3.3.7. (a) Let m be an index such that $W_m = \max\{W_i\colon 1 \leq i \leq n\}$. A chain decomposition of P of size W_m can be obtained by splicing together the edges of the partial matching between the adjacent levels. By Dilworth's theorem, an antichain has size at most W_m and hence $\mathcal{W}_m$ is a maximum-size antichain in P.

(b) Start at the middle level and proceed upward and downward in a consistent way. This can be done using the common transversal theorem (2.4.4) and checking that the normalized matching property implies that Ford–Fulkerson condition is satisfied.

3.3.8. Since every family of subsets all having size $\lfloor n/2 \rfloor$ is an antichain and defines a monotone function, the lower bound is clear. To show the upper bound, we will construct all monotone functions f on $2^{\{1,2,\ldots,n\}}$. We will use the symmetric chain decomposition defined by brackets to construct f inductively. Order the chains so that $|C_i| \leq |C_{i+1}|$. Since the smallest chain is of size 1 or 2, there are at most three ways to define f on the smallest chain. Suppose that f has been defined on all chains before the ith chain and let $E_k, E_{k+1}, \ldots, E_{n-k}$ be the subsets in C_i. If $E_j \supset D$, D occurs in an earlier chain, and if $f(D) = 1$, then $f(E_j)$ must equal 1. Dually, if $E_j \subset D$, D occurs in an earlier chain, and if $f(D) = 0$, then $f(E_j)$ must equal 0. In both cases, we say that $f(E_j)$ *is determined*. Observe that if $r < s < t$, and E_s is determined, then at least one of the sets E_r or E_t is determined. Thus, the undetermined subsets form a consecutive segment $E_r, E_{r+1}, \ldots, E_s$ of C_i.

We show next that there can be at most two subsets in the segment of undetermined subsets. Suppose E_r, E_{r+1}, E_{r+2} is a length-3 segment. In the bracket representations of E_r, E_{r+1}, E_{r+2}, we have, at two coordinates a, b, the brackets

$$)),\ )(,\ ((.$$

Consider the subset D represented by the same brackets as E_{r+1}, with the exception that at coordinates a, b, the brackets are (). Then D has a larger set of closed brackets and occurs in a chain of smaller length. Thus, $f(D)$ has already been chosen and at least one of E_r or E_{r+2} is determined.

We have shown that at most two subsets in C_i can be undetermined. Thus, there are at most three ways to define a monotone functions on C_i consistent with the earlier choices. The upper bound now follows.

(b) The idea is that for "most" chains C_i, only one subset is undetermined. Implementing this idea requires a complicated argument (see Kleitman's paper). Further improvements have been made by Kleitman and Markowsky, and others. An approach using entropy can be found in Kahn's paper.

3.4.3. (a) By Exercise 1.3.8, elements in $F_{\mathbf{D}}(3)$ are in bijection with nonempty antichains not equal to $\{\emptyset\}$ in the Boolean algebra $2^{\{1,2,3\}}$. Thus, $F_{\mathbf{D}}(3)$ has 18 elements. Let x, y, and z be three generators for $F_{\mathbf{D}}(3)$. Then $x \wedge y \wedge z$ is the minimum $\hat{0}$, and the eight elements

$$\begin{gathered} x \wedge y \wedge z, x \wedge y, y \wedge z, x \wedge z, \\ (x \wedge y) \vee (y \wedge z), (x \wedge y) \vee (x \wedge z), (x \wedge z) \vee (y \wedge z), \\ (x \wedge y) \vee (x \wedge z) \vee (y \wedge z) \end{gathered}$$

form a lower interval $[\hat{0}, u]$, where $u = (x \wedge y) \vee (x \wedge z) \vee (y \wedge z)$. This interval is a Boolean algebra. Looking upside-down, $x \vee y \vee z$ is the maximum $\hat{1}$, and the order-dual of the eight elements forms an upper interval. By distributivity (see Exercise 1.3.3),

$$(x \wedge y) \vee (z \wedge z) \vee (y \wedge z) = u = (x \vee y) \wedge (z \vee z) \wedge (y \vee z).$$

Thus the upper interval is $[u, \hat{1}]$ and u is in both intervals. Finally, the generator x fits in between $(x \wedge y) \vee (x \wedge z)$ and $(x \vee y) \wedge (x \vee z)$; analogously,

$$(x \wedge y) \vee (y \wedge z) < y < (x \vee y) \wedge (y \vee z),$$
$$(x \wedge z) \vee (y \wedge z) < z < (x \vee z) \wedge (y \vee z).$$

(b) Having practiced, we are ready to do the calculations for $F_{\mathbf{M}}(3)$. We use the same notation as in the text. As for $F_{\mathbf{D}}(3)$, we obtain the lower interval $[\hat{0}, u]$ and the upper interval $[v, \hat{1}]$. Since we are not assuming distributivity, $u < v$. Indeed, $x < x_1 < y$, $x < y_1 < y$, and $x < z_1 < y$. In addition, we have

$$(x \wedge y) \vee (x \wedge z) < x \wedge v < x < x \vee u < (x \vee y) \wedge (x \vee z),$$

and two analogous inequalities for y and z. Thus, we get 28 elements. The question now is whether there are more. This requires checking that the set of 28 elements is closed under meets and joins. Most of these cases are easy. The most complicated case, to decide whether $x_1 \wedge y_1$ equals one of the elements already constructed, is settled by the following calculation: using modularity twice, we have

$$\begin{aligned} x_1 \wedge y_1 &= ((x \wedge v) \vee u) \wedge ((y \wedge v) \vee u) \\ &= [(x \wedge v) \wedge ((y \wedge v) \vee u)] \vee u \\ &= [(x \wedge v) \wedge (y \vee u) \wedge v] \vee u = [(x \wedge v) \wedge (y \vee u)] \vee u. \end{aligned}$$

Now

$$x \wedge v = x \wedge (x \vee y) \wedge (x \vee z) \wedge (y \vee z) = x \wedge (y \vee z) = (y \vee z) \wedge x.$$

Similarly, $y \vee u = y \vee (x \wedge z)$. Thus, using modularity twice,

$$\begin{aligned} [(x \wedge v) \wedge (y \vee u)] \vee u &= [(y \vee z) \wedge x \wedge (y \vee (x \wedge z))] \vee u \\ &= [(y \vee z) \wedge ((x \wedge y) \vee (x \wedge z))] \vee u \\ &= ((y \vee z) \vee u) \wedge u \;=\; u. \end{aligned}$$

To finish, we check that these 28 elements are distinct. This is done by showing that the abstract elements can be represented as subspaces using Dedekind's representation. This will also verify that $F_{\mathbf{M}}(3)$ is modular. Briefly, if x, y, and z are the three subspaces of Dedekind, then $x \wedge y = \langle e_2 \rangle$, $x \wedge z = \langle e_4 \rangle$, and $y \wedge z = \langle e_6 \rangle$. These subspaces generate the lower interval $[0, u]$, where 0 is the zero subspace and $u = \langle e_2, e_4, e_6 \rangle$. Let $\widetilde{e_j}$ be the set $\{e_1, e_2, e_3, e_4, e_5, e_6, e_7, e_8\} \backslash \{e_j\}$. Then, $x \vee y = \langle \widetilde{e_7} \rangle$, $x \vee z = \langle \widetilde{e_3} \rangle$, and $y \vee z = \langle \widetilde{e_5} \rangle$, and these subspaces generate the upper interval $[v, H]$, where $v = \langle e_2, e_4, e_6, e_7, e_8 \rangle$. The subspaces v and u differ in dimension by 2, and between them are three subspaces of dimension 4, x_1, y_1, and z_1, given by

$$x_1 = (x \wedge v) \vee u = \langle e_2, e_4, e_6, e_8\rangle, \; y_1 = \langle e_2, e_4, e_6, e_7\rangle, \quad \text{and}$$
$$z_1 = \langle e_2, e_4, e_6, e_7 + e_8\rangle.$$

It should now be clear how to do the rest of the calculation.

3.4.4. (c) We construct the free lattice using lattice polynomials (see Exercise 1.3.8). Using the modular law and the fact that the variables x_i are comparable and the variables y_j are comparable, any lattice polynomial in x_i and y_j can be put into the form $\bigwedge(x_i \vee y_j)$ or $\bigvee(x_i \wedge y_j)$.

Let $u(i, j) = x_i \wedge y_j$ and $v(i, j) = x_i \vee y_j$. From the observation that $i_1 \geq i_2$ and $j_1 \geq j_2$ imply $u(i_1, j_1) \geq u(i_2, j_2)$ and $u(i_1, j_1) \vee u(i_2, j_2) = u(i_1, j_1)$, any join of the elements $u(i, j)$ can be written *irredundantly* in the form

$$u(i_1, j_1) \vee u(i_2, j_2) \vee \cdots \vee u(i_r, j_r),$$

where $i_1 > i_2 > \cdots > i_r$ and $j_1 < j_2 < \cdots < j_r$.

Next, we show the following *rewriting lemma*: if $x_1 \geq x_2 \geq \cdots \geq x_r$ and $y_1 \leq y_2 \leq \cdots \leq y_r$ in a modular lattice, then

$$\begin{aligned}
&(x_1 \wedge y_1) \vee (x_2 \wedge y_2) \vee \cdots \vee (x_r \wedge y_r)\\
&\quad = x_1 \wedge (y_1 \vee x_2) \wedge (y_2 \vee x_3) \wedge \cdots \wedge (y_{r-1} \vee x_r) \wedge y_r,\\
&(x_1 \vee y_1) \wedge (x_2 \vee y_2) \wedge \cdots \wedge (x_r \vee y_r)\\
&\quad = x_1 \vee (y_1 \wedge x_2) \vee (y_2 \wedge x_3) \vee \cdots \vee (y_{r-1} \wedge x_r) \vee y_r.
\end{aligned}$$

The two identities are dual to each other. We will prove both identities together by induction on r. The case $r = 1$ holds obviously. By the modular law, applied at the two ends,

$$\begin{aligned}
&(x_1 \wedge y_1) \vee (x_2 \wedge y_2) \vee \cdots \vee (x_{r-1} \wedge y_{r-1}) \vee (x_r \wedge y_r)\\
&\quad = x_1 \wedge [y_1 \vee (x_2 \wedge y_2) \vee \cdots \vee (x_{r-1} \wedge y_{r-1}) \vee x_r] \wedge y_r.
\end{aligned}$$

Using the induction hypothesis, the term in the middle equals

$$[(y_1 \vee x_2) \wedge (y_2 \vee x_3) \wedge \cdots \wedge (y_{r-1} \vee x_r)].$$

We conclude that the first equality holds. The second equality is the dual of the first and is proved similarly.

From the rewriting lemma, we conclude that the set of joins of elements $u(i, j)$ is closed under meets and joins. Hence, every element of $F_{\mathbf{M}}(P(m, n))$ is expressible as a join of elements $u(i, j)$. It remains to show that the irredundant joins are distinct. We shall do this by representing joins, written irredundantly, as a sublattice of $Y(m, n)$: to the irredundant join $\bigvee_{t=1}^{r} u(i_t, j_t)$, associate the order-ideal with minimal elements $(i_1, j_1), (i_2, j_2), \ldots, (i_r, j_r)$. It is easy to show that meets and joins are preserved by this representation.

3.4.7. These are difficult theorems of Jónsson (see Crawley and Dilworth, 1973, chapter 12, for a lucid exposition). We give a proof that a lattice with a type-2 representation is modular. Let $R : L \to \Pi(S)$ be a type-2 representation of a lattice L and x, y, and z be elements in L such that $x \geq z$. Suppose that a and b are elements in S such that $a\ R(x \wedge (y \vee z))\, b$. Then, $a\ R(x)\, b$ and $a\ R(y \vee z)\, b$ and, since R is type-2, there exist c and d in S such that

$$a\ R(z)\, c, \quad c\ R(y)\, d, \quad d\ R(z)\, b.$$

Since $z \leq x$, $R(z) \subseteq R(x)$. Thus, we have

$$a\ R(x)\, b, \quad a\ R(x)\, c, \quad d\ R(x)\, b,$$

and by transitivity, $c\ R(x)\, d$. Since $c\ R(y)\, d$, we have $cR(x \wedge y)d$. Together with $a\ R(z)\, c$, $d\ R(z)\, b$, we obtain $aR((x \wedge y) \vee z)b$. Thus,

$$R(x \wedge (y \vee z)) \subseteq R((x \wedge y) \vee z).$$

3.5.1. Let P be the matroid defined on the set S of atoms of L and $\mathcal{L}$ be the collection of subsets defined by rank-2 elements. Then every pair of elements in S determines a unique rank-2 element, and by modularity, every two rank-2 elements intersect at an atom. Hence, the set S is the set of points and the set $\mathcal{L}$ is the set of lines of a projective space. We can now apply the fundamental theorem of projective geometry. It is surprisingly hard to find an appropriate reference for this theorem. We suggest Crawley and Dilworth (1973, chapter 13).

3.5.2. (a) Observe that V is isomorphic to the vector space $\mathrm{GF}(q)^n$ for some q and n. The number of elements of rank k in the lattice $L(V)$ of subspaces is the q-binomial coefficient, defined by

$$\binom{n}{k}_q = \frac{(q^n - 1)(q^{n-1} - 1) \cdots (q^{n-k+1} - 1)}{(q^k - 1)(q^{k-1} - 1) \cdots (q - 1)}.$$

The q-binomial coefficients behave in a similar way to the binomial coefficients. In particular, they are symmetric and unimodal. Hence, $L(V)$ satisfies an analog of the LYM equality and is Sperner (see Exercise 3.3.7).

3.5.4. One way is to choose L_1 and L_2 to be the lattice of flats of two non-Desarguesian projective planes and glue them along a rank-1 interval. Another way is to choose L_1 and L_2 to be lattices of subspaces of vector spaces over two fields of different characteristics, again along a rank-1 interval. More complicated gluings exist (see Freese and Day, 1990).

3.5.7. (a) The main question is how to define $x^\dagger$. If x is not a meet-irreducible, then the interval $[x, x^*]$ has two or more atoms. If $\mu(x, x*) \neq 0$, then let $x^\dagger = x^*$. Otherwise,

$$0 = \mu(x, x^*) = - \sum_{y: x \leq y < x^*} \mu(x, y),$$

and there exists an element y, not an atom of $[x, x*]$, such that $\mu(x, y) \neq 0$. Let $x^\dagger = y$. Since $\mu(x, y) \neq 0$, y is not a join-irreducible (by, say, Theorem 3.1.7) and CS2 is satisfied.

3.6.3. We will prove the upper-crosscut version. Observe that

$$\hat{1} - x = \sum_{a: a \leq \hat{1}} e_a - \sum_{a: a \leq x} e_a = \sum_{a: a \not\leq x} e_a.$$

Thus,

$$\prod_{x: x \in C} (\hat{1} - x) = \prod_{x: x \in C} \left(\sum_{a: a \not\leq x} e_a \right).$$

This product on the right equals $e_{\hat{1}}$, because unless an idempotent e_a occurs in all the factors in the product, its contribution is zero by orthogonality. Since C is an upper crosscut, $e_{\hat{1}}$ is the only idempotent occurring in all the factors. We conclude that $\prod_{x: x \in C} (\hat{1} - x) = e_{\hat{1}}$. Equating the coefficients of $\hat{0}$, we have

$$\mu(\hat{0}, \hat{1}) = \sum_{X: \bigwedge X = \hat{0}} (-1)^{|X|}.$$

3.6.4. (a) Suppose $a \leq b$ in P. Then the inverse image $\varphi^{-1}(F(\varphi(a))$ of the principal filter generated by $\varphi(a)$ is a principal filter $F(c)$ for some c in P. Since $a \in F(c)$, $b \in F(c)$ and hence; $\varphi(b) \in \varphi(F(c)) = F(\varphi(a))$; that is, $\varphi(b) \geq \varphi(a)$.

(b) $a \leq b$ in Q $\Rightarrow$ $F(a) \supseteq F(b)$ $\Rightarrow$ $\varphi^{-1}(F(a)) \supseteq \varphi^{-1}(F(b))$ $\Rightarrow$ $F(\varphi^\Delta(a)) \supseteq F(\varphi^\Delta(b)) \Rightarrow \varphi^\Delta(a) \leq \varphi^\Delta(b)$.

(c) We need to check Axiom GC2 in Section 1.5. Let $x \in P$. Since $a \geq x$ implies $\varphi(a) \geq \varphi(x)$, $\varphi(F(x)) \subseteq F(\varphi(x))$ in Q. Pulling this back to P, we obtain

$$F(x) \subseteq \varphi^{-1}(F(\varphi(x))) \subseteq F(\varphi^\Delta(\varphi(x)));$$

that is, $\varphi^\Delta(\varphi(x))) \geq_\downarrow x$ in the order-dual $P^\downarrow$.

Next, let $y \in Q$ and $x = \varphi^\Delta(y)$. Since $\varphi^{-1}(F(y)) = F(x)$, $\varphi(x) \in F(y)$, and $\varphi(x) \geq y$, we conclude that $\varphi(\varphi^\Delta(y)) \geq y$.

(d) Let $Q' = \{q: q \in Q, q \le \varphi(p)$ for some p in $P\}$. Then Q' is in the image of Q, φ is defined as a function from P to Q', and $\varphi : P \to Q'$ is a residuated map. Let $x \mapsto \overline{x}$ be the closure operator defined by the Galois coconnection φ and φ^{Δ}.

Define $M(\varphi) : M(P, \mathbb{A}) \to M(Q', \mathbb{A})$ by

$$M(\varphi)(e_p) = \sum_{q:\overline{q}=\varphi(p)} e_q$$

if p is closed $M(\varphi)(e_p) = 0$ if p is not closed, and linearity.

Note that Q' is partitioned into (disjoint) subsets $\{q: \overline{q} = z\}$, where z ranges over all closed subsets of Q'. Thus, if $r \ne s$, $M(\varphi)(e_r)$ and $M(\varphi)(e_s)$ are sums over disjoint sets of orthogonal idempotents. This implies that $M(\varphi)$ is an $\mathbb{A}$-algebra homomorphism. Since $q \le \overline{q}$,

$$\begin{aligned} M(\varphi)(x) &= \sum_{y:y\le x} M(\varphi)e_y \\ &= \sum_{y:\, y \text{ is closed},\, y\le x} \left(\sum_{z:\,\overline{z}=\varphi(y)} e_z \right) \\ &= \sum_{z:z\le\varphi(x)} e_z = \varphi(x); \end{aligned}$$

that is, $M(\varphi)$ extends φ.

Now suppose that a function $\varphi : P \to Q$ extends to a homomorphism $M(P, \mathbb{A}) \to M(Q, \mathbb{A})$. Since $x \le y$ in P if and only if $xy = x$ in $M(P, \mathbb{A})$, φ is order-preserving. To show that φ is residuated, we use the equation

$$\sum_{a:\,a\le x} \varphi(e_a) = \varphi(x) = \sum_{b:\,b\le\varphi(x)} e_b. \tag{I}$$

Let $q \in Q$ such that $\varphi^{-1}(F(q))$ is nonempty. Then there exists $p \in P$ such that $\varphi(p) \in F(q)$. In Equation (I) with $x = p$, e_q occurs on the right side. Thus, e_q is a summand in $\varphi(e_r)$ for some element $r \in P$. It is not hard to check that r is unique, r is minimum in $\varphi^{-1}(F(q))$, and $q \le \varphi(r)$. Together, this proves that $\varphi^{-1}(F(q)) = F(r)$.

Chapter 4

4.1.2. An *inversion* in a permutation γ of $\{1, 2, \ldots, n\}$ is a pair (i, j) such that $i < j$ but $\gamma(i) > \gamma(j)$. Let $\mathrm{Inv}(\gamma)$ be the number of inversions of γ. Then it is straightforward to prove by induction that

$$\sum_{\gamma:\gamma\in\mathfrak{S}_n} q^{\mathrm{Inv}(\gamma)} = (1+q)(1+q+q^2)\cdots(1+q+q^2+\cdots+q^{n-1}),$$

the sum ranging over all permutations γ of $\{1, 2, \ldots, n\}$. Potter's formula can be proved in the same way.

4.1.4. Formally,

$$\sum_{n=1}^{\infty} a_n \left(\sum_{m=1}^{\infty} l_m z^{mn} \right) = \sum_{s=1}^{\infty} b_s z^s.$$

Grouping terms, we have $b_s = \sum_{n:\, mn=s} a_n l_m$.

4.1.5. Take the logarithm formally. Then

$$\begin{aligned}\sum_{s=1}^{\infty} b_s z^s &= \log C(z) \\ &= \sum_{n=1}^{\infty} a_n \log(1 - z^n) = \sum_{n=1}^{\infty} a_n \left(\sum_{m=1}^{\infty} -z^{mn}/m \right).\end{aligned}$$

Hence,

$$b_s = - \sum_{m,n:\, mn=s} \frac{a_n}{m} = -\frac{1}{s} \sum_{n:n|s} n a_n.$$

4.2.2. We use the same notation as in Theorem 4.1.2. Let α be an umbra with $L(\alpha^n) = a_n$ and γ be an umbra with $L(\gamma^k) = |C_k|$. Then it is easy to show combinatorially that

$$L(\alpha^{n+1}) = L(\gamma(\alpha + \gamma)^n).$$

Once this is done, we can imitate the proof in the text.

4.3.2. We use the delta operator $D(I - D)$, where I is the identity operator. The basic sequence $(p_n(x))$ is given by

$$p_n(x) = x(I - D)^{-n} x^{n-1}.$$

By the binomial theorem,

$$(I - D)^{-n} = \sum_{k=0}^{\infty} (-1)^k \binom{-n}{k} D^k = \sum_{k=0}^{\infty} \binom{n+k-1}{k} D^k.$$

Hence,

$$p_n(x) = \sum_{k=0}^{n-1} \binom{n+k-1}{k} (n-1)_{(k)} x^{n-k}.$$

Hence, by BT3 and $\sin xt = \frac{1}{2i}(e^{ixt} - e^{-ixt})$, where $i = \sqrt{-1}$,

$$e^{xt} = \sum_{n=0}^{\infty} \frac{p_n(x)}{n!}[(t(1-t)]^n, \quad \text{and} \quad \sin xt = \sum_{n=0}^{\infty} \frac{p_n(x) - p_n(ix)}{2i\, n!}[(t(1-t)]^n.$$

4.3.4. (a) Use Lemma 4.3.12. Another way is to differentiate both sides of

$$1 + \sum_{n=1}^{\infty} \left(\sum_{k=1}^{n} c_{nk} x^k \right) \frac{t^n}{n!} = \exp(xf(t))$$

with respect to x, obtaining

$$\sum_{n=1}^{\infty} \left(\sum_{k=1}^{n} c_{nk} k x^{k-1} \right) \frac{t^n}{n!} = f(t)\exp(xf(t)).$$

Setting $x = 0$, we obtain the required formula.

(b) Let S be an n-set and l_{nk} be the number of k-tuples $(L_1, L_2, \ldots, L_k)$, where L_i are subsets of S forming a partition of S and each subset L_i is linearly ordered. Given such a k-tuple and a function $\rho : \{1, 2, \ldots, k\} \to X$, we can build a reluctant function in $\mathcal{L}(S, X)$ by sending the minimum element in L_i to $\rho(i)$. All reluctant functions in $\mathcal{L}(S, X)$ can be uniquely constructed in this way. Thus, if $|S| \geq 1$,

$$|\mathcal{L}(S, X)| = \sum_{k=1}^{n} l_{nk} x^k.$$

Since there is a natural bijection from $\mathcal{L}(S, X \cup Y)$ to

$$\mathcal{L}(A, X) \times \mathcal{L}(S \backslash A, Y),$$

when X and Y are disjoint, the polynomials $\sum_{k=1}^{n} l_{nk} x^k$ form a sequence of binomial type. By part (a), this sequence is determined by l_{n1}. Since $l_{n1} = n!$ and $L_n'(0) = -n!$, we conclude that $\sum_{k=1}^{n} l_{nk} x^k = L_n(-x)$.

4.3.5. Formally,

$$\sum_{n=0}^{\infty} (-1)^n E^n = \frac{I}{I + E} = \frac{I}{2I + \Delta} = \frac{1}{2}\left(\frac{I}{I + \frac{1}{2}\Delta} \right).$$

4.4.1. Let $(s_n(x))$ be Sheffer with delta operator Q. Then by the isomorphism theorem (4.3.5), there exists an invertible shift-invariant operator R such that $Q = AR$. Then, we can start with the binomial identity and show that a recurrence holds with the sequence (a_n), where $a_0 = 0$ and $a_n = n[R^{-1} p_{n-1}(y)]_{y=0}$.

For the converse, suppose the recurrence holds. Let Q be the operator defined by $Qs_n(x) = ns_{n-1}(x)$, if $n \geq 1$, and $Qs_0(x) = 0$. Then applying Q to the recurrence, we have

$$QAs_n(x) = \sum_{k=0}^{n} \binom{n}{k} a_k(n-k)s_{n-k-1}(x) = nAs_{n-1}(x) = AQs_n(x).$$

Thus, $AQ = QA$ and $AQ^n = Q^nA$. As A is a delta operator, the first expansion theorem implies that Q is shift-invariant, and hence Q is a delta operator.

4.4.3. By Boole's formula, $\Delta = e^D - I$. Then formally,

$$\Delta - 1 = \frac{1}{DJ} = \frac{1}{D} \cdot \frac{D}{e^D - I} = \frac{B_0}{D} + \sum_{k=1}^{\infty} \frac{B_k}{k!} D^{k-1}.$$

4.5.3. The sequence $(L_n(ax))$ is basic for the delta operator

$$\frac{a^{-1}D}{a^{-1}D - I}.$$

Let

$$f(t) = \frac{t}{(1-a)t + a}.$$

The basic sequence $(p_n(x))$ for $f(D)$ is easily calculated:

$$p_n(x) = x((1-a)D + a)^n x^{n-1} = \sum_{k=0}^{n-1} \binom{n}{k} (1-a)^k a^{n-k} (n-1)_{(k)} x^{n-k}.$$

Changing the index of summation from k to $n - k$ and rearrranging the binomial coefficients and falling factorials, we have

$$p_n(x) = \sum_{k=1}^{n} \frac{n!}{k!} \binom{n-1}{k-1} (1-a)^{n-k} a^k x^k.$$

Since

$$\frac{a^{-1}t}{a^{-1}t - 1} = f\left(\frac{t}{t-1}\right),$$

the duplication formula follows from umbral composition.

Chapter 5

5.2.1. (a) Consider the partition consisting of a single block B. Then a placing in $\mathcal{H}_{\{B\}}$ can be obtained by choosing an ordinary function $f : B \to X$ and linearly ordering each nonempty inverse image $f^{-1}(x_i)$. We can obtain a placing f by choosing an integer partition $\tau_1, \tau_2, \ldots, \tau_s$ of $|B|$, a sequence $x_{i_1}, x_{i_2}, \ldots, x_{i_s}$ of distinct elements in X, and a permutation $d_1, d_2, \ldots, d_{|B|}$ of B, and then defining

$$f(d_1) = d_2,\ f(d_2) = d_3, \ldots, f(d_{\tau_1}) = x_{i_1},\ f(d_{\tau_1+1}) = d_{\tau_1+2},$$
$$f(d_{\tau_1+2}) = d_{\tau_1+3}, \quad \ldots, \quad f(d_{\tau_1+\tau_2}) = x_{i_2},$$

and so on. All placings in $\mathcal{H}_{\{B\}}$ can be obtained in this way We conclude that

$$\mathrm{Gen}\,(\mathcal{H}_{\{B\}}) = |B|! \sum_{\tau:\tau\vdash|B|} k_\tau = h_{|B|}.$$

Placings can be defined independently on each block of a partition. Hence if $\pi = \{B_1, B_2, \ldots, B_s\}$, $\mathrm{Gen}\,(\mathcal{H}_\pi) = \mathrm{Gen}\,(\mathcal{H}_{\{B_1\}})\mathrm{Gen}\,(\mathcal{H}_{\{B_2\}}) \cdots \mathrm{Gen}\,(\mathcal{H}_{\{B_s\}})$.

5.2.3. There seems to be no simple formula. The standard way is to derive a formula for the resultant of two polynomials and then specialize to the case of a polynomial and its derivative.

5.2.4. Observe that

$$\Phi(\sigma) = \sum_{f:\,\mathrm{coimage}(f)\geq\sigma} \prod_{i=1}^{n} x_{i,f(i)}.$$

Since a function is injective if and only if its coimage is the minimum partition $\hat{0}$, Crapo's formula follows by Möbius inversion.

5.2.5. Let π be a set partition having type λ. By Theorem 5.2.2,

$$a_\pi = \sum_{\sigma:\sigma\wedge\pi=\hat{0}} k_\sigma.$$

Sum over all such partitions, convert to integer partition indices using Theorem 5.2.3, and use Equation (N) in Section 5.2 to conclude that

$$\frac{1}{|\lambda|}\binom{n}{\lambda}\lambda! a_\lambda = \sum_{\pi,\sigma:\pi\wedge\sigma=\hat{0}} |\gamma| k_\gamma.$$

Hence,

$$c_{\gamma\delta} = \frac{|\gamma||\delta|}{n!} D(\gamma, \delta),$$

where $D(\gamma, \delta)$ is the number of pairs of set partitions π, σ such that π has type γ, σ has type δ, and $\pi \wedge \sigma = \hat{0}$.

There are many other proofs (see, for example, Stanley 1999, p. 290). In particular, it can be shown that the coefficients $c_{\gamma\delta}$ are integers. The solution to part (b) is similar.

5.2.6. This is one of two proofs given by Pólya. As is not uncommon, it is easier to prove a theorem with more variables. Let z, $x_1, x_2, \ldots, x_n$, $y_1, y_2, \ldots, y_n$ be variables, and

$$\begin{aligned} f(z) &= (z-x_1)(z-x_2)\cdots(z-x_n) = z^n - a_1 z^{n-1} + a_2 z^{n-2} - \cdots \pm a_n, \\ g(z) &= (z-y_1)(z-y_2)\cdots(z-y_n) = z^n - b_1 z^{n-1} + b_2 z^{n-2} - \cdots \pm b_n, \\ s_k &= x_1^k + x_2^k + \cdots + x_n^k, \qquad t_k = y_1^k + y_2^k + \cdots + y_n^k, \\ u_k &= s_k - t_k, \\ R &= \prod_{i,j=1}^{n} (x_i - y_j). \end{aligned}$$

Then

$$Ra_1, Ra_2, \ldots, Ra_n, Rb_1, Rb_2, \ldots, Rb_n$$

are polynomials in $u_1, u_2, \ldots, u_{2n-1}, u_{2n}$ and R is a polynomial in $u_1, u_2, \ldots, u_{2n-1}$.

To prove the more general theorem, observe that

$$\begin{aligned} \log \frac{g(z)}{f(z)} &= \sum_{i=1}^{n} \log(z-y_i) - \log(z-x_i) \\ &= \sum_{i=1}^{n} \left[\frac{x_i - y_i}{z} + \frac{x_i^2 - y_i^2}{2z^2} + \frac{x_i^3 - y_i^3}{3z^2} + \cdots \right] \\ &= \frac{u_1}{z} + \frac{u_2}{2z^2} + \frac{u_3}{3z^2} + \cdots . \end{aligned}$$

Let $U(z)$ be the finite sum

$$\frac{u_1}{z} + \frac{u_2}{2z^2} + \frac{u_3}{3z^2} + \cdots + \frac{u_{2n}}{2nz^{2n}}$$

and $\mathcal{O}(z^{-m})$ represents a series

$$\frac{c_m}{z^m} + \frac{c_{m+1}}{z^{m+1}} + \frac{c_{m+2}}{z^{m+2}} + \cdots .$$

In this notation,

$$\log \frac{g(z)}{f(z)} = U(z) + \mathcal{O}(z^{-(2n+1)}).$$

Exponentiating and rearranging terms,

$$f(z)e^{U(z)} = g(z) + \mathcal{O}(z^{-(n+1)}). \qquad \text{(P)}$$

Next, expand $e^{U(z)}$ in powers of z^{-1} to obtain

$$e^{U(z)} = 1 + \frac{v_1}{z} + \frac{v_2}{z^2} + \frac{v_3}{z^3} + \cdots,$$

where v_i is a polynomial in $u_1, u_2, \ldots, u_i$. For example,

$$v_1 = u_1, \; v_2 = \frac{1}{2}(u_1^2 + u_2), \; v_3 = \frac{1}{6}(u_1^3 + 3u_1u_2 + 2u_3).$$

On the right side of Equation (P), the coefficients of $z^{-1}, z^{-2}, \ldots, z^{-n}$ are zero. Hence, we obtain the equations

$$\begin{aligned} v_1a_n + v_2a_{n-1} + v_3a_{n-2} + \; \ldots + \; v_na_1 &= -v_{n+1} \\ v_2a_n + v_3a_{n-1} + v_4a_{n-2} + \ldots + v_{n+1}a_1 &= -v_{n+2} \\ &\vdots \\ v_na_n + v_{n+1}a_{n-1} + v_{n+2}a_{n-2} + \ldots + v_{2n-1}a_1 &= -v_{2n}. \end{aligned}$$

It is not hard to show that

$$\det[v_{i+j-1}]_{1 \le i,j \le n} = (-1)^{n(n-1)/2} R,$$

and hence, as a polynomial in $u_1, u_2, \ldots, u_{2n-1}$, the determinant is nonzero. Solving the equation using Cramer's rule, we obtain a_i as a quotient of two determinants with entries which are polynomials in $u_1, u_2, \ldots, u_{2n}$.

Finally, we can obtain Laguerre's theorem by setting $y_i = -x_i$. When this is done, $u_{2i-1} = 2s_{2i-1}$ and $u_{2i} = 0$ for $1 \le i \le n$.

5.2.9. Use the isomorphism sending $\ker(x_i - x_j)$ to the partition in which $\{i, j\}$ is a two-element block and all other blocks are one-element subsets.

5.3.1. The partitions that are not periods of subgroups of the alternating group are the rank-1 partitions (with one block of size 2 and all other blocks of size 1).

5.3.2. A related paper is Fendel (1967).

5.3.6. (a) If f is aperiodic, then for all permutations $\gamma \in G$, $f \neq f\gamma$. Hence, the orbits of aperiodic functions all have size $|G|$.

(b) Since $\mathcal{F}$ is assumed to be G-closed, if a function f contributes to a coefficient of a monomial in $A(G, \hat{0})$, then so does any function in its orbit. Hence, each coefficient in $A(G, \hat{0})$ is divisible by $|G|$.

5.4.1. By algebraic manipulations (assuming commutativity in $\mathcal{A}$),

$$\begin{aligned} &E(((1 - E)b)(Ea)) + E((1 - E)a)(Eb)) \\ &= (I - E)E(ab) - E((I - E)a(I - E)b). \end{aligned}$$

Operating on both sides by $(I - E)^{-1}$, we obtain

$$P(((1 - E)y)(Ex)) + P((1 - E)a)(Eb)) = E(ab) - P((I - E)a(I - E)b).$$

Now let $a = (I - E)^{-1}x$ and $(I - E)^{-1}y = b$ to obtain Baxter's identity.

5.4.3. Use induction. The shuffle identities are special cases of Cartier's identity (5.5.4).

5.6.1. Let $f(x) = x$. Then

$$\begin{aligned} Pf &= qx + q^2x + q^3x + \cdots = \left(\frac{q}{1-q}\right)x, \\ P(fPf) &= \left(\frac{q}{1-q}\right)P(x^2) \\ &= \left(\frac{q}{1-q}\right)((qx)^2 + (q^2x)^2 + (q^3x)^2 + \cdots) \\ &= \left(\frac{q}{1-q}\right)\left(\frac{q^2}{1-q^2}\right)x^2, \end{aligned}$$

and in general,

$$\underbrace{P(fP(f \cdots P(fP(f))))}_{n \text{ times}} = \frac{q^{1+2+\cdots+n}}{(1-q)(1-q^2)\cdots(1-q^n)}x^n.$$

On the other side,

$$P(f^k) = \left(\frac{q^k}{1-q^k}\right)x^k$$

and

$$\begin{aligned} \sum_{k=1}^{\infty} \frac{P(f^k)}{k} &= \sum_{k=1}^{\infty} \frac{x^k}{k}\left(\frac{q^k}{1-q^k}\right) \\ &= \sum_{k=1}^{\infty} \frac{x^k}{k}(1 + q^k + q^{2k} + q^{3k} + \cdots) \\ &= \sum_{j=0}^{\infty}\left(\sum_{k=1}^{\infty} \frac{x^k}{k} q^{jk}\right) \\ &= -\sum_{j=0}^{\infty} \log(1 + xq^j). \end{aligned}$$

5.6.4. (a) Let H be the step function, defined on the real numbers by $H(x) = 0$ if $x \le 0$ and 1 if $x > 0$. Observe that

$$\begin{aligned}\max_{0\le i\le k}\{s_i(\gamma(\underline{x}))\} - \max_{0\le i\le k-1}\{s_i(\gamma(\underline{x}))\} &= H(s_k(\gamma(\underline{x}))[x_{\gamma(1)} \\ &+ \max\{0, x_{\gamma(2)}, x_{\gamma(2)} + x_{\gamma(3)}, \ldots, x_{\gamma(2)} + x_{\gamma(3)} + \cdots + x_{\gamma(k)}\} \\ &- \max_{0\le i\le k-1}\{s_i(\gamma(\underline{x}))\}].\end{aligned}$$

Partition $\mathfrak{S}_n$ into $\binom{n}{k}$ disjoint subsets: for a k-element subset T of $\{1, 2, \ldots, n\}$, let $G(T) = \{\gamma\colon \gamma(\{1, 2, \ldots, k\}) = T\}$. Summing over $G(T)$, we obtain

$$\sum_{\gamma:\gamma\in G(T)} \max_{0\le i\le k}\{s_i(\gamma(\underline{x}))\} - \max_{0\le i\le k-1}\{s_i(\gamma(\underline{x}))\} = \sum_{\gamma:\gamma\in G(T)} x_{\gamma(1)} H(s_k(\gamma(\underline{x}))).$$

The two other sums one would expect on the right side cancel each other. To see this, regard a permutation γ as a rearrangement $i_1, i_2, \ldots, i_k, i_{k+1}, \ldots, i_n$ (so that $\gamma(j) = i_j$). Pair γ in $G(T)$ with γ' in $G(T)$, where γ' is the rearrangement $i_2, i_3, \ldots, i_k, i_1, i_{k+1}, \ldots, i_n$. Then

$$\begin{aligned}&\max\{0, x_{\gamma(2)}, x_{\gamma(2)} + x_{\gamma(3)}, \ldots, x_{\gamma(2)} + x_{\gamma(3)} + \cdots + x_{\gamma(k)}\} \\ &\quad = \max_{0\le i\le k-1}\{s_i(\gamma'(\underline{x}))\},\end{aligned}$$

and hence, when one sums over $G(T)$, the sum vanishes. Summing over all k-element subsets T, we obtain

$$\sum_{\gamma:\gamma\in\mathfrak{S}_n} \max_{0\le i\le k}\{s_i(\gamma(\underline{x}))\} - \max_{0\le i\le k-1}\{s_i(\gamma(\underline{x}))\} = \sum_{\gamma:\gamma\in\mathfrak{S}_n} x_{\gamma(1)} H(s_k(\gamma(\underline{x}))).$$

To finish the proof, sum over k and note that the left side is a telescoping sum to obtain

$$\sum_{\gamma:\gamma\in\mathfrak{S}_n} \max_{0\le i\le n}\{s_i(\gamma(\underline{x}))\} = \sum_{\gamma:\gamma\in\mathfrak{S}_n} x_{\gamma(1)} \left(\sum_{k=1}^{n} H(s_k(\gamma(\underline{x})))\right).$$

(c) We need a theorem of Cauchy in integral geometry. Let A be a compact convex set in the plane. Then the perimeter L of A equals

$$\int_0^{\pi} D(\theta)d\theta,$$

where $D(\theta)$ is the length of the projection of A onto the line through the origin at an angle of θ to the x-axis. Observe that

$$D(\theta) = \max_{(x,y)}(x\cos\theta + y\sin\theta) - \min_{(x,y)}(x\cos\theta + y\sin\theta),$$

where the maximum and minimum are taken over all points (x, y) in A.

5.7.2. When $p = 3$,

$$(1+X)^{\alpha_1}(1-X)^{\alpha_{-1}} = 1 + c_1X^1 + c_2X^2 + c_1c_2(c_1-c_2)X^3 \\ + c_1c_2(c_1-1)(c_2-1)X^4.$$

Unlike the case $p = 2$, there seems to be no nice formulas for a_m when $m \notin L_3$. The first two cases are

$$a_4 = a_3a_1 - a_2^2a_1^2 - a_2^2a_1 - a_2a_1^2, \quad \text{and} \quad a_5 = a_3a_2 - a_2^2a_1^2 + a_2a_1.$$

5.7.5. By Fermat's little theorem, the elements of GF(q) are the roots of the polynomial $X^q - X$. Apply Newton's identities (5.2.8), with $a_i = 0$, if $1 \le i \le q-2$, $a_{q-1} = -1$. Is there a purely field-theoretic proof?

5.7.6. (b) Condition H1 is clearly necessary. Let

$$f(X)^t = \sum_{i=0}^{q-1} b_{t,i}X^i, \quad \mod x^q - x.$$

Then, by Exercise 5.7.5,

$$\sum_{e:e\in \mathrm{GF}(q)} f(e)^t$$

$$= qb_{t,0} + b_{t,1}\left(\sum_{e:\, e\in \mathrm{GF}(q)} e\right) + \cdots + b_{t,q-2}\left(\sum_{e:\, e\in \mathrm{GF}(q)} e^{q-2}\right)$$

$$+ b_{t,q-1}\left(\sum_{e:\, e\in \mathrm{GF}(q)} e^{q-1}\right) = -b_{t,q-1}.$$

Further, if $f(X)$ is a permutation polynomial, then $f(e)$ ranges over GF(q) as e ranges over GF(q). Hence, if $t \le q-2$, Exercise 5.7.2 implies that $\sum_e f(e)^t = 0$ and $b_{t,q-1} = 0$. We conclude that Condition H2 is necessary.

To show sufficiency, we show that

$$\prod_{e:\, e\in \mathrm{GF}(q)} (Y - f(e)) = Y^q - Y = \prod_{e:\, e\in \mathrm{GF}(q)} (Y - e).$$

This will show that the image $\{f(e): e \in \mathrm{GF}(q)\}$ equals GF(q); that is, f is surjective and hence a permutation.

Let

$$\prod_{e:\, e\in \mathrm{GF}(q)} (Y - f(e)) = \sum_{i=0}^{q-1} a_iY^{q-1-i},$$

where a_i are elementary symmetric functions in the roots $f(e)$. Condition H1 implies that $a_q = 0$. Condition H2 allows us to calculate easily some of power sums in the roots. Indeed, as in the proof of necessity,

$$s_t = \sum_{e:\, e \in \mathrm{GF}(q)} f(e)^t = -b_{t,q-1}.$$

By Condition H2, $s_t = 0$ whenever $1 \le t \le q-2$ and $t \not\equiv 0 \bmod p$. Since $x^{q-1} = 1$ if $x \ne 0$, we have $s_{q-1} = q - 1 = -1$. Using Newton's identities (5.2.8), we conclude that

$$a_m = 0 \text{ if } 1 \le m \le q-2 \text{ and } m \not\equiv 0 \bmod p. \tag{A}$$

Knowing (A), we can also conclude from Newton's identities that

$$s_t = 0 \text{ if } t = kp,\ 1 \le k \le p^{a-1}. \tag{B}$$

It remains to determine the value of $a_p, a_{2p}, a_{3p}, \ldots$. To do so, we apply Newton's identities for $m \ge q$. Since we know (A) and $a_q = 0$ by H1, Newton's identity for m can be written

$$s_m + (-1)^p a_p s_{m-p} + a_{2p} s_{m-2p} + \cdots + (-1)^{q-p} a_{q-p} s_{m-q+p} \\ + (-1)^{q-1} a_{q-1} s_{m-q+1} = 0.$$

Since $e^q = e$ for all $e \in \mathrm{GF}(q)$, $s_{r+q-1} = s_r$. Applying this to the case $m = q + p - 1$ and using (B), almost all the terms in Newton's identity vanish, giving the identity $a_p s_{q-1} = 0$. We conclude that $a_p = 0$. We can now finish by letting $m = q + lp - 1$ and $l = 2, 3, \ldots, p^{a-1} - 1$, obtaining, one by one, $a_{2p} = 0$, $a_{3p} = 0$, $\ldots$, $a_{p(p^{a-1}-1)} = 0$.

(c) By the binomial theorem modulo p,

$$\bar{f}(x_1) - \bar{f}(x_2) = \bar{f}(x_1 - x_2).$$

Hence, if $X = 0$ is the only root of $\bar{f}(X)$, $\bar{f}(x_1) = \bar{f}(x_2)$ if and only if $x_1 = x_2$.

(d) The function $f(x)$ is a permutation if and only if the equation $e = f(x)$ can be solved for every element e in GF(q). This equation is equivalent to the system of m linear equations

$$e^{p^{ai}} = f(x)^{p^{ai}}$$

in the unknowns $x, x^{p^a}, x^{p^{2a}}, \ldots, x^{p^{m-1(a)}}$.

5.7.8. (a) Let $\ell(x_1, x_2, \ldots, x_n) = a_1x_1 + a_2x_2 + \cdots + a_nx_n$ and (a_{ij}) be a matrix in GL(n, p) with first row equal to $(a_1, a_2, \ldots, a_n)$. The linear form

x_1 divides $[\alpha_1, \alpha_2, \ldots, \alpha_n]$. Applying (a_{ij}), we conclude that ℓ divides $[\alpha_1, \alpha_2, \ldots, \alpha_n]$.

(b) This seems to be somewhat difficult (see Dickson, 1911, and Ore, 1933).

5.7.9. (a) By the characteristic-p binomial theorem and Fermat's little theorem, if $\alpha, \beta \in \mathbb{H}$,

$$(\alpha x_1 + \beta x_2)^{p^i} = \alpha x_1^{p^i} + \beta x_2^{p^i}.$$

In other words, the map $y \mapsto y^{p^i}$ is $\mathbb{H}$-linear. Hence, if $f(x)$ is an Ore polynomial,

$$f(\alpha x_1 + \beta x_2) = \alpha f(x_1) + \beta f(x_2).$$

(b) Use the characteristic-p binomial theorem.

5.7.10. (a) This seems to be a difficult problem, requiring the machinery of straightening algorithms over letter-place algebras.

(b) Modify the second proof of Theorem 5.2.7.

(c) Differential analogs of the Jacobi–Trudi and Nagelsbach identities for Schur functions have been obtained (see Kung, 2000, and Kung and Rota, 1984a).

Chapter 6

6.1.4. Proceed by induction on n. Then we may assume that $p_{n-1}(x)$ has $n-1$ distinct real zeros, $\lambda_1, \lambda_2, \ldots, \lambda_{n-1}$ (written in increasing order). Then by the recurrence and the intermediate value theorem in calculus, it is easy to show that the zeros ν_i of $p_n(x)$ are real and *interlace* the zeros λ_i:

$$\nu_1 < \lambda_1 < \nu_2 < \lambda_2 < \cdots < \nu_{n-1} < \lambda_{n-1} < \nu_n.$$

6.2.2. Let $w_{ij} = B(x_0^i x_1^{n-i}, x_0^j x_1^{n-j})$. Then the bilinear form B is determined by the $(n+1) \times (n+1)$ matrix $(w_{ij})_{0 \le i,j \le n}$. Consider the action of the following matrices on $B(f, g)$:

$$\begin{pmatrix} c & 0 \\ 0 & 1 \end{pmatrix}, \quad \begin{pmatrix} 1 & 1 \\ 0 & 1 \end{pmatrix}.$$

Applying the left matrix, we have $c^n B(x_0^i x_1^{n-i}, x_0^j x_1^{n-j}) = B((cx_0)^i x_1^{n-i}, (cx_0)^j x_1^{n-j})$; that is, $c^n w_{ij} = c^{i+j} w_{ij}$. Since c can be arbitrary chosen, $w_{ij} = 0$

if $i + j \neq n$. Next, apply the right matrix to $B(x_0^i x_1^{n-i}, x_0^{n-k+1} x_1^{k-1})$. This yields the recursion

$$(n + i + 1)w_{i,n-i} + i w_{i-1,n-i+1} = w_{i,n-i+1} = 0.$$

The recursion has the unique solution

$$w_{i,n-i} = (-1)^i w_{0n} \Big/ \binom{n}{i}.$$

This implies that $B(f, g) = w_{0n}\{f, g\}$.

6.3.3. Use the fact that if $p(z)$ has degree n,

$$\int_{\xi_1}^{\xi_2} p'(z)dz = p(\xi_2) - p(\xi_1).$$

6.4.3. Recall from Exercise 4.4.2 that the double factorials $(2j + 1)!!$ occur as normalized coefficients of Hermite polynomials. The Appell polynomials $g_n(x)$ of the sequence are related by simple transformations to Hermite polynomials. Using this and the fact that Hermite polynomials, being orthogonal polynomials, have only real zeros, we conclude that $g_n(x)$ has only real zeros. Now apply Theorem 6.4.4.

6.4.4. This problem requires two esoteric facts. To show that $\sum_{k=0}^{m} a_k b_k x^k$ has only real zeros, we need to know that all the zeros of the polynomials $Q_n(x)$, defined by

$$Q_n(x) = 1 + \binom{n}{1}^2 x + \binom{n}{2}^2 x^2 + \cdots + \binom{n}{n-1}^2 x^{n-1} + x^n,$$

are real and negative. To see this, use the identity

$$Q_n(x) = (1 - x)^n P_n\left(\frac{1 + x}{1 - x}\right),$$

where $P_n(x)$ is the degree-n Legendre polynomial, and the fact that all the zeros of $P_n(x)$ lie in the interval $[-1, 1]$.

For the second polynomial, we need to know that all the zeros of the polynomial

$$\sum_{j=0}^{n} \binom{n}{j} \frac{x^j}{j!}$$

are real. This can be done by looking at generalized Laguerre polynomials.

6.6.3. Since the random variables U_i are independent and identically distributed and expectation is linear,

$$\begin{aligned}
&\frac{1}{k!} E[\det(f_i(U_j)) \det(g_i(U_j))] \\
&= \frac{1}{k!} E\left[\sum_{\pi}\sum_{\sigma} \operatorname{sign}(\pi)\operatorname{sign}(\sigma) \prod_{i=1}^{k} f_{\pi(i)}(U_i) g_{\sigma(i)}(U_i)\right] \\
&= \frac{1}{k!} \sum_{\pi}\sum_{\sigma} \operatorname{sign}(\pi)\operatorname{sign}(\sigma) \prod_{i=1}^{k} E\left[f_{\pi(i)}(U) g_{\sigma(i)}(U)\right] \\
&= \sum_{\tau} \operatorname{sign}(\tau) \prod_{i=1}^{k} E[f_i(U) g_{\tau(i)}(U)] \\
&= \det(E[f_i(U) g_j(U)]).
\end{aligned}$$

In the second last step, we use the fact that for a given permutation τ of $\{1, 2, \ldots, k\}$, the product $\prod_{i=1}^{k} a_{i,\tau(i)}$ equals $\prod_{i=1}^{k} a_{\pi(i),\sigma(i)}$ whenever $\tau = \sigma\pi^{-1}$, and hence, $\prod_{i=1}^{k} a_{i,\tau(i)}$ occurs exactly $k!$ times in the multiset $\{\prod_{i=1}^{k} a_{\pi(i),\sigma(i)}\}$ as π and σ range over all possible permutations.

Let $U, U_1, U_2, \ldots, U_k$ be independent uniformly distributed random variables on $\{1, 2, \ldots, k\}$. Using the same notation in Formula 6.6.5, let $f_i(s) = f_{is}$, where $(f_{i1}, f_{i2}, \ldots, f_{in})$ is the ith row in the matrix F and $g_j(s) = g_{sj}$, where $(g_{1j}, g_{2j}, \ldots, g_{nj})^T$ is the jth column of G. Then

$$\frac{1}{k!} E[\det(f_i(U_j)) \det(g_i(U_j))] = \frac{1}{k! n^k} \sum_{(a_1, a_2, \ldots, a_k)} \det(f_i(a_j)) \det(g_i(a_j)),$$

where the sum ranges over all k-tuples $(a_1, a_2, \ldots, a_k)$ with $a_i \in \{1, 2, \ldots, n\}$. If two coordinates a_i and a_j are the same, then the determinant is zero. Therefore, we can assume that the coordinates α_i are distinct. If we exchange two coordinates, then the two determinants change sign simultaneously, and hence their product remains the same. Hence, we can rearrange each k-tuple in increasing order, with each k-tuple $(a_1, a_2, \ldots, a_k)$ such that $a_1 < a_2 < \cdots < a_k$

occurring $k!$ times. When this is done, we have

$$\begin{aligned}
&\frac{1}{k!}E[\det(f_i(U_j))\det(g_i(U_j))]\\
&\quad= \frac{1}{n^k}\sum_{(a_1,a_2,\ldots,a_k)} \det(f_i(a_j))\det(g_i(a_j))\\
&\quad= \frac{1}{n^k}\sum_{(a_1,a_2,\ldots,a_k)} \det(f_{i,a_j})\det(g_{a_j,i})\\
&\quad= \frac{1}{n^k}\sum_{K:\,K=\{a_1,a_2,\ldots,a_k\}} \det F[\{1,2,\ldots,k\}|K]\det G[K|\{1,2,\ldots,k\}]).
\end{aligned}$$

On the right side, we have

$$\begin{aligned}
\det(E[f_i(U)g_j(U)]) &= \det\left(\frac{1}{n}\sum_{a=1}^{n} f_i(a)g_i(a)\right)\\
&= \frac{1}{n^k}\det\left(\sum_{a=1}^{n} f_{ia}g_{ai}\right) = \frac{1}{n^k}\det FG.
\end{aligned}$$

Bibliography

The bibliography is in three parts. The first part is a list of papers of Rota that provide the framework for this book. We include all the papers in the Foundations series for completeness. The second part consists of books of special relevance. The last is a list of papers or books cited in this book.

I. Papers of Gian-Carlo Rota and coworkers

Foundations

I. G.-C. Rota, On the foundations of combinatorial theory. I. *Z. Wahrscheinlichkeitstheo. Verwandte Geb.* 2 (1964) 340–368.

II. H.H. Crapo and G.-C. Rota, On the foundations of combinatorial theory. II. Combinatorial geometries, *Stud. Appl. Math.* 49 (1970) 109–133; a longer book version in typescript was published by M.I.T. Press, Cambridge, MA, 1970.

III. R. Mullin and G.-C. Rota, On the foundations of combinatorial theory. III. Theory of binomial enumeration, in B. Harris, ed., *Graph Theory and Its Applications*, Academic Press, New York, 1970, pp. 167–213.

IV. J.R. Goldman and G.-C. Rota, On the foundations of combinatorial theory. IV. Finite vector spaces and Eulerian generating functions, *Stud. Appl. Math.* 49 (1970) 239–258.

V. G.E. Andrews, On the foundations of combinatorial theory. V. Finite vector spaces and Eulerian generating functions, *Stud. Appl. Math.* 50 (1971) 345–375.

VI. P. Doubilet, G.-C. Rota, and R. Stanley, On the foundations of combinatorial theory. VI. The idea of generating function, in *Proceedings of the Sixth Berkeley Symposium on Mathematical Statistics and Probability (University of California, Berkeley, California, 1970/1971), Vol. II: Probability Theory*, University of California Press, Berkeley, CA, 1972, pp. 267–318.

VII. P. Doubilet, On the foundations of combinatorial theory. VII. Symmetric functions through the theory of distribution and occupancy, *Stud. Appl. Math.* 51 (1972) 377–396.

VIII. D. Kahaner, G.-C. Rota, and A. Oldyzko, On the foundations of combinatorial theory. VIII. Finite operator calculus, *J. Math. Anal. Appl.* 42 (1973) 684–760.

IX. P. Doubilet, G.-C. Rota, and J. Stein, On the foundations of combinatorial theory. IX. Combinatorical methods invariant theory, *Stud. Appl. Math.* 53 (1974) 185–216.

X. F. Bonetti, G.-C. Rota, D. Senato, and A.M. Venezia, On the foundations of combinatorial theory. X. A categorical setting for symmetric functions, *Stud. Appl. Math.* 86 (1992) 1–29.

Matching theory

L.H. Harper and G.-C. Rota, Matching theory: An introduction, in P. Ney, ed., *Advances in Probability and Related Topics*, Vol. 1, Dekker, New York, 1971, pp. 169–215.

Twelve problems

G.-C. Rota, Twelve problems in probability no one likes to bring up, in H. Crapo and D. Senato, eds., *Algebraic Combinatorics and Computer Science*, Springer Italia, Milan, 2001, pp. 57–93.

Selected papers

I. Gian-Carlo Rota on Combinatorics. Introductory Papers and Commentaries, J.P.S. Kung, ed., Birkhäuser, Boston and Basel, 1995.

II. Gian-Carlo Rota on Analysis and Probability, J. Dhombres, J.P.S. Kung, and N. Starr, eds., Birkhäuser, Boston and Basel, 2003.

II. Books for further reading

M. Aigner, *Combinatorial Theory*, Springer, Berlin, 1979; reprinted Springer, Berlin, 1997.

I. Anderson, *Combinatorics of Finite Sets*, Clarendon Press, Oxford, 1987. [3.2, 3.3]

G.E. Andrews, R. Askey, and R. Roy, *Special Functions*, Cambridge University Press, Cambridge, 1999. [6.1, 6.4]

R. Balbes and P. Dwinger, *Distributive Lattices*, University of Missouri Press, Columbia, MI, 1974. [1.3]

C. Berge, *Principles of Combinatorics*, Academic Press, New York, 1971. [5.3]

G. Birkhoff, *Lattice Theory*, 1st edition, American Mathematical Society, New York, 1940; 2nd edition, Providence, RI, 1948; 3rd edition, Providence, RI, 1967. [Chapters 1 and 3]

R. Brualdi and H.J. Ryser, *Combinatorial Matrix Theory*, Cambridge University Press, Cambridge, 1991. [2.3, 2.6]

P.M. Cohn, *Universal Algebra*, 2nd edition, Reidel, Dordrecht, 1981. [1.3, 5.4]

P. Crawley and R.P. Dilworth, *Algebraic Theory of Lattices*, Prentice-Hall, Englewood Cliffs, NJ, 1973. [Chapters 1 and 3]

B.A. Davey and H.A. Priestley, *Introduction to Lattices and Order*, 2nd edition, Cambridge University Press, New York, 2002. [Chapter 1]

R.P. Dilworth, *The Dilworth Theorems*, K.P. Bogart, R. Freese, and J.P.S. Kung, eds., Birkhäuser, Boston and Basel, 1990. [Chapters 1 and 3]

K. Engel, *Sperner Theory*, Cambridge University Press, Cambridge, 1997. [3.3]

R. Freese, J. Ježek, and J.B. Nation, *Free Lattices*, American Mathematical Society, Providence, RI, 1995. [1.3, 3.4]

R.L. Graham, D.E. Knuth, and O. Patashnik, *Concrete Mathematics: A Foundation for Computer Science*, Addison-Wesley, Reading, MA, 1988. [4.1]

G. Grätzer, *General Lattice Theory*, 2nd edition, Birkhäuser, Basel, 2003. [Chapters 1 and 3]

G.H. Hardy, J.E. Littlewood, and G. Pólya, *Inequalities*, 2nd edition, Cambridge University Press, Cambridge, 1952. [2.6]

S. Karlin, *Total Positivity*, Stanford University Press, Stanford, CA, 1968. [Chapter 6]

W. Ledermann, *Introduction to Group Characters*, 2nd edition, Cambridge University Press, Cambridge, 1987. [Chapter 5]

L. Lovász and M.D. Plummer, *Matching Theory*, North-Holland, Amsterdam and New York, 1986. [Chapter 2]

I.G. MacDonald, *Symmetric Functions and Hall Polynomials*, 2nd edition, Oxford University Press, New York, 1995. [Chapter 5]

P.A. MacMahon, *Collected Papers, Vol. I, Combinatorics, Vol. 2, Number Theory, Invariants and Applications*, G.E. Andrews, ed., M.I.T. Press, Cambridge, MA, 1978, 1986. [4.3, 4.4, 4.5, 6.1, 6.3]

M. Marden, *Geometry of Polynomials*, American Mathematical Society, Providence, RI, 1949. [6.1, 6.3]

A.W. Marshall and I. Olkins, *Inequalities: Theory of Majorization and Its Applications*, Academic Press, New York and London, 1979. [2.6]

H. Minc, *Permanents*, Addison-Wesley, Reading, MA, 1978. [Chapter 2]

H. Minc, *Nonnegative Matrices*, Wiley, New York, 1988. [Chapter 2]

L. Mirsky, *Transversal Theory*, Academic Press, New York, 1971. [Chapter 2]

T. Muir and W.H. Metzler, *A Treatise on the Theory of Determinants*, Longmans, London, 1933; reprinted Dover, New York, 1960. [5.2]

O. Ore, *Theory of Graphs*, American Mathematical Society, Providence, RI, 1962. [1.4, 2.2]

J.G. Oxley, *Matroid Theory*, Oxford University Press, Oxford 1992. [2.4, 3.5]

M. Petkovšek, H.S. Wilf, and D. Zeilberger, $A = B$, AK Peters, Wellesley, MA, 1996. [Chapter 4]

G. Pólya and G. Szegö, *Problems and Theorems in Analysis II*, revised English edition, Springer, Berlin and New York, 1976. [Chapter 6]

A. Rényi, *A Diary on Information Theory*, Académiai Kiadó, Budapest, 1984; reprinted Wiley, New York, 1987. [1.4]

B.E. Sagan, *The Symmetric Group*, 2nd edition, Springer, New York, 2001. [Chapter 5]

V.N. Saliĭ, *Lattices with Unique Complements*, Translations of Mathematical Monographs, Vol. 69, American Mathematical Society, Providence, RI, 1988. [1.1, 1.3]

R. Stanley, *Enumerative Combinatorics, Volume I*, Wadsworth and Brooks/Cole, Monterey CA, 1986; reprinted, Cambridge University Press, Cambridge, 1997. [Chapters 3 and 4]

R. Stanley, *Enumerative Combinatorics, Volume II*, Cambridge University Press, Cambridge, 1999. [Chapters 4 and 5]

M. Stern, *Semimodular Lattices, Theory and Applications*, Cambridge University Press, Cambridge, 1999. [3.5]

W.T. Trotter, *Combinatorics and Partially Ordered Sets, Dimension Theory*, Johns Hopkins University Press, Baltimore, MD, 1992. [1.2, 3.2]

H.S. Wilf, *generatingfunctionology*, 3rd edition, AK Peters, Wellesley, MA, 2006. [4.1]

III. References

The sections where the book or paper is discussed or referenced are indicated in square brackets. Thus, the paper of Aberth is discussed in Section 5.7.

O. Aberth, The elementary symmetric functions in a finite field of prime order, *Illinois J. Math.* 8 (1964) 132–138. [5.7]

M. Aguiar and W. Moreira, Combinatorics of the free Baxter algebra, *Electron. J. Comb.* 13 (2006) #R17. [5.5]

M. Aigner, *Combinatorial Theory*, Springer, Berlin, 1979; reprinted Springer, Berlin, 1997. [3.1]

M. Aissen, I.J. Schoenberg, and A. Whitney, On the generating functions of totally positive sequences, *J. Anal. Math.* 2 (1952) 93–103. [6.9]

G.E. Andrews, R. Askey, and R. Roy, *Special Functions*, Cambridge University Press, Cambridge, 1999. [6.1]

L. Alhfors, *Complex Analysis*, 2nd edition, McGraw-Hill, New York, 1966. [6.3]

S. Altmann and E.L. Ortiz, eds., *Mathematics and Social Utopia in France, Olinde Rodrigues and His Times,* American Mathematical Society, Providence, RI, 2005. [4.3]

M. Atiyah and I.G. MacDonald, *Introduction to Commutative Algebra*, Addison-Wesley, Reading, MA, 1969. [1.1]

R. Balbes and P. Dwinger, *Distributive Lattices*, University of Missouri Press, Columbia, MI, 1974. [1.3]

P. Appell, Sur une classe de polynômes, *Ann. Sci. École Norm. Super.* 9 (1880) 119–144. [4.4]

K. Ball, Completely monotonic rational functions and Hall's marriage theorem, *J. Comb. Theory Ser. B* 61 (1994) 118–124. [2.6]

R.B. Bapat, König's theorem and bimatroids, *Linear Algebr. Appl.* 212/213 (1994) 353–365. [2.3]

M. Barnabei, A. Brini, and G.-C. Rota, Systems of section coefficients. I, II. *Rend. Circ. Mat. Palermo (2)* 29 (1980) 457–484, 30 (1981) 161–198. [4.3]

M. Barnabei, A. Brini and G.-C. Rota, On the exterior calculus of invariant theory, *J. Algebr.* 96 (1985) 120–160. [3.4]

O. Barndorff-Nielsen and G. Baxter, Combinatorial lemmas in higher dimensions, *Trans. Am. Math. Soc.* 106 (1963) 313–325. [5.6]

L.D. Baumert, R.J. McEliece, E.R. Rodemich, and H. Rumsey, A probabilistic version of Sperner's lemma, *Ars Comb.* 9 (1980) 91–100. [3.3]

G. Baxter, An analytic problem whose solution follows from a simple algebraic identity, *Pac. J. Math.* 10 (1960) 731–742. [5.4]

E.T. Bell, Exponential polynomials, *Ann. Math.* 35 (1934) 258–277. [4.1]

E.T. Bell, Exponential numbers, *Trans. Am. Math. Soc.* 41 (1934) 411–419. [4.1]

E.T. Bell, Postulational bases for the umbral calculus, *Am. J. Math.* 62 (1940) 717–724. [4.2]

C. Berge, *Principles of Combinatorics*, Academic Press, New York, 1971. [5.3]

F. Bergeron, G. Labelle, and P. Leroux, *Combinatorial Species and Tree-like Structures*, Cambridge University Press, Cambridge, 1998. [4.3]

G. Birkhoff, On the combination of subalgebras, *Proc. Camb. Phil. Soc.* 29 (1933) 441–464. [1.3, 3.4]

G. Birkhoff, Applications of lattice algebras, *Proc. Camb. Phil. Soc.* 30 (1934) 115–122. [1.1]

G. Birkhoff, On the structure of abstract algebras, *Proc. Camb. Phil. Soc.* 31 (1935) 433–454. [3.4]

G. Birkhoff, *Lattice Theory*, 1st edition, American Mathematical Society, New York, 1940. [1.3]

G. Birkhoff, Tres observaciones sobre el algebra lineal (Three observations on linear algebra), *Univ. Nac. Tucumán Rev. A* 5 (1946) 147–151. [2.6]

G. Birkhoff, *Lattice Theory*, 2nd edition, American Mathematical Society, New York, 1948. [1.3]

G. Birkhoff, *Lattice Theory*, 3rd edition, American Mathematical Society, Providence, RI, 1967. [3.4]

G. Birkhoff and M. Ward, A characterization of Boolean algebras, *Ann. Math. (2)* 40 (1939) 609–610. [1.3]

A. Björner, Topological methods, in *Handbook of Combinatorics*, Elsevier, Amsterdam, 1995, pp. 1819–1872. [3.1]

R.P. Boas and R.C. Buck, *Polynomial Expansions of Entire Functions*. Springer, New York, 1964. [4.5]

K.P. Bogart, R. Freese, and J.P.S. Kung, eds., *The Dilworth Theorems*, Birkhäuser, Boston and Basel, 1991. [1.3, 3.2]

K.P. Bogart, C. Greene, and J.P.S. Kung, The impact of the chain decomposition theorem on classical combinatorics, in *The Dilworth Theorems*, Birkhäuser, Boston, 1990, pp. 19–29. [3.3]

G. Boole, *An Investigation into the Laws of Thought*, Macmillan, London, 1854. [1.1]

G. Boole, *Calculus of Finite Differences*, 2nd edition, London, 1872; reprinted, Chelsea, New York, 1970. [4.2]

F. Brenti, *Unimodal, Log-Concave and Pólya Frequency Sequences in Combinatorics*, *Mem. Am. Math. Soc.* 81 (1989) no. 413, American Mathematical Society, Providence RI, 1989. [6.9]

R. Brualdi, Transversal theory and graphs, in D.R. Fulkerson, ed., *Studies in Graph Theory*, Mathematical Association of America, Washington, DC, 1975, pp. 23–88. [Chapter 2]

R. Brualdi and H.J. Ryser, *Combinatorial Matrix Theory*, Cambridge University Press, Cambridge, 1991. [2.7]

R. Brualdi and B.L. Shader, *Matrices of Sign-Solvable Linear Systems*, Cambridge University Press, Cambridge, 1995. [2.3]

R. Canfield, On a problem of Rota, *Adv. Math.* 29 (1978) 1–10. [3.5]

L. Carlitz, On Schur's expansion of $\sin \pi x$, *Bull. Un. Mat. Ital.* 21 (1966) 353–357. [4.3]

P. Cartier, On the structure of Baxter algebras, *Adv. Math.* 9 (1972) 253–265. [5.5]

C. Cercignani, *Ludwig Boltzmann: The Man Who Trusted Atoms*, Oxford University Press, Oxford, 1998. [1.4]

K.-T. Chen, *Collected Papers of K.-T. Chen*, P. Tondeur, ed., Birkhäuser, Boston, 2001. [5.4]

C. Chevalley, Invariants of finite groups generated by reflections, *Am. J. Math.* 77 (1955) 778–782 [5.2]

P.M. Cohn, *Universal Algebra*, 2nd edition, Reidel, Dordrecht, 1981. [1.3]

G. Converse and M. Katz, Symmetric matrices with given row sums, *J. Comb. Theory Ser. A* 18 (1975) 171–176. [2.6]

D. Cox, J. Little, and D. O'Shea, *Ideals, Varieties, and Algorithms*, Springer, New York, 1992. [1.2]

H.H. Crapo, The Möbius function of a lattice, *J. Comb. Theory* 1 (1966) 126–131. [3.1]

H.H. Crapo, Möbius inversion in lattices, *Arch. Math.* 19 (1968) 595–607. [3.1]

H.H. Crapo, Permanents by Möbius inversion, *J. Comb. Theory* 4 (1968) 198–200. [3.1]

H.H. Crapo, Erecting geometries, in *Proceedings of the Second Chapel Hill Conference on Combinatorics and Its Applications, 1970*, University of North Carolina, Chapel Hill, NC, 1970, pp. 74–99. [1.5]

H. Crapo, Unities and negations: On the representations of finite lattices, *J. Pure Appl. Algebr.* 23 (1982) 109–135. [1.5]

H.H. Crapo and G.-C. Rota, *On the Foundations of Combinatorial Theory. Combinatorial Geometries*, Preliminary edition, MIT Press, Cambridge, MA, 1970. [2.4]

P. Crawley and R.P. Dilworth, *Algebraic Theory of Lattices*, Prentice-Hall, Englewood Cliffs, NJ, 1973. [1.3, 3.4, 3.5]

G.B. Dantzig, *Linear Programming and Extensions*, Princeton University Press, Princeton, NJ, 1963. [2.6]

R.L. Davies, Order algebras, *Bull. Am. Math. Soc.* 76 (1970) 83–87. [3.6]

A.C. Davis, A characterization of complete lattices, *Pac. J. Math.* 5 (1955) 311–319. [1.3]

A. Day, Geometric applications in modular lattices, in *Universal Algebra and Lattice Theory (Puebla, 1982)*, Lecture Notes in Mathematics 1004, Springer, Berlin and New York, 1983, pp. 111–141. [3.4]

A. Day and R. Freese, The role of gluing constructions in modular lattice theory, in *The Dilworth Theorems*, Birkhäuser, Boston and Basel, 1990, pp. 251–260. [3.5]

A. Day and D. Pickering, The coordinatization of Arguesian lattices, *Trans. Am. Math. Soc.* 278 (1983) 507–522. [3.4]

R. Dedekind, Über Zerlegungen von Zahlen durch ihre grössten gemeinsamen, *Festschrift Technische Hochschule Braunschweig*, 1897; reprinted *Gesammelte Werke*, Vol. 2, Vieweg, Braunschweig, 1931, pp. 103–148. [3.4]

R. Dedekind, Über die von drei Moduln erzeugte Dualgruppe, *Math. Ann.* 53 (1900) 371–403; reprinted *Gesammelte Werke*, Vol. 2, Vieweg, Braunschweig, 1931, pp. 236–271. [1.3, 3.4, 3.5]

N.G. de Bruijn, C. van Ebbenhorst Tengbergen, and D. Kruyswijk, On the set of divisors of a number, *Nieuw Arch. Wiskd.* (2) 23 (1949) 191–193. [3.3]

A. De Morgan, On the Syllogism: IV, and on the logic of relations, *Trans. Camb. Phil. Soc.* 10 (1864) 331–358. [1.5]

E. Deutsch and B.E. Sagan, Congruences for Catalan and Motzkin numbers and related sequences, *J. Number Theory* 117 (2006) 191–215. [5.3]

L.E. Dickson, *Linear Groups: With an Exposition of the Galois Field Theory*, Teubner, Leipzig 1901; reprinted Dover, New York, 1958. [5.7]

L.E. Dickson, A fundamental system of invariants of the general modular linear group with a solution of the form problem, *Trans. Am. Math. Soc.* 12 (1911) 75–98. [5.7]

L.E. Dickson, Finiteness of the odd perfect and primitive abundant numbers with n distinct prime factors, *Am. J. Math.* 35 (1913) 413–422. [1.2]

R.P. Dilworth, Lattices with unique complements, *Trans. Am. Math. Soc.* 57 (1945) 123–154. [1.3]

R.P. Dilworth, A decomposition theorem for partially ordered sets, *Ann. Math.* 51 (1950) 161–166. [1.2, 3.2]

R.P. Dilworth, Some combinatorial problems on partially ordered sets, in *Combinatorial Analysis, Proceedings of the Tenth Symposium on Applied Mathematics*, American Mathematical Society, Providence RI, 1960, pp. 85–90. [3.2, 3.3]

R.P. Dilworth, *The Dilworth Theorems*, K.P. Bogart, R. Freese, and J.P.S. Kung, eds., Birkhäuser, Boston and Basel, 1990. [3.2]

R.P. Dilworth and C. Greene, A counterexample to the generalization of Sperner's theorem, *J. Comb. Theory* 10 (1971) 18–21. [3.5]

S.J. Dow and P.M. Gibson, Permanents of d-dimensional matrices, *Linear Algebr. Appl.* 90 (1987) 133–145. [2.8]

T.A. Dowling, A class of geometric lattices based on finite groups, *J. Comb. Theory Ser. B* 14 (1973) 61–86. [5.2]

T.A. Dowling, Complementing permutations in finite lattices, *J. Comb. Theory Ser. B* 23 (1977) 223–226. [3.1]

T.A. Dowling and R.M. Wilson, Whitney number inequalities for geometric lattices, *Proc. Am. Math. Soc.* 47 (1975) 504–512. [3.5]

M. Downs, The Möbius function of $PSL_2(q)$, with application to the maximal normal subgroups of the modular group, *J. Lond. Math. Soc. (2)* 43 (1991) 61–75. [3.1]

P. Dubreil and M.-L. Dubreil-Jacotin, Théorie algébrique des relations d'équivalence, *J. Math. Pures Appl.* 18 (1939) 63–95. [3.4]

B. Dushnik and E.W. Miller, Partially ordered sets, *Am. J. Math.* 63 (1941) 600–610. [1.2]

T.E. Easterfield, A combinatorial algorithm, *J. Lond. Math. Soc.* 21 (1946) 219–226. [2.2]

K. Ebrahimi-Fard and L. Guo, Rota-Baxter algebras and dendriform algebras, *J. Pure Appl. Algebr.* 212 (2008) 320–339. [5.5]

P.H. Edelman, Zeta polynomials and the Möbius function, *Eur. J. Comb.* 1 (1980) 335–340. [3.1]

J. Edmonds, Systems of distinct representatives and linear algebra, *J. Res. Natl. Bur. Stand.* 71B (1967) 241–245. [2.3]

J. Edmonds, Submodular functions, matroids, and certain polyhedra, in *Combinatorial Structures and Their Applications*, R. Guy, H. Hanani, N. Sauer, and J. Schönheim, eds., Gordon and Breach, New York, 1970, pp. 69–87. [2.4]

A. Edrei, On the generating functions of totally positive sequences II, *J. Anal. Math.* 2 (1952) 104–109. [6.9]

G.P. Egorychev, Van der Waerden conjecture and applications, in *Handbook of Algebra*, Vol. 1, North-Holland, Amsterdam, 1996, pp. 3–26. [2.6]

R. Ellerman and G.-C. Rota, A measure-theoretic approach to logical quantification, *Rend. Sem. Mat. Univ. Padova* 59 (1978) 227–246. [3.6]

P. Erdős, On a lemma of Littlewood and Offord, *Bull. Am. Math. Soc.* 51 (1945) 898–902. [3.3]

P. Erdős and G. Szekeres, A combinatorial problem in geometry, *Compos. Math.* 2 (1935) 464–470. [3.2]

C.J. Everett, Closure operators and Galois theory in lattices, *Trans. Am. Math. Soc.* 55 (1944) 514–525. [3.6]

C.J. Everett and G. Whaples, Representations of sequences of sets, *Am. J. Math.* 71 (1949) 287–293. [2.2]

H.K. Farahat and L. Mirsky, Permutation endomorphisms and refinement of a theorem of Birkhoff, *Proc. Camb. Phil. Soc.* 56 (1960) 322–328. [2.6]

R.B. Feinberg, Polynomial identities of incidence algebras, *Proc. Am. Math. Soc.* 55 (1976) 25–28. [3.1]

R.B. Feinberg, Characterization of incidence algebras, *Discrete Math.* 17 (1977) 47–70. [3.1]

D. Fendel, The number of classes of linearly equivalent functions, *J. Comb. Theory* 3 (1967) 48–53. [5.3]

J.V. Field and J.J. Gray, *The Geometric Work of Girard Desargues*, Springer, Berlin and New York, 1987. [3.4]

N.J. Fine, On the asymptotic distribution of the elementary symmetric function (mod *p*), *Trans. Am. Math. Soc.* 69 (1950) 109–129. [5.7]

D. Foata, A combinatorial proof of the Mehler formula, *J. Comb. Theory Ser. A* 24 (1978) 367–376. [4.4]

D. Foata and M.-P. Schützenberger, *Théorie Géométrique des Polynômes Eulériens*, Lecture Notes in Mathematics, Vol. 138, Springer, Berlin and New York, 1970. [4.1]

J. Folkman, The homology groups of a lattice, *J. Math. Mech.* 15 (1966) 631–636. [3.1]

S.V. Fomin, Finite, partially ordered sets and Young diagrams, *Dokl. Akad. Nauk SSSR* 243 (1978) 1144–1147. [3.2]

L.R. Ford and D.R. Fulkerson, Network flows and systems of representatives, *Can. J. Math.* 10 (1958, 78–85. [2.3]

A. Frank, On chain and antichain families of a partially ordered set, *J. Comb. Theory Ser. B* 29 (1980) 176–184. [3.2]

R. Freese, An application of Dilworth's lattice of maximal antichains, *Discrete Math.* 7 (1974) 107–109. [3.2, 3.3]

R. Freese, J. Ježek, and J.B. Nation, *Free Lattices*, American Mathematical Society, Providence, RI, 1995. [1.3, 3.4]

G. Frobenius, Über zerlegbare determinanten, *Sitzungsber. Preuss. Akad. Wiss. Berl.* (1917) 274–277. [2.3]

R. Frucht and G.-C. Rota, La función de Möbius para el retículo di particiones de un conjunto finito, *Scientia (Valparaiso)* 122 (1963) 111–115. [5.2]

D.R. Fulkerson, Note on Dilworth's decomposition theory for partially ordered sets, *Proc. Am. Math. Soc.* 7 (1956) 701–702. [3.2]

D. Gale, A theorem on flows in networks, *Pac. J. Math.* 7 (1957) 1073–1082. [2.7]

F. Galvin, A proof of Dilworth's chain decomposition theorem, *Am. Math. Mon.* 101 (1994) 353–353. [3.2]

F.R. Gantmacher and M.G. Krein, Sur les matrices complétement non-négative et oscillatoires, *Compos. Math.* 4 (1937) 445–476. [6.7]

W. Gaschütz, Die Eulersche Funktion endlicher auflösbarer Gruppen, *Illinois J. Math.* 3 (1959) 469–476. [3.1]

J. Geelen, B. Gerards, and G. Whittle, Towards a structure theory for matrices and matroids, *International Congress of Mathematicians*, Vol. III, European Mathematical Society, Zürich, 2006, pp. 827–842. [1.2]

L. Geissinger, Valuations of distributive lattices. I. *Arch. Math. (Basel)* 24 (1973) 230–239; II. 24 (1973) 337–345; III. 24 (1973) 475-481. [3.6]

I.M. Gelfand, M.M. Kapranov, and A.V. Zelevinsky, *Discriminants, Resultants and Multidimensional Determinants*, Birkhäuser, Boston, 1994. [2.8]

I.M. Gelfand and V.A. Ponomarev, Problems of linear algebra and classification of quadruples of subspaces in a finite-dimensional vector space, in *Hilbert Space Operators and Operator Algebras (Proceedings of the International Conference, Tihany, 1970)*, Vol. 5, Colloquia Mathematica Societatis János Bolyai, North-Holland, Amsterdam, 1972, pp. 163–237. [3.4]

I.M. Gelfand and V.A. Ponomarev, Free modular lattices, and their representations, *Uspehi Mat. Nauk* 29 (1974) 3–58. [3.4]

P.M. Gibson, Conversion of the permanent into the determinant, *Proc. Am. Math. Soc.* 27 (1971) 471–476. [2.3]

P. Gordan, *Vorlesungen über Invariantentheorie*, G. Kershensteiner, ed., Vols. I, II, Teubner, Leipzig, 1885, 1887. [1.2]

H.W. Gould, A binomial identity of Greenwood and Gleason, *Math. Stud.* 29 (1961) 53–57. [4.5]

J.H. Grace, The zeros of a polynomial, *Proc. Camb. Phil. Soc.* 11 (1902) 242–247. [6.3]

K.M. Gragg and J.P.S. Kung, Consistent dually semimodular lattices, *J. Comb. Theory Ser. A* 60 (1992) 246–263. [1.3, 3.5]

R.L. Graham, D.E. Knuth, and O. Patashnik, *Concrete Mathematics: A Foundation for Computer Science*, Addison-Wesley, Reading, MA, 1988. [4.1]

H. Grassmann, *Ausdehnungslehre*, Enslin, Berlin, 1862; English translation by L.C. Kannenberg, *Extension Theory*, American Mathematical Society, Providence, RI, 2000. [3.4]

G, Grätzer, *General Lattice Theory*, 2nd edition, Birkhäuser, Basel, 2003. [1.3]

G. Grätzer, Two problems that shaped a century of lattice theory, *Not. Am. Math. Soc.* 54 (2007) 696–707. [1.3]

C. Greene, On the Möbius algebra of a partially ordered set, *Adv. Math.* 10 (1973) 177–187. [3.6]

C. Greene, An extension of Schensted's theorem, *Adv. Math.* 14 (1974) 254–265. [3.2]

C. Greene, Some partitions associated with a partially ordered set, *J. Comb. Theory Series A* 20 (1976) 69–79. [3.2]

C. Greene, A class of lattices with Möbius function $\pm 1, 0$, *Eur. J. Comb.* 9 (1988) 225–240. [3.1]

C. Greene and D.J. Kleitman, The structure of Sperner k-families, *J. Comb. Theory Ser. A* 20 (1976) 41–68. [3.2]

C. Greene and D.J. Kleitman, Proof techniques in the theory of finite sets, in G.-C. Rota, ed., *Topics in Combinatorics*, Mathematical Association of America, Washington, DC, 1978, pp. 22–79. [3.3]

J.R. Griggs, Sufficient conditions for a symmetric chain order, *SIAM J. Appl. Math.* 32 (1977) 807–809. [3.3]

F. Grosshans, The work of Gian-Carlo Rota in invariant theory, *Algebr. Univ.* 49 (2003) 213–258. [2.8]

S. Gundelfinger, Zur Theorie der binären Formen, *J. Reine Angew. Math.* 100 (1886) 413–424. [6.2]

T. Hailperin, *Boole's Logic and Probability*, 2nd edition, North-Holland, Amsterdam, 1986. [1.1]

M. Haiman, Proof theory for linear lattices, *Adv. Math.* 58 (1985) 209–242. [3.4]

M. Haiman, Two notes on the Arguesian identity, *Algebr. Univ.* 21 (1985) 167–171. [3.4]

M. Haiman, Arguesian lattices which are not type-1, *Algebr. Univ.* 28 (1991) 128–137. [3.4]

M. Haiman and W. Schmitt, Antipodes, incidence coalgebras and Lagrange inversion in one and several variables, *J. Comb. Theory Ser. A* 50 (1989) 172–185. [4.3]

L.H. Haines, On free monoids partially ordered by embedding, *J. Comb. Theory* 6 (1968) 94–98. [1.2]

M. Hall, Distinct representatives of subsets, *Bull. Am. Math. Soc.* 54 (1948) 922–926. [2.2]

M. Hall and R.P. Dilworth, The imbedding problem for modular lattices, *Ann. Math.* 45 (1944) 450–456. [3.4, 3.5]

P. Hall, On representatives of subsets, *J. Lond. Math. Soc.* 10 (1935) 26–20. [2.2]

P. Hall, The Eulerian functions of a group, *Q. J. Math. (Oxford)* 7 (1936) 134–151. [3.1]

P. Hall, The algebra of partitions, in *Proceedings of the 4th Canadian Mathematical Congress, Banff*, University of Toronto Press, Toronto, Ontario, 1959, pp. 147–59; reprinted in *The Collected Works of Philip Hall*, Oxford University Press, Oxford, 1988. [5.1]

P.R. Halmos, *Lectures on Boolean Algebras*, van Nostrand, New York, 1963; reprinted, Springer, New York, 1974. [1.1]

P.R. Halmos and H.E. Vaughan, The marriage problem, *Am. J. Math.* 72 (1950) 214–215. [2.2]

G. Hansel, Sur le nombre des fonctions booléennes monotones de n variables, *C.R. Acad. Sci., Paris* 262 (1966) 1088–1090. [3.3]

F. Harary and E. Palmer, *Graphical Enumeration*, Academic Press, New York, 1973. [5.3]

G.H. Hardy, J.E. Littlewood, and G. Pólya, *Inequalities*, 2nd edition, Cambridge University Press, Cambridge, 1952. [2.6]

G.H. Hardy and E.M. Wright, *An Introduction to the Theory of Numbers*, 4th edition, Oxford University Press, Oxford, 1960, p. 257. [4.1]

J. Hartmanis, Two embedding theorems for finite lattices, *Proc. Am. Math. Soc.* 7 (1956) 571–577. [1.5]

J. Hartmanis, A note on the lattice of geometries, *Proc. Am. Math. Soc.* 8 (1957) 560–562. [1.5]

M. J. Hawrylycz, Geometric identities, invariant theory, and a theorem of Bricard, *J. Algebr.* 169 (1994) 287–297. [3.4]

M. J. Hawrylycz, Arguesian identities in invariant theory, *Adv. Math.* 122 (1996) 1–48. [3.4]

M. Henle, Binomial enumeration on dissects, *Trans. Am. Math. Soc.* 202 (1975) 1–39. [4.1]

P.J. Higgins, Disjoint transversals of subsets, *Can. J. Math.* 11 (1959) 280–285. [2.7]

G. Higman, Ordering by divisibility in abstract algebras, *Proc. Lond. Math. Soc.* 2 (1952) 326–336. [1.2]

A.J. Hoffman and D.E. Schwartz, On partitions of a partially ordered sets, *J. Comb. Theory Ser. B* 20 (1977) 3–13. [3.2]

A.J. Hoffman and H.W. Wielandt, The variation of the spectrum of a normal matrix, *Duke Math. J.* 20 (1953) 37–39. [2.6]

R. Huang and G.-C. Rota, On the relations of various conjectures on Latin squares and straightening coefficients, *Discrete Math.* 128 (1994) 225–236. [2.4]

E.V. Huntington, A new set of independent postulates for the algebra of logic with special reference to Whitehead and Russell's Principia Mathematica, *Trans. Am. Math. Soc.* 35 (1933) 274–304. [1.1]

A. Hurwitz, Über den Vergleich des arithmetischen und des geometrischen Mittels, *J. Reine Angew. Math.* 108 (1891) 266–268. [2.6]

A.W. Ingleton, Representation of matroids, in D.J.A. Welsh, ed., *Combinatorial Mathematics and Its Applications*, Academic Press, London and New York, 1971, pp. 149–167. [2.8]

J.R. Isbell, Birkhoff's Problem 111, *Proc. Am. Math. Soc.* 6 (1955) 217–218. [2.6]

N. Jacobson, *Algebra I*, W.H. Freeman, New York, 1985. [6.1]

M.L. Jacotin-Dubreil, Quelques propriétés des applications multiformes, *C.R. Acad. Sci., Paris* 230 (1950) 806–808. [1.5, 3.4]

K.L. Johnson, Real representations of finite directed graphs, Ph.D. thesis, University of Alabama, 1971. [1.2]

P.T. Johnstone, *Stone Spaces*, Cambridge University Press, Cambridge, 1982. [1.1]

S.A. Joni and G.-C. Rota, Coalgebras and bialgebras in combinatorics, *Stud. Appl. Math.* 61 (1979) 93–139. [4.3]

B. Jónsson, On the representation of lattices, *Math. Scand.* 1 (1953) 193–206. [1.5, 3.4]

M. Kac, Toeplitz matrices, translation kernels, and a related problem in probability theory, *Duke Math. J.* 21 (1954) 501–509. [5.6]

J. Kahn, Some non-Sperner paving matroids, *Bull. Lond. Math. Soc.* 12 (1980) 268. [3.5]

J. Kahn, On lattices with Möbius function $\pm 1, 0$, *Discrete Comput. Geom.* 2 (1987) 1–8. [3.1]

J. Kahn, Entropy, independent sets and antichains: A new approach to Dedekind's problem, *Proc. Am. Math. Soc.* 130 (2002) 371–378. [3.3]

J. Kahn and J.P.S. Kung, A classification of modularly complemented geometric lattices, *Eur. J. Comb.* 7 (1986) 243–248. [5.2]

S. Karlin, *Total Positivity*, Stanford University Press, Stanford, CA, 1968. [6.5]

P.W. Kasteleyn, Dimer statistics and phase transitions, *J. Math. Phys.* 4 (1963) 287–293. [2.3]

G. Katona, on a conjecture of Erdős and a stronger form of Sperner's theorem, *Stud. Sci. Math. Hung.* 1 (1966) 59–63. [3.3]

M. Katz, On the extreme points of a certain convex polytope, *J. Comb. Theory* 8 (1970) 417–423. [2.6]

M. Katz, On the extreme points of the set of substochastic and symmetric matrices, *J. Math. Anal. Appl.* 37 (1972) 576–579. [2.6]

D.G. Kendall, On infinite doubly-stochastic matrices and Birkhoff's Problem 111, *J. Lond. Math. Soc.* 35 (1960) 81–84. [2.6]

D. Klain and G.-C. Rota, A continuous analogue of Sperner's theorem, *Comm. Pure Appl. Math.* 50 (1997) 205–223,

D.J. Kleitman, On a lemma of Littlewood and Offord on the distribution of certain sums, *Math. Z.* 90 (1965) 251–259. [3.3]

D.J. Kleitman, On Dedekind's problem: The number of monotone Boolean functions, *Proc. Am. Math. Soc.* 21 (1969) 677–682. [3.3]

D.J. Kleitman, On a lemma of Littlewood and Offord on the distribution of linear combinations of vectors, *Adv. Math.* 5 (1970) 155–157. [3.3]

D.J. Kleitman, On an extremal property of antichains in partial orders. The LYM property and some of its implications and applications, in *Combinatorics (Proceedings NATO Advanced Study Institute, Breukelen, 1974), Part 2*, Mathematical Centre Tracts 55, Mathematisch Centrum, Amsterdam, 1974, pp. 77–90. [3.3]

D.J. Kleitman, Some new results on the Littlewood-Offord problem, *J. Comb. Theory Ser. A* 20 (1976) 89–113. [3.3]

D.J. Kleitman, Extremal hypergraph problems, in *Surveys in Combinatorics (Proceedings of the Seventh British Combinatorial Conference, Cambridge, 1979)*, Cambridge University Press, Cambridge, 1979. [Preface]

D.J. Kleitman, On the future of combinatorics, in S. S. Hecker and G.-C. Rota, eds., *Essays on the Future in Honor of Nick Metropolis*, Birkhäuser, Boston, 2000, pp. 123–134. [Preface]

D.J. Kleitman, M. Edelberg, and D. Lubell, Maximal sized antichains in partial orders, *Discrete Math.* 1 (1971) 47–53. [3.2]

D.J. Kleitman and G. Markowsky, On Dedekind's problem: The number of monotone Boolean functions II, *Trans. Am. Math. Soc.* 45 (1974) 373–389. [3.3]

D. Kleitman and B.L. Rothschild, The number of finite topologies, *Proc. Am. Math. Soc.* 25 (1970) 276–282. [1.2]

D. Kleitman and B.L. Rothschild, Asymptotic enumeration of partial orders on a finite set, *Trans. Am. Math. Soc.* 205 (1975) 205–220. [1.2]

B. Knaster, Un théorème sur les fonctions d'ensembles, *Ann. Soc. Polon. Math.* 6 (1928) 133–134. [1.3]

D.E. Knuth, *The Art of Computer Programming*, Vol. 2, 2nd edition, Addison-Wesley, Reading, MA, 1969. [3.1]

D. König, Über Graphen und ihre Anwendung auf Determinantentheorie und Mengenlehre, *Math. Ann.* 77 (1916) 453–465. [2.1, 2.2, 2.3]

J.B. Kruskal, Well-quasi-ordering, the tree theorem and Vázsonyi's conjecture, *Trans. Am. Math. Soc.* 95 (1960) 210–225. [1.2]

J.B. Kruskal, The theory of well-quasi-ordering: A frequently discovered concept, *J. Comb. Theory Ser. A* 13 (1972) 297–305. [1.2]

H.W. Kuhn, The Hungarian method for the assignment problem, *Nav. Res. Logist. Quart.* 2 (1955) 83–97. [2.2]

J.P.S. Kung, Bimatroids and invariants, *Adv. Math.* 30 (1978) 238–249. [2.8]

J.P.S. Kung, On algebraic structures associated with the Poisson process, *Algebr. Univ.* 13 (1981) 137–147. [3.4]

J.P.S. Kung, Matchings and Radon transforms in lattices. I. Consistent lattices, *Order* 2 (1985) 105–112. [1.3, 3.5]

J.P.S. Kung, Gundelfinger's theorem on binary forms, *Stud. Appl. Math.* 75 (1986) 163–170. [6.2]

J.P.S. Kung, Matchings and Radon transforms in lattices. II. Concordant sets, *Math. Proc. Camb. Phil. Soc.* 101 (1987) 221–231. [3.5]

J.P.S. Kung, Matroid theory, in M. Hazewinkel, ed., *Handbook of Algebra,* Vol. 1, North-Holland, Amsterdam, 1996, pp. 157–184. [2.4]

J.P.S. Kung, Critical problems, in J.E. Bonin, J.G. Oxley, and B. Servatius, eds., *Matroid Theory*, Contemporary Mathematics, Vol. 197, American Mathematical Society, Providence, RI, 1996, pp. 1–127. [3.1]

J.P.S. Kung, Differential symmetric functions, *Ann. Comb.* 4 (2000) 285–297. [5.7]

J.P.S. Kung and G.-C. Rota, On the differential invariants of an ordinary differential equation, *Proc. Roy. Soc. Edinburgh* 89A (1984) 111–123. [5.7]

J.P.S. Kung and G.-C. Rota, The invariant theory of binary forms, *Bull. Am. Math. Soc. (N.S.)* 10 (1984) 27–85. [6.1, 6.2]

J.P.S. Kung and C. Yan, Gončarov polynomials and parking functions, *J. Comb. Theory Ser. A* 102 (2003) 16–37. [4.5]

A. Kurosch, Durchschnittsdarstellungen mit irreduziblen Komponenten in Ringen und in sogenannten Dualgruppen, *Mat. Sb.* 42 (1935) 613–616. [1.3]

P.S. Laplace, *Philosophical Essay on Probabilities*, translation by A.I. Dale of *Essai Philosophique sur les Probabilités*, 5th edition, Paris, 1825, Springer, New York and Berlin, 1995. [4.1]

B. Lindström, Determinants on semilattices, *Proc. Am. Math. Soc.* 20 (1969) 207–208. [3.1]

L. Lovász, Three short proofs in graph theory, *J. Comb. Theory Ser. B* 19 (1975) 95–98. [2.2]

L. Lovász, Perfect graphs, in L.W. Beineke and R.J. Wilson, eds., *Selected Topics in Graph Theory 2*, Academic Press, London and New York, 1983, pp. 55–87. [3.2]

D. Lubell, A short proof of Sperner's lemma, *J. Comb. Theory* 1 (1966) 299. [3.3]

F. Lucas, Sur une applications de la méchanique rationelle á la théorie des équationes, *C.R. Acad. Sci., Paris* 89 (1879) 224–226. [6.3]

P.A. MacMahon, The law of symmetry and other theorems in symmetric functions, *Q. J. Math.* 22 (1887) 74–81. [5.2]

H.M. MacNeille, Partially ordered sets, *Trans. Am. Math. Soc.* 42 (1937) 416–460. [1.5]

R.D. Maddux, *Relation Algebras*, Elsevier, Amsterdam, 2006. [1.5]

M. Mainetti and C.H. Yan, Geometric identities in lattice theory, *J. Comb. Theory Ser. A* 91 (2000) 411–450. [3.4]

M. Marcus and M. Newman, On the minimum of the permanent of a doubly stochastic matrix, *Duke Math. J.* 26 (1959) 61–72. [2.6]

M. Marden, *Geometry of Polynomials*, American Mathematical Society, Providence, RI, 1949. [6.1]

A.W. Marshall and I. Olkins, *Inequalities: Theory of Majorization and Its Applications*, Academic Press, New York and London, 1979. [2.6]

J.C.C. McKinsey and A. Tarski, The algebra of topology, *Ann. Math. (2)* 45 (1944) 141–191. [1.3]

J.E. McLaughlin, Atomic lattices with unique comparable complements, *Proc. Am. Math. Soc.* 7 (1956) 864–866. [1.3, 3.4]

P. McMullen, Valuations and dissections, in P.M. Gruber and J.M. Wills, eds., *Handbook of Convex Geometry*, Elsevier, Amsterdam, 1993, pp. 933–988. [6.1]

N. Metropolis and G.-C. Rota, Witt vectors and the algebra of necklaces, *Adv. Math.* 50 (1983) 359–389. [4.1]

L.D. Mešalkin, A generalization of Sperner's theorem on the number of subsets of a finite set (Russian), *Teor. Verojatnost. i Primen.* 8 (1963) 219–220. [3.3]

L. Mirsky, Proofs of two theorems on doubly-stochastic matrices, *Proc. Am. Math. Soc.* 9 (1958) 371–374. [2.6]

L. Mirsky, On a convex set of matrices, *Arch. Math.* 10 (1959) 88–92. [2.6]

L. Mirsky, Even doubly-stochastic matrices, *Math. Ann.* 144 (1961) 418–421. [2.6]

L. Mirsky, Results and problems in the theory of doubly-stochastic matrices, *Z. Wahrscheinlichkeittheor. Verwandte Geb.* 1 (1962/1963) 319–334. [2.6]

L. Mirsky, A dual of Dilworth's decomposition theorem, *Am. Math. Mon.* 78 (1971) 876–877. [3.2]

L. Mirsky, *Transversal Theory*, Academic Press, New York, 1971. [Chapter 2]

L. Mirsky and H. Perfect, Applications of the notion of independence to problems of combinatorial analysis, *J. Comb. Theory* 2 (1967) 327–357. [2.3]

G. Markowsky, The factorization and representation of lattices, *Trans. Am. Math. Soc.* 203 (1975) 185–200. [1.5]

T. Motzkin, Contributions to the theory of linear inequalities, *RAND Corporation Translations 22*, Santa Monica CA, 86 pp., translation of *Beiträge zur Theorie der linearen Ungleichungen*, doctoral thesis, Basel, 1934, printed Azriel, Jerusalem, 1936, 73 pp. Translation reprinted in T.S. Motzkin, *Selected Papers*, D. Cantor, B. Gordon, and B. Rothschild, eds., Birkhäuser, Boston and Basel, 1983. [6.8]

T. Muir and W.H. Metzler, *A Treatise on the Theory of Determinants*, Longmans, London, 1933; reprinted Dover, New York, 1960. [2.8, 5.2]

R.F. Muirhead, Inequalities relating to some algebraic means, *Proc. Edinburgh Math. Soc.* 19 (1901) 36–45. [2.6]

M.R. Murty and Y.-R. Liu, Sieve methods in combinatorics, *J. Comb. Theory Ser. A* 111 (2005) 1–23. [2.2]

C.St.J.A. Nash-Williams, On well-quasi-ordering infinite trees, *Proc. Camb. Phil. Soc.*, 61 (1965) 697–720. [1.2]

C.St.J.A. Nash-Williams, On well-quasi-ordering trees, in F. Harary, ed., *A Seminar on Graph Theory*, Holt, Rinehart and Winston, New York, 1967, pp. 79–82. [1.2]

E. Netto, *The Theory of Substitutions and Its Applications to Algebra*, 2nd edition, English translation by F.N. Cole, A.A. Register, 1892; reprinted Chelsea, Bronx, NY, 1964. [5.7]

T. Ogasawara and U. Sasaki, On a theorem in lattice theory, *J. Sci. Hiroshima Univ. Ser. A* 14 (1949) 13. [1.3]

O. Ore, On a special class of polynomials, *Trans. Am. Math. Soc.* 35 (1933) 559–584. [5.7]

O. Ore, On the foundations of abstract algebra. II, *Ann. Math. (2)* 37 (1936) 265–292. [1.3]
O. Ore, Galois connexions, *Trans. Am. Math. Soc.* 55 (1944) 493–513. [1.5]
O. Ore, Graphs and matching theorems, *Duke Math. J.* 22 (1955), 625–639. [2.2, 2.4]
O. Ore, On coset representatives in groups, *Proc. Am. Math. Soc.* 9 (1958) 665–670. [2.2]
O. Ore, *Theory of Graphs*, American Mathematical Society, Providence, RI, 1962. [1.5]
P. Orlik and L. Solomon, Combinatorics and topology of complements of hyperplanes, *Invent. Math.* 56 (1980) 167–189. [6.6]
P. Orlik and H. Terao, *Arrangements of Hyperplanes*, Springer, New York and Berlin, 1992. [5.2]
P.A. Ostrand, Systems of distinct representatives. II, *J. Math. Anal. Appl.* 32 (1970) 1–4. [2.2]
J.G. Oxley, *Matroid Theory*, Oxford University Press, Oxford, 1992. [2.4]
M.A. Perles, A proof of Dilworth's decomposition theorem for partially ordered sets, *Isr. J. Math.* 1 (1963) 105–107. [3.2]
M. Petkovšek, H.S. Wilf, and D. Zeilberger, $A = B$, AK Peters, Wellesley, MA, 1996. [4.1]
S. Pincherle, *Lo Operazioni Distributive e le loro Applicazioni all'Analisi*, Zanichelli, Bologna, 1901. [3.3]
S. Pincherle, Operatori lineari e coefficienti di fattoriali, *Atti Accad. Naz. Lincei, Rend. Cl. Fis. Mat. Nat. (6)* 18 (1933) 417–519. [3.3]
G. Pólya, Remarques sur une problème d'algebre étudié par Laguerre, *J. Math. Pures Appl.* 31 (9) (1952) 37–47. [5.2]
G. Pólya, On picture writing, *Am. Math. Mon.* 75 (1969) 330–334. [4.1]
G. Pólya and R.C. Read, *Combinatorial Enumeration of Groups, Graphs, and Chemical Compounds*, Springer, New York, 1987. [5.3]
G. Pólya and G. Szegö, *Problems and Theorems in Analysis*, Vol. II, Springer, Berlin, 1976. [3.1]
E.L. Post, *Solvability, Provability, Definability: The Collected Works of Emil L. Post*, M. Davis, ed., Birkhäuser, Boston, 1994. [1.1]
A. Postnikov and B.E. Sagan, What power of two divides a weighted Catalan number? *J. Comb. Theory Ser. A* 114 (2007) 970–977. [5.3]
H.S.A. Potter, On the latent roots of quasi-commutative matrices, *Am. Math. Mon.* 57 (1950) 321–322. [4.1]
O. Pretzel, Another Proof of Dilworth's decomposition theorem, *Discrete Math.* 25 (1979) 91–92. [3.2]
C. Procesi, Positive symmetric functions, *Adv. Math.* 29 (1978) 219–225. [2.6]
P. Pudlák and J. Tůma, Every finite lattice can be embedded in a finite partition lattice, *Algebr. Univ.* 10 (1980) 74–95. [1.5]
R. Rado, A theorem on general measure functions, *Proc. Lond. Math. Soc. (2)* 44 (1938) 61–91. [2.5]
R. Rado, A theorem on independence relations, *Q. J. Math. Oxford* 13 (1942) 83–89. [2.4]
R. Rado, An inequality, *J. Lond. Math. Soc.* 27 (1952) 1–6. [2.6]
R. Rado, On the number of systems of distinct representatives of sets, *J. Lond. Math. Soc.* 42 (1967) 107–109.

R. Rado, Note on the transfinite case of Hall's theorem on representatives, *J. Lond. Math. Soc.* 42 (1967) 321–324. [2.2]

R.M. Redheffer, Eine explizit lösbare Optimierungsaufgabe, *Int. Schriftenreihe Numer. Math.* 36 (1977) 213–216. [3.1]

A. Rényi, *A Diary of Information Theory*, Académiai Kiadó, Budapest, 1984; reprinted Wiley, New York, 1987. [1.4]

K. Reuter, The Kurosh-Ore exchange property, *Acta Math. Hung.* 53 (1989) 119–127. [1.3]

J. Riguet, Quelques propriétés des relations difonctionelles, *C.R. Acad. Sci., Paris* 230 (1950) 1999–2000. [1.5]

J. Riguet, Les relations de Ferrers, *C.R. Acad. Sci., Paris* 232 (1951) 1729–1730. [1.5]

N. Robertson and P.D. Seymour, Graph minors. XX. Wagner's conjecture, *J. Comb. Theory Ser. B* 92 (2004) 325–357. [1.2]

G.B. Robison and E.S. Wolk, The imbedding operators on a partially ordered set, *Proc. Am. Math. Soc.* 8 (1957) 551–559. [1.5]

V.A. Rohlin, On the fundamental ideas of measure theory, *Mat. Sbornik (N.S.)* 25 (67) (1949) 107–150. [3.5]

S. Roman and G.-C. Rota, The umbral calculus, *Adv. Math.* 27 (1978) 95–188. [4.3]

G.-C. Rota, The number of partitions of a set, *Am. Math. Mon.* 71 (1964) 498–504. [4.2]

G.-C. Rota, Combinatorial analysis, in *The Mathematical Sciences: A Collection of Essays*, M.I.T. Press, Cambridge, MA, 1969. [1.2]

G.-C. Rota, Baxter algebras and combinatorial identities. I, II, *Bull. Am. Math. Soc.* 75 (1969) 325–329, 330–334. [5.4, 5.5, 5.6]

G.-C. Rota, On the combinatorics of the Euler characteristic, in L. Mirsky, ed., *Studies in Pure Mathematics*, Academic Press, London, 1971, pp. 221–233. [3.6]

G.-C. Rota, The valuation ring of a distributive lattice, in *Proceedings University of Houston Lattice Theory Conference (Houston, Texas 1973)*, Department of Mathematics, University of Houston, Houston, TX, 1973, pp. 574–628. [3.6]

G.-C. Rota, Hopf algebra methods in combinatorics, in *Problémes Combinatoires et Théorie des Graphes (Colloques Internationaux CNRS, Université Orsay, Orsay, 1976)*, CNRS, Paris, 1978, pp. 363–365. [4.3]

G.-C. Rota, Baxter operators, an introduction, in *Gian-Carlo Rota on Combinatorics*, Birkhäuser, Boston and Basel, 1995, pp. 504–512. [5.4]

G.-C. Rota, Combinatorics, representation theory and invariant theory: The story of a ménage à trois, *Discrete Math.* 193 (1998) 5–16. [5.1]

G.-C. Rota, Ten mathematics problems I will never solve, *Mitt. Dtsch. Math.-Ver.* (1998), 45–52. [5.5]

G.-C. Rota, What is invariant theory? in H. Crapo and D. Senato, eds., *Algebraic Combinatorics and Computer Science*, Springer-Italia, Milan, 2002, pp. 41–56. [2.8]

G.-C. Rota, An example of profinite combinatorics, in *Gian-Carlo Rota on Analysis and Probability*, Birkhäuser, Boston and Basel, 2003, pp. 286–289. [3.6]

G.-C. Rota and B. Sagan, Congruences derived from group action, *Eur. J. Comb.* 1 (1980) 67–76. [5.3]

G.-C. Rota and J. Shen, On the combinatorics of cumulants, *J. Comb. Theory Ser. A* 91 (2000) 283–304. [4.2]

G.-C. Rota and D.A. Smith, Fluctuation theory and Baxter algebras, in *Symposia Mathematica, Vol. IX (Convegno di Calcolo delle Probabilità, INDAM, Rome, 1971)*, Academic Press, London, 1972, pp. 179–201. [5.4, 5.5, 5.6]

G.-C. Rota and D.A. Smith, Enumeration under group action, *Ann. Scuola Norm. Super. Pisa* 4 (1977) 637–646. [5.3]

L. Rowen, *Polynomial Identities in Ring Theory*, Academic Press, New York, 1980. [3.1]

H.J. Ryser, A combinatorial theorem with an application to Latin rectangles, *Proc. Am. Math. Soc.* 2 (1951) 550–552. [2.2]

H.J. Ryser, Combinatorial properties of matrices of 0's and 1's, *Can. J. Math.* 9 (1957) 371–377. [2.7]

H.J. Ryser, *Combinatorial Mathematics*, Mathematical Association of America, Washington, DC, 1963. [3.1]

M. Saks, A short proof of the existence of k-saturated partitions of partially ordered sets, *Adv. Math.* 33 (1979) 19–22. [3.2]

V.N. Saliĭ, *Lattices with Unique Complements*, Translations of Mathematical Monographs, Vol. 69, American Mathematical Society, Providence, RI, 1988. [1.3]

J.R. Sangwine-Yager, Mixed volumes, in P.M. Gruber and J.M. Wills, eds., *Handbook of Convex Geometry*, Elsevier, Amsterdam, 1993, pp. 43–71. [6.1]

W. Schmitt, Hopf algebra methods in graph theory, *J. Pure Appl. Algebr.* 101 (1995) 77–90. [4.3]

H. Schneider, The concepts of irreducibility and full indecomposability of a matrix in the works of Frobenius, König and Markov, *Linear Algebr. Appl.* 18 (1977) 139–162. [2.3]

I.J. Schoenberg, On the zeros of the generating functions of multiply positive sequences and functions, *Ann. Math.* (2) 62 (1955) 447–471. [6.9]

O. Schreier, Über den J-H'schen Satz, *Abh. Math. Sem. Univ. Hamburg* 6 (1928) 300–302. [3.4]

A. Schrijver, Matroids and linking systems, *J. Comb. Theory Ser. B* 26 (1979) 349–369. [2.8]

I. Schur, Zwei Sätze über algebraische Gleichungen mit lauter reellen Wurzeln, *J. Reine Angew. Math.* 144 (1914) 75–88. [6.4]

M.P. Schützenberger, Sur certains axioms de la théorie des structures, *C.R. Acad. Sci., Paris* 221 (1945) 218–220. [3.4]

M.-P. Schützenberger, Une interprétation de certaines solutions de l'equation fonctionelle $F(x + y) = F(x)F(y)$, *C.R. Acad. Sci., Paris* 236 (1953) 352–353. [4.1]

C.E. Shannon, A mathematical theory of communication, *Bell Syst. Tech. J.* 22 (1948) 379–423, 623–656; reprinted, *Claude Elwood Shannon. Collected Papers*, N.J.A. Sloane and A.D. Wyner, eds., IEEE Press, New York, 1993. [1.4]

J. Shareshian, On the Möbius number of the subgroup lattice of the symmetric group, *J. Comb. Theory Ser. A* 78 (1997) 236–267. [3.1]

H.M. Sheffer, A set of five independent postulates for Boolean algebras, *Trans. Am. Math. Soc.* 14 (1913) 481–488. [1.1]

I.M. Sheffer, Some properties of polynomials of type zero, *Duke Math. J.* 5 (1939) 590–622. [4.4]

G.C. Smith, *The Boole-De Morgan Correspondences, 1842–1864*, Oxford University Press, Oxford, 1982. [1.1]

L. Solomon, The Burnside algebra of a finite group, *J. Comb. Theory* 2 (1967) 603–615. [3.5]

E. Sperner, Ein Satz über Untermengen einer endlichen Menge, *Math. Z.* 27 (1928) 544–548. [3.3]

E. Spiegel and C.J. O'Donnell, *Incidence Algebras*, Marcel Dekker, New York, 1997. [3.1]

F. Spitzer, A combinatorial lemma with applications to probability theory, *Trans. Am. Math. Soc.* 82 (1956) 323–339. [5.6]

F. Spitzer and H. Widom, The circumference of a convex polygon, *Proc. Am. Math. Soc.* 12 (1961) 506–509. [5.6]

R.P. Stanley, Structure of incidence algebras and their automorphism groups, *Bull. Am. Math. Soc.* 76 (1970) 1236–1239. [3.1]

R.P. Stanley, Theory and application of plane partitions, I and II, *Stud. Appl. Math.* 50 (1971) 167–188, 259–279. [5.1, 5.2]

R.P. Stanley, A Brylawski decomposition for finite ordered sets, *Discrete Math.* 4 (1973) 77–82. [1.2]

R.P. Stanley, Combinatorial reciprocity theorems, *Adv. Math.* 14 (1974) 194–253. [1.2]

R.P. Stanley, Exponential structures, *Stud. Appl. Math.* 59 (1978) 73–82. [4.1]

R.P. Stanley, *Enumerative Combinatorics*, Vol I, Cambridge University Press, Cambridge, 1997. [3.1, 4.1]

R.P. Stanley, *Enumerative Combinatorics*, Vol II, Cambridge University Press, Cambridge, 1999. [2.6]

G.P. Steck, The Smirnov two-sample tests as rank tests, *Ann. Math. Stat.* 40 (1969) 1449–1466. [4.5]

R. Steinberg, On Dickson's theorem on invariants, *J. Fac. Sci. Univ. Tokyo Sect. IA Math.* 34 (1987) 699–707. [5.7]

M.H. Stone, The theory of representations for Boolean algebras, *Trans. Am. Math. Soc.* 40 (1936) 37–111. [1.1]

R.G. Swan, An application of graph theory to algebra, *Proc. Am. Math. Soc.* 14 (1963) 367–373 and 21 (1969) 379–380. [3.1]

J.J. Sylvester, An essay on canonical forms, supplement to a sketch of a memoir on elimination, George Bell, Fleet Street, London, 1851. *Collected Mathematical Papers*, Vol. I, Paper 34. [6.2]

J.J. Sylvester, On a remarkable discovery in the theory of canonical forms and of hyperdeterminants, *Math. Mag.* 2 (1851) 391–410. *Collected Mathematical Papers*, Vol. I, Paper 41. [6.2]

J.J. Sylvester, *Collected Mathematical Papers*, Vols. I–IV, Cambridge University Press, Cambridge, 1904–1912. [6.2]

G. Szegö, Bemerkungen zu einen Satz von J. H. Grace über die Wurzeln algebraischer Gleichungen, *Math. Z.* 13 (1922) 28–55. [6.4]

G. Szegö, *Orthogonal Polynomials*, 4th edition, American Mathematical Society, Providence RI, 1975. [6.1]

E. Szpilrajn, Sur l'extension de l'ordre partiel, *Fundam. Math.* 16 (1930) 386–389. [1.2]

T. Tao and V. Vu, *Additive Combinatorics*, Cambridge University Press, Cambridge, 2006. [3.3]

A. Tarski, Sur les classes closes par rapport á certaines opérations élémentaires, *Fundam. Math.* 16 (1929) 181–304. [1.1]

A. Tarski, A lattice-theoretical fixpoint theorem and its applications, *Pac. J. Math.* 5 (1955) 285–309. [1.3]
A. Tarski and S.R. Givant, *A Formalization of Set Theory Without Variables*, American Mathematical Society, Providence, RI, 1987. [1.5]
E.C. Titchmarsh, *Introduction to the Theory of Fourier Integrals*, 3rd edition, Chelsea, New York, 1986. [5.5]
J. Touchard, Nombres exponentiels et nombres de Bernoulli, *Can. J. Math.* 8 (1956) 305–320. [4.5]
W.T. Trotter, Inequalities in dimension theory for posets, *Proc. Am. Math. Soc.* 47 (1975) 311–316. [3.2]
W.T. Trotter, *Combinatorics and Partially Ordered Sets. Dimension Theory*, Johns Hopkins University Press, Baltimore, MD, 1992. [1.2, 3.2]
H.W. Turnbull, The invariant theory of a general bilinear form, *Proc. Lond. Math. Soc. (2)* 33 (1930) 3–21. [6.1]
W.T. Tutte, On dichromatic polynomials, *J. Comb. Theory* 2 (1967) 301–320. [4.3]
W.T. Tutte, On chromatic polynomials and the golden ratio, *J. Comb. Theory* 9 (1970) 289–296. [6.1]
H.A. Tverberg, A proof of Dilworth's decomposition theorem, *J. Comb. Theory Ser. A* 17 (1967) 305–306. [3.2]
B.L. van der Waerden, Ein Satz Über Klasseneinteilungen von endlichen Mengen, *Abh. Math. Sem. Univ. Hamburg* 5 (1927) 185–188. [2.6]
J. von Neumann, On regular rings, *Proc. Natl. Acad. Sci. USA* 23 (1936) 707–713. [3.4]
J. von Neumann, A certain zero-sum two-person game equivalent to the optimal assignment problem, in *Contributions to the Theory of Games*, Vol. II, Princeton University Press, Princeton, NJ, 1953, pp. 5–12. [2.6]
J. von Neumann, *Continuous Geometry*, Princeton University Press, Princeton, NJ, 1960. [3.7]
F. Vogt and B. Voigt, Symmetric chain decompositions of linear lattices, *Combin., Probab. Comput.* 6 (1997) 231–245. [3.5]
M. Ward, The algebra of lattice functions, *Duke Math. J.* 5 (1939) 357–371. [3.1]
R. Webster, *Convexity*, Oxford University Press, Oxford, 1994. [2.6]
L. Weisner, Abstract theory of inversion of finite series, *Trans. Am. Math. Soc.* 38 (1935) 474–484. [3.1]
D.J.A. Welsh, Generalized versions of Hall's theorem, *J. Comb. Theory Ser. B* 10 (1971) 95–101. [2.4]
P.M. Whitman, Free lattices, *Ann. Math.* (2) 42 (1941) 325–329. [1.3]
P.M. Whitman, Free lattices II, *Ann. Math.* (2) 43 (1942) 104–105. [1.3]
P.M. Whitman, Lattices, equivalence relations, and subgroups, *Bull. Am. Math. Soc.* 52 (1946) 507–522. [1.5]
H. Whitney, On the abstract properties of linear dependence, *Am. J. Math.* 57 (1935) 509–533. [2.4]
H. Wielandt, Eine Verallgemeinerung der invarianten Untergruppen, *Math. Z.* 45 (1939) 209–244. [3.5]
H.S. Wilf, Hadamard determinants, Möbius functions, and the chromatic number of a graph, *Bull. Am. Math. Soc.* 74 (1968) 960–964. [3.1]
H.S. Wilf, The Redheffer matrix of a partially ordered set, *Electron. J. Comb.* 11 (2004/2005), Research Paper 10. [3.1]

H.S. Wilf, *generatingfunctionology*, 3rd edition, AK Peters, Wellesley, MA, 2006. [4.1]

P. Winkler, *Mathematical Puzzles*, AK Peters, Wellesley MA, 2004. [2.1]

K. Yamamoto, Logarithmic order of free distributive lattice, *J. Math. Soc. Japan* 6 (1954) 343–353. [3.3]

C.H. Yan, The theory of commuting Boolean sigma-algebra, *Adv. Math.* 144 (1999) 94–116. [3.4]

H. Zassenhaus, Zum Satz von Jordan-Hölder-Schreier, *Abh. Math. Sem. Univ. Hamburg* 10 (1934) 106–108. [3.4]

Subject Index

Printed in the United States
By Bookmasters